Lecture Notes in Networks and S

AF869408

Volume 349

The series "Lecture Notes in Networks and Systems" publishes the latest developments in Networks and Systems—quickly, informally and with high quality. Original research reported in proceedings and post-proceedings represents the core of LNNS.

Volumes published in LNNS embrace all aspects and subfields of, as well as new challenges in, Networks and Systems.

The series contains proceedings and edited volumes in systems and networks, spanning the areas of Cyber-Physical Systems, Autonomous Systems, Sensor Networks, Control Systems, Energy Systems, Automotive Systems, Biological Systems, Vehicular Networking and Connected Vehicles, Aerospace Systems, Automation, Manufacturing, Smart Grids, Nonlinear Systems, Power Systems, Robotics, Social Systems, Economic Systems and other. Of particular value to both the contributors and the readership are the short publication timeframe and the world-wide distribution and exposure which enable both a wide and rapid dissemination of research output.

The series covers the theory, applications, and perspectives on the state of the art and future developments relevant to systems and networks, decision making, control, complex processes and related areas, as embedded in the fields of interdisciplinary and applied sciences, engineering, computer science, physics, economics, social, and life sciences, as well as the paradigms and methodologies behind them.

Indexed by SCOPUS, INSPEC, WTI Frankfurt eG, zbMATH, SCImago.

All books published in the series are submitted for consideration in Web of Science.

More information about this series at http://www.springer.com/series/15179

David Guralnick · Michael E. Auer ·
Antonella Poce
Editors

Innovations in Learning and Technology for the Workplace and Higher Education

Proceedings of 'The Learning Ideas Conference' 2021

Springer

Editors
David Guralnick
Kaleidoscope Learning
International E-Learning Association
New York, NY, USA

Michael E. Auer
Carinthia University of Applied Sciences
St. Magdalen, Austria

Antonella Poce
Università degli Studi di Modena e Reggio
Emilia
Rome, Italy

ISSN 2367-3370 ISSN 2367-3389 (electronic)
Lecture Notes in Networks and Systems
ISBN 978-3-030-90676-4 ISBN 978-3-030-90677-1 (eBook)
https://doi.org/10.1007/978-3-030-90677-1

This Springer imprint is published by the registered company Springer Nature Switzerland AG
The registered company address is: Gewerbestrasse 11, 6330 Cham, Switzerland

Preface

The Learning Ideas Conference began life in 2008 as the International Conference on E-Learning in the Workplace (ICELW), which focused on new and interesting work relating to workplace learning. As ICELW evolved, it began to seem more and more critical to integrate work from higher education along with workplace learning, broadening the focus to adult learning in various forms, and to expand beyond the term "e-learning." The conference integrated one of its sister conferences, the International Conference on Interactive Collaborative and Blended Learning (ICBL), added the more technical Adaptive Learning via Interactive, Collaborative and Emotional Approaches (ALICE) workshop as a special track, and was reborn and rebranded as the Learning Ideas Conference.

With the subtitle "Innovations in Learning and Technology for the Workplace and Higher Education," the conference's goal is to bring together people from around the world to help reimagine what learning can be, particularly using, and inventing, new technologies.

The Learning Ideas Conference 2021 was held as a fully online event due to the COVID-19 pandemic and was the first conference under the new name. The conference featured 5 fantastic keynote speakers:

Prof. Dr. Ilona Buchem, Professor of Media and Communication, Faculty of Economics and Social Sciences, Beuth University of Applied Sciences Berlin, Berlin, Germany. *"Wearable Enhanced Learning (WELL): Trends, Opportunities, and Challenges."*

Deborah Howes, Professor, Museum Studies Program, Johns Hopkins University, and President of Howes Studio Inc, New York, NY, USA. *"Museums as Catalysts for Digital Learning."*

Michael Kanaan, Director of Operations, Department of the Air Force/MIT Artificial Intelligence, Cambridge, Massachusetts, USA. *"Artificial Intelligence: Reaching the End of the Rainbow."*

David Kelly, EVP and Executive Director, The Learning Guild, New York, NY, USA. *"A Look Ahead: The Now and the Next of Learning and Technology within Organizations."*

Alicia Sanchez, Ph.D., Director of Innovation, Defense Acquisition University, Department of Defense, Fort Belvoir, Virginia, USA. *"Learning on the Seam: The Intersection between Learning Science and User-Centered Design."*

The conference featured a panel discussion entitled "The Future of Learning: Reimagining Education and its Impact on Society," along with dozens of other sessions, over a 5-day period, from June 14 to 18, 2021. All papers were double-blind peer reviewed.

I very much appreciate all of the work it took to make The Learning Ideas Conference 2021 a success, from our keynotes, our Executive Committee and Program Committee members, our reviewers, and of course our conference organizing team.

I am looking forward to The Learning Ideas Conference 2022, to be held as a hybrid event, both in New York and online.

David Guralnick
Conference Chair, The Learning Ideas Conference 2021

Organization

Committees

Conference Chair

David Guralnick	Kaleidoscope Learning and Columbia University, New York, NY, USA

Executive Committee Chairs

Michael E. Auer	CTI, Frankfurt, Germany
Antonella Poce	Università degli Studi di Modena e Reggio Emilia, Rome, Italy

Publication Chair

Lara Ramsey	Kaleidoscope Learning, New York, NY, USA

Executive Committee

Mohammed Ali Akour	A'Sharqiyah University, Ibra, Oman
Kostas Apostolou	McMaster University, Hamilton, Ontario, Canada
Ryan Baker	University of Pennsylvania, Philadelphia, Pennsylvania, USA
Patricia Behar	Federal University of Rio Grande do Sul, Brazil
John Black	Teachers College, Columbia University, New York, USA
Patrick Blum	Inside Business Group, Aachen, Germany
Santi Caballé	Open University of Catalonia, Barcelona, Spain
Nicola Capuano	University of Basilicata, Potenza, Italy
Imogen Casebourne	University of Oxford, Oxford, UK

Veronica Chehtman	AySA Water and Sanitation Argentina, Buenos Aires, Argentina
Hal Christensen	QuickCompetence, New York, USA
Samir El-Seoud	The British University in Egypt (BUE), Egypt
Kai Erenli	University of Applied Sciences BFI Vienna, Vienna, Austria
Christian Guetl	Graz University of Technology, Graz, Austria
Alexander Kist	University of Southern Queensland, Queensland, Australia
Gila Kurtz	Holon Institute of Technology, Holon, Israel
Mark J. W. Lee	Charles Sturt University, Bathurst, Australia
Christy Levy	Kaleidoscope Learning, Chicago, Illinois, USA
Matthea Marquart	Columbia University, New York, USA
Bruce McLaren	Carnegie Mellon University, Pittsburgh, Pennsylvania, USA
Jorge Membrillo Hernández	Tecnológico de Monterrey, México
Dominik May	University of Georgia, Athens, Georgia, USA
Christina Merl	TalkShop/2CG®, Vienna, Austria
Barbara Oakley	Oakland University, Oakland, Michigan, USA
Andreas Pester	Carinthia Tech Institute, Villach, Austria
Teresa Restivo	University of Porto, Porto, Portugal
Fernando Salvetti	Logosnet, Turin, Italy
Alicia Sanchez	Czarina Games, Alexandria, Virginia, USA
Sabine Seufert	Universität St.Gallen (HSG), St. Gallen, Switzerland
Thrasyvoulos Tsiatsos	Aristotle University of Thessaloniki, Thessaloniki, Greece
James Uhomoibhi	Ulster University, Newtownabbey, UK
Matthias Utesch	Technical University of Munich, Munich, Germany
Ellen Wagner	North Coast EduVisory Services, Sonoma, California; and University of Central Florida, Orlando, Florida, USA
Sarah Wang	Xi'an Jiaotong-Liverpool University, Suzhou, Jiangsu, China
Xiao-Guang Yue	European University Cyprus, Nicosia, Cyprus

Program Committee

Carme Anguera Iglesias	Open University of Catalonia, Barcelona, Spain
Fahriye Altinay Aksal	Near East University, Nicosia, Cyprus
Zehra Altinay Gazi	Near East University, Nicosia, Cyprus
Sarah Appleby	Online Learning International, New York, USA
Anabel Bugallo	Kaleidoscope Learning, New York, USA

Martha Burkle	Southern Alberta Institute of Technology, Calgary, Canada
Mihai Caramihai	University Politehnica Bucharest, Bucharest, Romania
Nunzio Casalino	Guglielmo Marconi University and LUISS Business School, Rome, Italy
Mark Cassetta	Pfizer, New York, USA
Gary J. Dickelman	EPSScentral, Annandale, Virginia, USA
David Foster	ExecOnline, Chapel Hill, North Carolina, USA
Sarah Frame	University of East London, London, England
Marga Franco i Casamitjana	Open University of Catalonia, Barcelona, Spain
Genevieve Gallant	GG Consultants Limited, St. John's, Newfoundland, Canada
Manir Abdullahi Kamba	Bayero University, Kano, Nigeria
Okba Kazar	Biskra University, Biskra, Algeria
J. C. Kinnamon	Practising Law Institute, New York, USA
Adamantios Koumpis	Bern University of Applied Sciences, Bern, Switzerland
Molly Koenen	Pioneer Management Consulting, Minneapolis, Minnesota, USA
Maria Lambrou	University of the Aegean, Greece
Stacy Lindenberg	Talent Seed Consulting, Columbia, South Carolina, USA
Allison Littlejohn	Glasgow Caledonian University, Glasgow, Scotland
Luis Ochoa Siguencia	Jerzy Kukuczka Academy of Physical Education, Katowice, Poland
Grace O'Malley	National College of Ireland, Dublin, Ireland
Rikke Orngreen	Aalborg University, Copenhagen, Denmark
Michael Paraskevas	Technological Educational Institute of Western Greece, Antirrio, Greece
Iina Paarma	United Nations Development Programme, Copenhagen, Denmark
Kinga Petrovai	The Art & Science of Learning, Ottawa, Ontario, Canada
Stefanie Quade	Berlin School of Economics and Law, Berlin, Germany
Maria Rosaria Re	Roma Tre University, Rome, Italy
Laura Ricci	MIA Digital University, Barcelona, Spain
Gina Ann Richter	St. Charles Consulting Group, New York, USA
Andree Roy	University of Moncton, Moncton, Canada
John Sandler	Telstra Corporation, Melbourne, Australia
Steven Schmidt	East Carolina University, Greenville, North Carolina, USA

Barbara Schwartz-Bechet	Misericordia University, Elkins Park, Pennsylvania, USA
Julie-Ann Sime	Lancaster University, Lancashire, UK
Amando P. Singun, Jr.	Higher College of Technology, Muscat Sultanate of Oman
Anelise Spyer	Docta, São Paulo, Brazil
Christian Stracke	eLC Institute for eLearning, Bonn, Germany
Kyla L. Tennin	University of Phoenix, Tempe, Arizona, USA
Terrie Lynn Thompson	University of Alberta, Edmonton, Alberta, Canada
Caryn Tilton	MyPlaceToLearn, Welches, Oregon, USA
Leyla Y. Tokman	Anadolu University, Eskisehir, Turkey
Christos Troussas	University of West Attica, Athens, Greece
Chris Turner	University of Winchester, Winchester, UK
Karin Tweddell Levinsen	Danish University of Education, Denmark
Maggie M. Wang	The University of Hong Kong, Hong Kong
Steve Wheeler	University of Plymouth, Plymouth, UK
Annika Wiklund-Engblom	Abo Akademi University, Vasa, Finland
Rusen Yamacli	Anadolu University, Eskisehir, Turkey

ALICE Special Track Co-chairs

Santi Caballé	Open University of Catalonia, Barcelona, Spain
Nicola Capuano	University of Basilicata, Potenza, Italy

ALICE Track Program Committee

Mohammad Al-Smadi	Jordan University of Science and Technology (JUST), Jordan
Joan Casas	Open University of Catalonia, Spain
Jordi Conesa	Open University of Catalonia, Spain
Thanasis Daradoumis	University of the Aegean, Greece
Christian Gütl	Graz University of Technology, Austria
Giuseppina Rita Mangione	Institute of Educational Documentation, Innovation and Research, Italy
Agathe Merceron	Beuth University of Applied Sciences Berlin, Germany
Anna Pierri	University of Salerno, Italy
Antonio Sarasa	Universidad Complutense de Madrid, Spain
Krassen Stefanov	Sofia University, Bulgaria

Contents

Main Conference

Dynamic Plan Generation and Digital Storyboarding for the Professional Training of Accident Prevention with Time Travel Games

Oksana Arnold[1], Ronny Franke[2], Klaus P. Jantke[3](✉), and Hans-Holger Wache[4]

[1] Fachhochschule Erfurt, Altonaer Str. 25, 99085 Erfurt, Germany
oksana.arnold@fh-erfurt.de

[2] Fraunhofer Institut für Fabrikbetrieb und -Automatisierung IFF, Sandtorstr. 22, 39106 Magdeburg, Germany

[3] ADICOM Software, Frauentorstr. 11, 99423 Weimar, Germany
klaus.p.jantke@adicom-group.de

[4] Berufsgenossenschaft Rohstoffe und Chemische Industrie, Präventionszentrum Berlin, Innsbrucker Str. 26/27, 10825 Berlin, Germany

Abstract. Time travel games are a recent form of edutainment media having high potential in areas such as environmental education and prevention training. Time travel prevention games for purposes such as accident prevention in the industries are advantageous due to their conservation of resources including human health and lives. They are affective by allowing for unprecedented learner/player/trainee experiences, and they are effective due to the fascination of application-oriented game play including opportunities to influence the fate, the latter being less close to reality — but the more attractive and worth telling. Digital storyboarding is the ultimate design methodology allowing for properly dovetailing pedagogical and game design. It works simultaneously bottom-up, top-down, or both at once in interdisciplinary teams flexibly in space and time. Storyboarding is the organization of future learner/player/trainee experiences. The expressive power and the reach of digital storyboarding is due to its roots in dynamic plan generation for the mastery of disturbances in the industries. Storyboards are finite hierarchically structured families of graphs small in size encoding patterns of game design, principles of didactics, and their interference. The technology particularly supports the design of time travel adaptive to the players' needs and goals aiming at affective experiences and effective learning.

Keywords: Time travel game · Time travel prevention game · Game design · Didactic design · Design of experience · Dynamic plan generation · Adaptivity · Accident prevention training · Patterns of game design · Didactic principles

D. Guralnick et al. (Eds.): TLIC 2021, LNNS 349, pp. 3–18, 2022.
https://doi.org/10.1007/978-3-030-90677-1_1

1 Introduction

It makes a difference whether a human learner or trainee has experienced an accident, especially a self-induced one. However, this shall not be misinterpreted as a call for a didactic principle of deliberately damaging industrial installations or injuring humans at their workplace for more affective and effective learning about accident prevention.

Apparently, there is an alternative — virtual training environments in which humans can act and interact without fear and without the danger of real damages and injuries. The authors' approach is to offer affecting experience to learn from — preferably human experience so touching and exciting that learners and trainees find it worth telling.

This includes that damage happens, that accidents take place — of course, virtually. But how to learn effectively? Just making a mistake does not imply that you learn from it. The key idea advocated by means of this paper is, roughly, virtual time travel.

The term *time travel prevention game* has been coined at the conference and expo Deutscher Präventionstag in 2015 with a focus on crime prevention [18]. Currently, we consider so-called *time travel games*, in general, and classify according to didactic concepts into *time travel exploratory games* [6] and *time travel prevention games* as dealt with in the present contribution.

Before we introduce the technicalities in as much detail as necessary and before we demonstrate how to wield the tools, this section shall be completed by an illustration.

Imagine a training session in which a trainee is striving hard to solve a certain problem in the virtual world of the training environment. Training is particularly relevant for tasks of high complexity and risk. Consequently, incorrect operations and inaccuracies are likely. If a virtual accident occurs in the virtual world, it is impressively animated to clearly articulate the problem. But trainees should not get stuck at any point of no return. Therefore, they get offered time travel opportunities indicated by iconic visualizations, such as the Tardis-like one in Fig. 1. This is accompanied by an information of what to do next.

Fig. 1. Entrance point to time travel.

Travelling back in time brings with it a chance to think about what went wrong and to act more successfully the next time. Even repeated journeys in time may take place (Fig. 2).

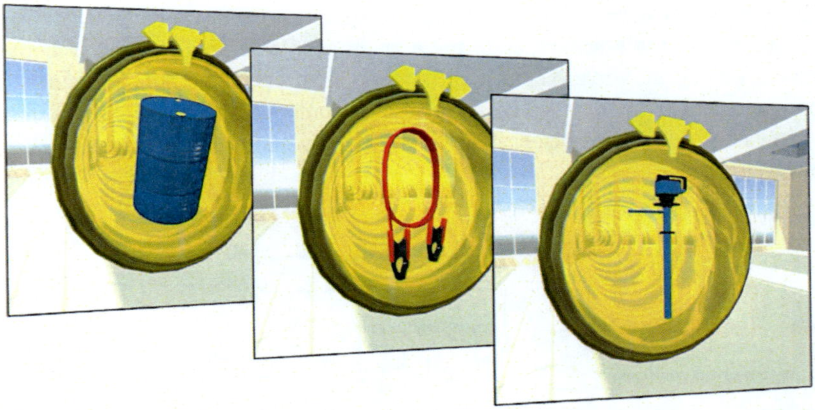

Fig. 2. These three screenshots visualize one of the authors' implementations of a time tunnel. The objects in the tunnel represent a trainee's activities and move backwards or forwards, resp., in dependence on pressing the upper left or right button. A click to the central button selects the object on display and starts a journey in time back to the point of handling the current object.

2 Dynamic Plan Generation

In complex dynamic environments — the application domain of reference [1] is the therapy of disturbed industrial installations that shall be brought back to a normal mode of operation — it is undecidable at the time of plan generation whether a plan under construction will ultimately succeed or not.

Therefore, plan generation does not result in a single plan, but in a space of plans. At execution time, this space is dynamically traversed. Decisions are made at the time of execution in dependence on dynamic data. As a result, there emerges a successful conventional plan in the very end — planning is learning [5].

The aim of this section is to survey the essentials of dynamic plan generation with technicalities as few as possible. The focus is on those concepts that are important to the authors' storyboarding approach and to the development of time travel prevention games and their deployment in practical training for accident prevention.

2.1 Concepts of Dynamic Plan Generation

Finite directed graphs with entry nodes and exit nodes — graph concepts that are called pin graphs [14] — are the building blocks of plans. If one keeps these graphs small and simple in structure, this will lead to explainable plans and allows for the negotiation of planning details in interdisciplinary teams of stakeholders. Later on, the authors' storyboarding approach will substantially rely on this simplicity and its relevance for the co-operation of interdisciplinary teams of educators, psychologists, game designers, AR and VR specialists, and others.

All graphs may have two types of nodes — compound nodes and elementary nodes. The latter are having a semantics in the domain such as a repair action or cooling a unit for a certain duration. There may be alternative meanings having conditions of execution to control their respective deployment. Compound nodes are placeholders for graph substitution. At this place, it becomes obvious that we need a whole set of pin graphs.

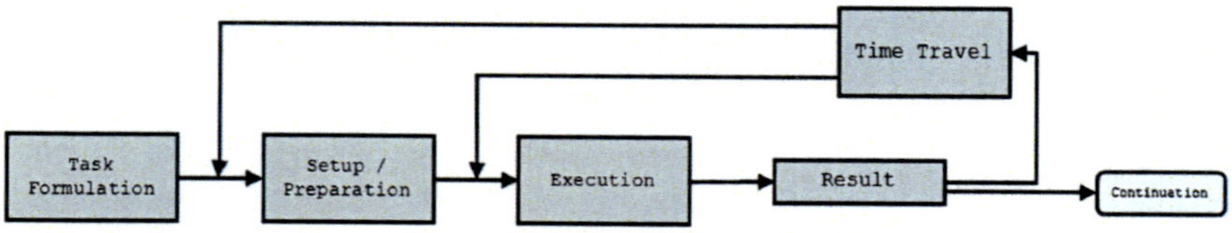

Fig. 3. By way of illustration, we present some pin graph, not from the process therapy domain, but from the domain of time travel prevention games to be studied in more detail subsequently.

The collapsed boxes in Fig. 3 denote elementary nodes, whereas the larger boxes are compound nodes. The semantics of the ‘Result’ node depends on some data from the actions performed before and may be a cutscene displaying the successful completion of the task or, alternatively, an animated accident. There may be varying substitutions for compound nodes, such as a dialogue to prepare for time travel. The ‘Execution’ compound node may be replaced by varying sequences of actions.

Following reference [1], a plan $\boldsymbol{P}$ is a finite hierarchically structured family $\boldsymbol{F}$ of pin graphs together with a distinguished graph $\boldsymbol{G_1}$ in $\boldsymbol{F}$, which is considered top level. Furthermore, every compound node in any graph in $\boldsymbol{F}$ gets assigned the graphs that may be used to expand this node. In addition, every graph has a substitution condition that determines the dynamic conditions in which this graph may be used.

The hierarchical structure of $\boldsymbol{F}$ is due to the regulations of substituting pin graphs into one another. There are a few more technicalities that may remain under the hood.

2.2 Dynamic Plan Generation as Graph Expansion

At planning time, the values of variables that occur in logical conditions like those for the replacement of a node by a graph may be unknown and substitution is suspended.

At execution time, current values of variables determine admissible replacements. Sometimes, substitutions take place immediately before the execution of the actions.

The authors are aware of the problem that the presentation of the plan generation algorithm on display in Fig. 4 is a bit dense, so to speak. For the purpose of the present contribution, some essentials of dynamic planning that are relevant to the storyboarding approach shall be discussed in more detail. In doing so, we will use a slightly simplified font.

algorithm: *planner1*
input: $\mathcal{G}_i\ ,\ t_0\ ,\ \mathcal{R} = [\mathcal{F}', \mathcal{G}_i]$
output: $\mathcal{P} = [\mathcal{F}''', \mathcal{G}_i]$

1. $\mathcal{G} := \mathcal{G}_i$
2. $\mathcal{P} := [\mathcal{F}''', \mathcal{G}_i]$ with $\mathcal{F}''' := \{\mathcal{G}_i\}$
3. **while** $C \neq \emptyset$ **do**
   - 3.1 $A := \{v \mid v \textit{ active in } \mathcal{G}\}$
   - 3.2 **forall** $v \in A$ **do**
     $sub_{curr}(v) := sub_{t_0}^{cons}(v)$
   - 3.3 **while** $A \neq \emptyset$ **do**
     - 3.3.1 **find** $v \in A$
     - 3.3.2 **repeat**
       - i $j := \max sub_{curr}(v)$
       - ii $sub_{curr}(v) := sub_{curr}(v) \setminus \{j\}$
       - iii $\mathcal{F}'_{part} := \{\mathcal{G} \mid \mathcal{G}_i \preceq^{+}_{\mathcal{F}'} \mathcal{G}\}$
       - iv $\mathcal{P}_{part} = [\mathcal{F}'''_{part}, \mathcal{G}_i] := planner1(\ \mathcal{G}_j, \tau(v), [\mathcal{F}'_{part}, \mathcal{G}_i]\ )$

       until $sub_{curr}(v) = \emptyset \vee \mathcal{P}'''_{part} \neq \emptyset$
     - 3.3.3 **if** $\mathcal{P}_{part} \neq \emptyset$
       then
       - i **find** $\mathcal{G}_{part}$ with $\mathcal{G}_i \Rightarrow^{+}_{\mathcal{F}'''} \mathcal{G}_{part} \wedge \mathcal{G}_{part} \downarrow$
       - ii $\mathcal{G} := \mathcal{G}[v \leftarrow \mathcal{G}_{part}]$
       - iii $\mathcal{F}''' := \mathcal{F}''' \cup \mathcal{F}'''_{part}$
       - iv $\mathcal{P} := [\mathcal{F}''', \mathcal{G}_i]$

       else
       - i $\mathcal{P} := [\mathcal{F}''', \mathcal{G}_i]$ with $\mathcal{F}''' = \emptyset$
4. **return** $\mathcal{P}$

Fig. 4. This plan generation algorithm from reference [3], Figure 4 on page 23 (see also reference [1], Figure 3.12 on page 149), demonstrates that the process of planning effectively is graph processing [4]. The decisive plan generation steps are graph rewriting — see step 3.3.3.i in this figure — and graph substitution — see step 3.3.3.ii.

The top-level graph of a plan above is named $\boldsymbol{G_i}$. A family of graphs ***F'*** contains all the available alternative substitutions for compound nodes. By way of illustration, readers may have a look at Fig. 3 with the compound node 'Time Travel.' Fig. 5 shows an expansion of this node by substitution of a whole graph.

The expansion begins at the entry nodes of $\boldsymbol{G_i}$ as a leftmost depth-first rewriting. The flat nodes of a graph like 'Result' in Fig. 3 are called elementary. Elementary nodes are no subject to expansion, as they have an operational meaning in the domain. This motivates the introduction of the term *active*. A node is active — see step 3.1 in the description of the planning algorithm — if it is a compound node and all preceding nodes are elementary.

Intuitively, active nodes are compound nodes that are 'reached' in the planning process. All preceding substitution decisions have been made. It is now time to decide about the active node. It may happen that this takes place as late as at execution time. At earlier time points, data to resolve conflicts of graph substitution may be missing.

Substitution of a compound node ***v*** by a graph of ***F'*** brings in new nodes. These nodes may be elementary or compound. In case there is at least one new compound node, the algorithm needs to find an active one — intuitively, one that is leftmost. Therefore, we speak about leftmost rewriting. Those compound nodes are replaced next. When newly introduced compound nodes are replaced, further compound nodes may occur. The planner is called recursively (see step 3.3.2.iv of the planning algorithm). Finally, if the substitution of the node v including all newly introduced lower level compounds is completed and only elementary nodes are left over, the process of plan generation proceeds, so to speak, to the right. The rewriting is leftmost depth-first.

Notice that the process of plan generation is — to some extent — non-deterministic. There is a set of active nodes A. The selection of a node v from A is not controlled. Furthermore, $sub(v)$ is a set of all the graphs — more precisely, the names of the graphs — that are foreseen as potential replacements for the compound node named v. When t_0 is the time point of expanding the node v, the planner checks the current data of the system, the process, and the environment (see the lower and upper indices t_0 and $cons$, resp., in step 3.2). Names of graphs that are admissible for substitution — i.e., their substitution conditions are currently consistent with the data, are collected in some set named $sub_{curr}(v)$. From these graphs in F', the algorithm selects one for the replacement of the compound node v. When dealing with severely disturbed industrial processes, the occurrence of non-determinism is a lucky case. It means that there is more than only one repair action – therapy action in reference [1] – executable.

In edutainment and, in particular, in game-based learning, the non-determinism in the process of unfolding player/learner/trainee experience at the time of learning, playing, or training brings with it variety — an important criterion of didactic quality.

The planning methodology sketched above will be carried over to didactic and game design subsequently called digital storyboarding.

3 Digital Storyboarding

Digital storyboards are highly modular design documents that specify on hierarchical layers of granularity — see the concept of layered languages of ludology introduced by reference [15] and applied to complex game studies by reference [26] – even large-scale applications [2].

The advantage of the present digital storyboard concept over earlier conventional approaches is that the digital storyboards developed on the basis of dynamic planning are *executable*. With an appropriate interpreter at your fingertips you can run a digital storyboard — the storyboard itself is the program. This innovative methodology that unfortunately goes far beyond the limits of the present conference paper is called *storyboard interpretation technology* [11].

Essentials of the authors' concepts of storyboards and storyboarding as introduced by reference [20] are carried over from the dynamic planning approach as sketched in the preceding section.

Basically, a storyboard is a hierarchically structured family of directed pin graphs with compound nodes and/or elementary nodes. The compound nodes are named *episodes* and the elementary nodes are *scenes*.

Logical conditions that control the expansion of episodes and that determine the scenes' semantics, respectively, refer to the interaction history — i.e. to game play and

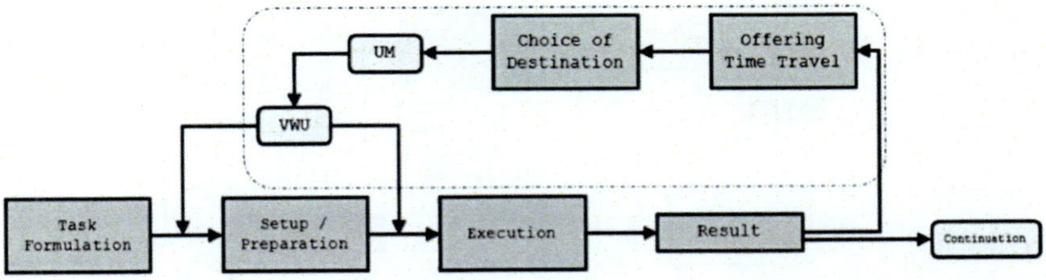

Fig. 5. Replacement of the episode 'Time Travel' by some graph (surrounded by a dashed line) that consists of 2 episodes 'Offering Time Travel' and 'Choice of Destination' and of 2 scenes.

its effects, to the player/learner/trainee profile, and to environmental data. Consider, by way of illustration, the episode 'Time Travel' in the graph on display in Fig. 3.

Figure 5 shows one alternative result of replacing the episode — compound node, in terms of dynamic plan generation — 'Time Travel' in the storyboard graph of Fig. 3. A potential expansion of the new episode 'Offering Time Travel' will be presented in Sect. 5 below. A variant of how to perform a 'Choice of Destination' is sketched in Fig. 2 via the authors' time tunnel with the opportunity to select objects on display.

The scenes 'UM' and 'VWU' mean user modeling and virtual world updates, resp. All edges in a storyboard graph have their execution conditions. The edges out of the 'VWU' scene have complementary conditions that refer to the destination selected.

Storyboarding means the design of experience, as reference [20] put it. The design process is iterative and interdisciplinary. Take the design of user models that evolves throughout negotiations of where, when, and why to use these data.

4 Patterns of Didactics and Game Design

Storyboarding allows for a much more precise and in-depth treatment of patterns in game design than some previous approaches like reference [8]. The authors keep the present section on patterns short and refer to another very recent discussion in reference [6], Sect. 8.

Very roughly speaking, storyboard graphs represent ideas, principles, and concepts of pedagogy, of game design, and of the interference of them. Those concepts that are implemented become structurally visible. From this particular perspective of design, digital storyboarding is the methodology of interdisciplinary didactic and game design, and storyboard graphs are the carriers that bring these ideas into the media system.

4.1 Patterns and Varying Instances

A particular aspect not to be underestimated when dealing with time travel games shall be briefly illustrated and discussed — repeated time travel.

Figure 6 shows a refinement of the graph in Fig. 5. There are two edges outgoing from the episode node 'Offering Time Travel' that need complementary conditions. When it may happen that players fail to master a certain task repeatedly and, therefore, need several attempts to try it again, an adaptive system assistance is required [22, 25]. For

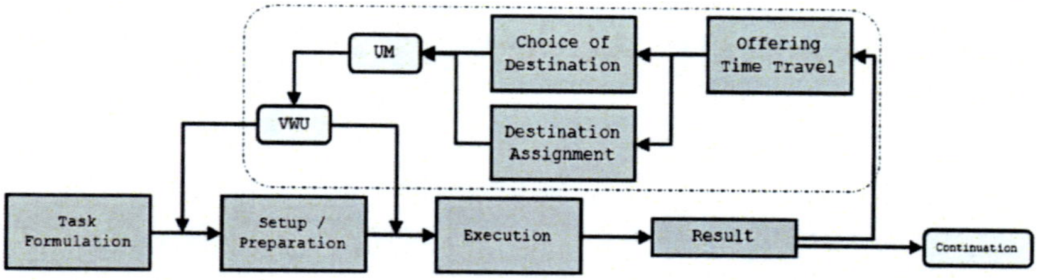

Fig. 6. Refinement of the pattern of time travel to deal with some repeatedly occurring accident.

this purpose, a time travel counter is inserted into the user model. Designers, especially educators and domain specialists, decide how many times they want to allow the trainees to choose their destinations when traveling backward in time by themselves. If a limit is set, the actual value is checked, and if the limit is exceeded, the system can take over and determines the point in time to travel to. Setting varying limits leads to varying training experiences.

4.2 Occurring Instances and Their Patterns

Which instances of a pattern really occur when a time travel prevention game is used may inform the designers about the effectiveness of their design decisions.

The occurrence of instances can be used as indicators of mastery [16]. Hence, patterns and their occurring instances open access to a systematic assessment. Needless to say, this brings with it several issues of data protection and privacy.

4.3 Patterns of Game Characteristics

Taxonomies are key to the understanding and to the mastery of digital games, especially for their exploitation aiming at affective and effective learning [19]. Reference [16] explicitly relate taxonomies to storyboard patterns, whereas references [10, 12], and others, do not identify any connection like this.

By way of illustration, let us have a closer look at pervasive games in which human players act in the real world that is virtually interwoven with a digital game world. What players may experience and, thus, potentially learn or train does essentially depend on characteristics of the dovetailing of these two worlds. In pervasive games [27], storyboards explicitly represent features decisive to learning [2, 7, 21]. Storyboards — more precisely, patterns in storyboards — explicate the way in which data flow from one world in the other and how and why virtual objects establish meaning in the real world.

As discussed above and to be demonstrated in the next section, storyboarding is an appealing methodology of interdisciplinary communication in game design superior to tables of parallel text proposed elsewhere [13].

By way of illustration, let us briefly analyze the question for forward time travel in prevention games. Recall the representation of backward time travel as specified by storyboard graphs on display in the Figs. 4 and 6 and think of our time tunnel.

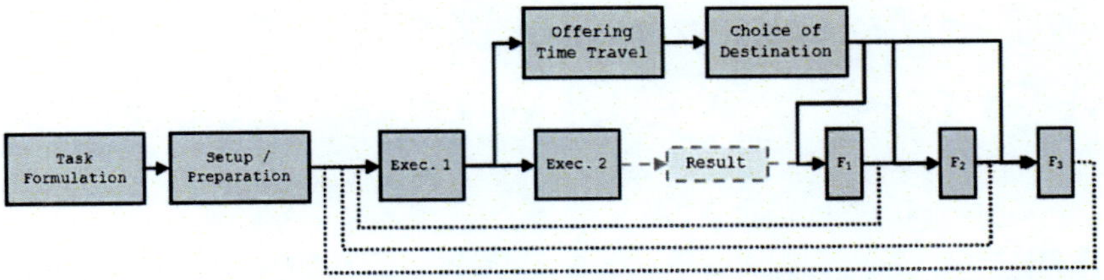

Fig. 7. Virtual forward time travel that leads to completely new phenomena visualized; dotted and dashed lines are not new features of storyboards; they shall only simplify references.

In a time travel prevention game, backward time travel is initiated by an accident or by any other undesirable event. Forward time travel is intended to provoke the human player's imagination. "Have you any idea of the implications of your doing? Let's have a look into the future." Something like that may be the form of offering a virtual time travel.

But to what point in time to travel forward ...? There are no objects familiar to the trainee that could be visualized in the time tunnel. How to name future events $\mathbf{F_i}$ ...?

Assume that a human trainee has been traveling forward in time to $\mathbf{F_1}$, $\mathbf{F_2}$, or $\mathbf{F_3}$. What is an appropriate point in time to return to afterwards (dotted lines in Fig. 7)? Definitely, this should not be the time point of departure, because at this point, it may be too late to correct earlier erroneous behavior. Is repeated time travel still useful?

There are other designs than the one on display in Fig. 7. However, a storyboard graph like the one above is an ideal tool for communication among different experts. It makes certain characteristics of the ideas under discussion explicit.

5 Time Travel Prevention Games at Work

This section is intended to deepen the discussion of the authors' methodology and technology by way of illustration. Several details may be supplemented en passant. Because of space limitations, not all storyboard graphs are presented in every detail. Sometimes, it is sufficient to show how graphs unfold, so to speak, at execution time.

Fig. 8. These 4 screenshots illustrate the authors' implementation developed at Fraunhofer IFF and deployed by industrial co-operation partners, as well as within the fourth author's BG RCI prevention center on varying hardware and software platforms incl. mobile 3D kiosk terminals.

The last screenshot of Fig. 8 in the lower right position shows a certain workplace. Subsequently, we will discuss a training session at this workplace dealing with the task of decanting a flammable liquid from a larger source container to a barrel.

5.1 Main Training Session Case Study

Assume a design of training, game play, and interaction as laid out in Fig. 6 before. During the execution of the 'Task Formulation' episode, trainees get detailed advice describing the task to be performed. The system performs the actions of this episode.

When the 'Task Formulation' episode is completed, the system initiates the next episode by a scene placing the trainee in a working environment such as the virtual locations displayed in Fig. 8. The game is experienced in a first person perspective, very much like a conventional first person shooter. The only difference lies in the trainee's weapons and, naturally, in the tasks to be performed.

Next, it is the trainee's turn to interact with the environment. Usually, the episode offers varying alternatives. A trainee selects one, then possibly another one and so on.

What is shown abstractly in Fig. 9 will be illustrated by means of the authors' game for training accident prevention in practice. Three subsequent figures are presenting subsequent screenshots of game play from a single training session.

The first screenshot in Fig. 10 shows the personal protective equipment. After dressing, the trainee has several opportunities. One is to turn on the de-aeration as on display

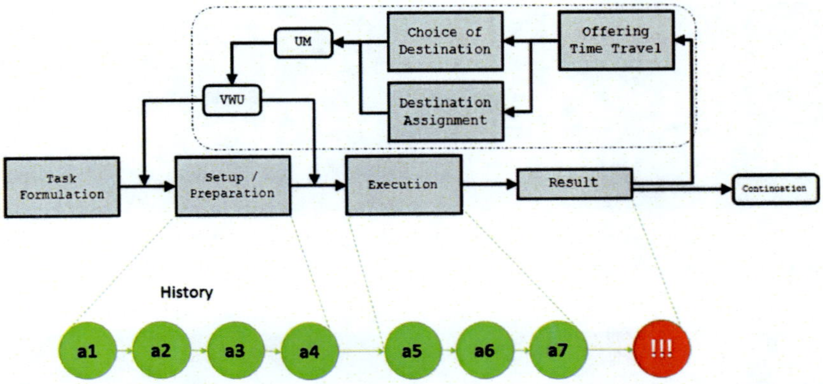

Fig. 9. A history of game play that unfolds at execution time; the actions abstractly presented are both actions of the system and the trainee's actions. The whole action sequence is recorded.

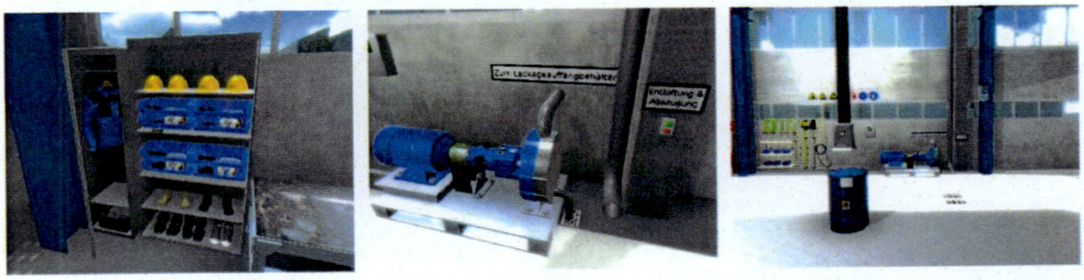

Fig. 10. Three screenshots from a training session show the unfolding experience of game play.

in the middle. In Fig. 10, the third screenshot illustrates that the trainee has provided the target barrel.

Notice that, in Sect. 5.2 below, we will discuss an extension that, if applicable, might change the unfolding game play history prior to the rightmost screenshot above. Recall that a digital storyboard being a hierarchically structured family of pin graphs, does not represent a single action sequence, but a large — even potentially infinite — space of experiences one of which unfolds at execution time.

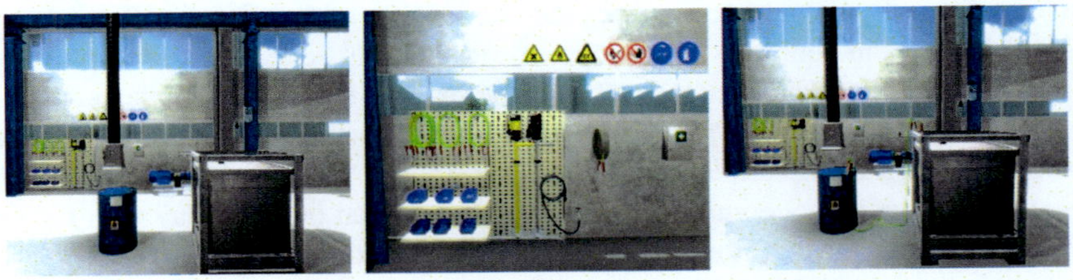

Fig. 11. Three screenshots continuing the survey of the human's unfolding training experience.

Fig. 12. The pump is inserted into the source container (left), the counter is installed and the pump nozzle is connected to the target container (middle), and the suction device is in place (right).

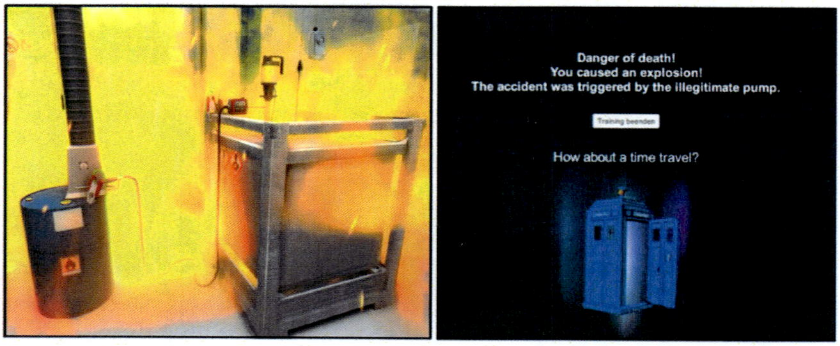

Fig. 13. The trainee's last action was to turn on the pump. According to the logical conditions of execution, the following action is not the one decanting the fluid from the source container to the target barrel, but the accident action executed by the system and denoted by !!! in Fig. 9.

On the left in Fig. 11, the source container has been supplied. In the middle, the trainee inspects the available equipment and, on the right, has attached the grounding cables.

All preparations are completed and the decanter installation is fully set up. Let's go! Unfortunately, as on display in Fig. 12, an explosion-proof pump was not selected.

In this case, the 'Result' scene is implemented by an animated accident visualized on the left in Fig. 13. The 'Result' scene has other semantics available for execution. But their conditions of execution are, unfortunately, not satisfied. From the two edges outgoing from the 'Result' scene, only the one leading to the 'Offering Time Travel' episode has a valid condition of deployment (in the right screenshot; for the readers' convenience, we have changed the text from German to English, except the button).

If the accident that just happened — virtually only, fortunately — took place for the first time in this training session, the following episode is 'Choice of Destination' where the authors' time tunnel comes into play. The trainee can literally play with the time tunnel and ponder the question how far to travel back in time.

Otherwise, adaptivity as discussed in Sect. 4.1 comes into play and leads to the 'Destination Assignment' episode, if the limit set by the designers is exceeded.

5.2 Case Study Extensions

Experience shows that members of interdisciplinary designer teams frequently come up with more and more additional ideas to enrich a training session. Domain experts may be interested in integrating further hazard sources.

Fig. 14. The workplace needs to be tidied up prior to performing the main task.

In the first screenshots of Fig. 14, the working trolley at the workplace carries a mobile phone and a small radio that both are hazard sources because of their batteries. Integrating such an extension brings with it several more problems to be negotiated within the team of designers. Does it play a role in how far away the trolley is moved; how to integrate distances like that into the edutainment system's bookkeeping; which of the possibly following actions including, if applicable, the accident action executed by the system have execution conditions that refer to this distance value; or what formulas are needed to process this value appropriately?

Usually, educators bring in different desirable extensions in comparison to what domain experts (see above) or AR and VR technicians suggest. One may expect more adaptivity to the human trainee's needs and desires, more guidance, and, possibly, more variety of experience. This may be accommodated by further storyboard graphs with appropriate execution conditions of their edges, as well as substitution conditions. Stakeholder ideas result in emerging storyboards and novel player experiences.

6 Summary and Conclusions

Although there are many more aspects to be reported, many more questions to be asked and a few of them to be answered, the authors want to stay within the limits and — with a metaphorical tear in the eye — come to an end.

6.1 Related Work

There is time travel galore in commercial video games; hundreds of them are listed online under https://en.wikipedia.org/wiki/List_of_games_containing_time_travel.

In contrast, time travel games for purposes of education are rare. An ERASMUS+ project named "Game Based Learning to Alleviate Early School Leaving" running from 2015 to 2017 educed a guidebook tool [9] describing, in some detail, 38 games that might be useful in elementary school classes. Just one single game on the list draws benefits from time travel ideas — the game "The Silent Age." This is an extremely simple point and click adventure. The player's time travel does not explicitly serve a particular learning purpose and does not reflect any explicit pattern of didactics. This digital game is considered useful for the development of writing skills, when students are encouraged to take notes and to write simple reports [9, pp. 97/98].

For more affective virtual time travel with the potential of effective learning and/or training results, there is a need of more sophistication including a dovetailing of both game design principles and didactics [24].

6.2 State of the Art

To the authors' very best knowledge, their *learning idea* — a word chosen to refer to the conference title — of *time travel prevention games* is an original one. The digital game designed and developed — see Sect. 5 for a survey within given limits — is the first time travel prevention game for *training of accident prevention in the industries* world-wide.

Several of the concepts and results underlying the authors' present approach have their particular strength. The graph-based approach to dynamic plan generation has turned out to be provably more expressive than all earlier alternatives [23]. Pattern concepts that take the form of pin graphs in a storyboard are appropriate to clearly formalize patterns of pedagogy that are likely to remain blurred otherwise [17]. Furthermore, studies of occurring instances of patterns are suggesting new avenues to the assessment of learning and training [16].

Intuitively, time travel prevention games in practical training are advantageous over conventional approaches because they prevent trainees from getting stuck at points of no return, help to avoid frustration, and motivate pondering the alternatives for doing it better the next time.

Despite this good baseline, systematic studies of training effectiveness are pending, among others due to reduced trainings during the Sars-Cov-2 pandemic.

6.3 A Glance in the Crystal Ball

Encouraged by the current success of their time travel prevention game approach to accident prevention, the authors are planning further developments and applications in areas such as *machine safety*.

Furthermore, there are negotiations ongoing about time travel prevention games for *health care* topics addressing school children and young adults in settings such as the so-called flying classroom [6].

Acknowledgments. The authors gratefully acknowledge an anonymous reviewer's fruitful suggestions. Particularly exciting was his challenge to have a closer look at the distinction between virtual time travel for a training purpose back to the past and forward to the future. Taking up this challenge led the authors to a few ideas reflected in Sect. 4.3 above.

References

1. Arnold, O.: Die Therapiesteuerungskomponente einer wissensbasierten Systemarchitektur für Aufgaben der Prozeßführung. Vol. 130 of DISKI, infix, St. Augustin, Germany (1996)
2. Arnold, S., Fujima, J., Jantke, K.P.: Storyboarding serious games for large-scale training applications. In: Foley, O., Restivo, M., Uhomoibhi, J., Helfert, M. (eds.) Proceedings of the 5th Intl. Conference on Computer Supported Education, pp. 651–655. SciTePress, May 2013, Aachen, Germany (2013)
3. Arnold, O., Jantke, K.P.: Therapy plan generation in complex dynamic environments. ICSI Report TR-94-054. International Computer Science Institute, Berkeley (1994)
4. Arnold, O., Jantke, K.P.: Therapy plans as hierarchically structured graphs. In: Fifth Intl. Workshop on Graph Grammars and their Application to Computer Science. Williamsburg, VA, USA, Nov 1994 (1994)
5. Arnold, O., Jantke, K.P.: Planning is learning. In: Dilger, W., et al. (eds.) Machine Learning, 1996 Annual Meeting of the GI Special Interest Group, Chemnitzer Informatik-Berichte CSR-96-06, pp. 12–17. TU Chemnitz (1996)
6. Arnold, O., Jantke, K.P.: AI planning for unique learning experiences: the time travel exploratory games approach. In: Csapó, B., Uhomoibhi, J. (eds.) Proc. of the 13th Intl. Conference on Computer Supported Education, CSEDU, vol. 1, pp. 124–132. SciTePress (2021)
7. Arnold, O., Jantke, K.P., Spundflasch, S.: Hierarchies of pervasive games by storyboarding. In: Proc. 5th Intl. Games Innovation Conference, IGIC 2013, Vancouver, Canada, Sept. 2013, pp. 8–15. IEEE Consumer Electronics Society (2013)
8. Björk, S., Holopainen, J.: Patterns in Game Design. Charles River Media, Hingham (2005)
9. Camilleri, V., Dingli, A., Montebello, M.: Gamify your classroom: a guidebook tool for an alternative pedagogical approach to class-based teaching. ERASMUS+ project document. https://toolkit-gbl.com/downloads/guidebook (2017). Last Accessed 13 April 2021
10. Dai, D.Y., Wind, A.P.: Computer games and opportunity to learn: Implications for teaching students from low socioeconomic background. In: Tobias, S., Fletcher, J.D. (eds.) Computer Games and Instruction, pp. 477–500. Information Age Publ (2011)
11. Fujima, J., Jantke, K.P., Arnold, S.: Digital game playing as storyboard interpretation. In: Proceedings of the 5th Intl. Games Innovation Conference, Sept. 2013, Vancouver, BC, Canada, pp. 64–71. IEEE Consumer Electronics Society (2013)
12. Gee, J.P.: Reflection on empirical evidence in games and learning. In: Tobias, S., Fletcher, J.D. (eds.) Computer Games and Instruction, pp. 223–232. Information Age Publ (2011)
13. Hirumi, A., Stapleton, C.: Applying pedagogy during game development to enhance game-based learning. In: Miller, C.T. (ed.) Games: Purpose and Potential in Education, pp. 127–162. Springer Verlag (2008)
14. Höfting, F., Lengauer, T., Wanke, E.: Processing of hierarchically defined graphs and graph families. In: Monien, B., Ottmann, T. (eds.) Data structures and efficient algorithms. LNCS, vol. 594, pp. 44–69. Springer, Heidelberg (1992). https://doi.org/10.1007/3-540-55488-2_21
15. Jantke, K.P.: Layered languages of ludology: the core approach. Diskussionsbeiträge 25, Inst. Medien- und Kommunikationswissenschaften, TU Ilmenau (2006)
16. Jantke, K.P.: Patterns of game playing behavior as indicators of mastery. In: Ifenthaler, D., Eseryel, D., Gee, X. (eds.) Assessment in Game-Based Learning: Foundations, Innovations, and Perspectives, pp. 85–103. Springer Verlag, New York, Heidelberg Dordrecht, London (2012)
17. Jantke, K.P.: Pedagogical patterns and didactic memes for memetic design by educational storyboarding. In: Arnold, O., Spickermann, W., Spyratos, N., Tanaka, Y. (eds.) WWS 2013. CCIS, vol. 372, pp. 143–154. Springer, Heidelberg (2013). https://doi.org/10.1007/978-3-642-38836-1_12

18. Jantke, K.P.: https://www.praeventionstag.de/nano.cms/vortraege/begriff/Time-Travel-Prevention-Games?sb=Time+Travel+Prevention+Games (2015). Last Accessed 15 Mar 2021
19. Jantke, K.P., Gaudl, S.: Taxonomic contributions to digital games science. In: Bradbeer, R. Ahmadi, S. (eds.) Proc. 2nd Intl. Games Innovation Conference, IGIC 2010, pp. 27–34. IEEE Consumer Electronics Society (2010)
20. Jantke, K.P., Knauf, R.: Didactic design through storyboarding: standard concepts for standard tools. In: Baltes, B.R., et al. (eds.) First Intl. Workshop on E-Learning Technologies and Applications, Jan. 2005, Cape Town, South Africa, pp. 20–25. Computer Science Press, Trinity College, Dublin, Ireland (2005)
21. Jantke, K.P., Spundflasch, S.: Storyboarding pervasive learning games. In: Tan, D. (ed.), Proc. Intl. Conf. Advanced Information and Communication Technology for Education, ICAICTE 2013, Hainan, China, Sept. 2013, pp. 42–53. Atlantis Press (2013)
22. Kaschek, R.H.: Intelligent Assistant Systems: Concepts. Techniques and Technologies. Idea Group Publ, Hershey, London, Melbourne, Singapore (2007)
23. Kirsten, D.: Properties of formal languages of therapy plans created by graph grammars. In: Kutrib, M., Worsch, T. (eds.) 5. Theorietag der Fachgruppe 0.1.5 der Gesellschaft für Informatik, Schloss Rauischholzhausen, Sep. 1995, pp. 132–142, Bericht 9503, Dec. 1995. Justus-Liebig-Univ. Giessen, Germany (1995)
24. Krebs, J., Jantke, K.P.: Methods and technologies for wrapping educational theory into serious games. In: Zvacek, S., Restivo, M.T., Uhomoibhi, J., Helfert, M. (eds.) Proc. 6th Intl. Conf. Computer Supported Education, CSEDU 2014, Barcelona, Spain, May 2014, pp. 497–502. SciTePress (2014)
25. Kreuzberger, G., Lunzer, A., Kaschek, R.H.: Interdisciplinary Advances in Adaptive and Intelligent Systems: Concepts, Techniques, Applications, and Use. IGI Global (2011)
26. Lenerz, C.: Layered languages of ludology. In: Beyer, A., Kreuzberger, G. (eds.) Digitale Spiele – Herausforderung und Chance, pp. 39–52. VWH, Boitzenburg, Germany (2009)
27. Montola, M., Stenros, J., Wærn, A.: Pervasive Games: Theory and Design. Morgan Kaufmann Publ./Elsevier (2009)

Speech Analysis for Advanced Medical Simulation

Barbara Bertagni[1,2,3](✉), Fernando Salvetti[1,2,3], Roxane Gardner[4,5], and Rebecca Minehart[6,7]

[1] Centro Studi Logos, 10143 Turin, TO, Italy
{bertagni,salvetti}@logosnet.org
[2] Logosnet, 6900 Lugan, CH, Switzerland
[3] Logosnet, Houston, TX 77008, USA
[4] Center for Medical Simulation, Boston, MA 02129, USA
[5] Brigham and Women's Hospital/Children's Hospital/Massachusetts General Hospital and Harvard Medical School, Boston, MA 02114, USA
rgardner1@bwh.harvard.edu
[6] Massachusetts General Hospital and Harvard Medical School, Boston, MA 02114, USA
rminehart@mgh.harvard.edu
[7] Center for Medical Simulation, Boston, MA 02129, USA

Abstract. The e-REAL voice and speech analysis app helps medical simulation instructors to track learners individually with a focus on multiple factors such as tone of voice and spoken words, volume, rhythm, and other relevant aspects that are essential to provide a semantic and pragmatic overview of the interpersonal communication that is happening within a simulation setting. Functions and visual outputs include the following: 1. An integral transcript or a dialogue which can be visualized — audio clips, automatically divided phrase by phrase, are also available. 2. A word counter shows the number of spoken words per minute. 3. An internal search engine enables keyword search, highlighting the words in the transcript. 4. A word cloud tool visually summarizes the most spoken words. 5. A Voice Analysis tool is available in order to measure and visualize waveform (Decibel), perceived loudness (Hertz), and pitch.

Keywords: Speech analysis · Verbal communication · Medical simulation

1 An Innovative Voice and Speech Analysis App for Medical Simulation

1.1 The e-REAL Voice and Speech Analysis App

The e-REAL® voice and speech analysis app has been designed and developed by Centro Studi Logos (Turin, Italy) in collaboration with Logosnet (Lugan, Switzerland and Houston, Texas), the Polytechnic School of Turin (Italy), the Italian National Research Council (CNR, ISTI Dept., Pisa), the Tiny Bull Studios (Turin, Italy), and the Center for Medical Simulation (Boston, Massachusetts). It helps simulation instructors to track the

D. Guralnick et al. (Eds.): TLIC 2021, LNNS 349, pp. 19–27, 2022.
https://doi.org/10.1007/978-3-030-90677-1_2

learners individually with a focus on multiple factors such as tone of voice and spoken words, volume, rhythm, and other relevant aspects that are essential to provide a semantic and pragmatic overview of the interpersonal communication that is happening within a simulation setting.

The use of medical simulation is growing exponentially in academic and service settings. Simulation can enable both students and new graduates, as well as experienced professionals, to develop clinical competence and confidence in caring for patients in a learning environment that is cognitively and emotionally realistic [1]. Simulation is a technique that creates a situation or environment to allow persons to experience a representation of a real event for the purpose of practice, learning, evaluating, testing, or gaining an understanding of systems and human actions [2]. Usually, simulation replaces or amplifies real experiences with guided experiences, often immersive in nature, that evoke or replicate substantial aspects of the real world in a fully interactive fashion [3].

Reflective practice is an important strategy for learning through practice experiences and is triggered by a debriefing phase, which purpose is to assimilate new learning with the intent to transfer it to future clinical situations. During the debriefing phase, reflective thinking is encouraged, and feedback is given on the simulation performance. Participants are encouraged to question, explore emotions, and give feedback to each other [4], within a learning setting that is cognitively and emotionally demanding.

Despite the availability of some powerful audio-visual recording platforms, the existing solutions are not allowing for individual recording, speech-to-text transcription, and communicative analysis that are the specific features of the e-REAL speech analysis system. It uses a variety of techniques from communicative analysis, which not only focus on spoken words, but consider other features such as intonation, pitch, tempo, and nonverbal cues.

The e-REAL speech analysis functions include the following:

- An integral transcript
- Audio clips, automatically divided phrase by phrase and speaker by speaker
- A word counter that shows the number of spoken words per minute
- An internal search engine that enables keyword search, highlighting the words in the transcript
- A word cloud tool that summarizes the most spoken words
- A voice analysis tool that measures and visualizes waveform (Decibel), perceived loudness (Hertz), and pitch

Some of these features are visible in Fig. 1.

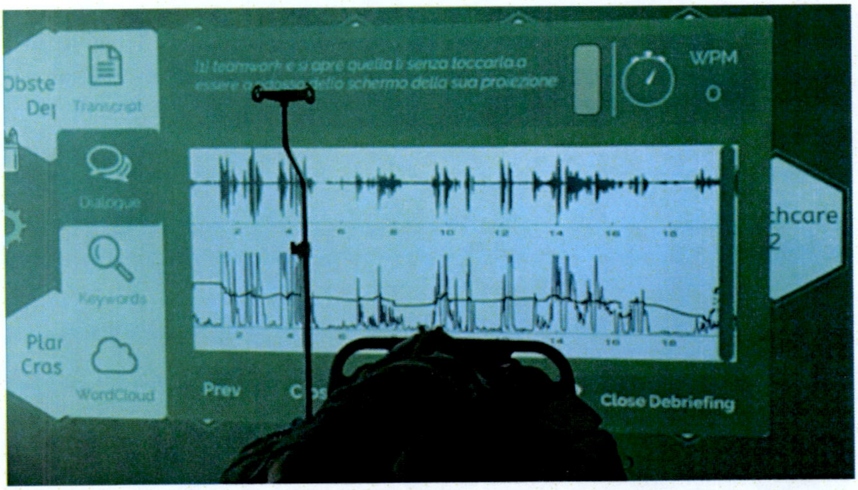

Fig. 1. The main features of the e-REAL® Voice and Speech Analysis App: speech to text transcript, audio clips automatically divided phrase by phrase and speaker by speaker, word counter displaying the number of spoken words per minute, search engine, word cloud to summarize the most spoken words, and voice analysis that measures and visualizes waveform (Decibel), perceived loudness (Hertz), and pitch.

A video introduction is available via the following URL:
https://youtu.be/3-hOdSYOmwg.

1.2 Debriefing and Communication Analysis

An effective debriefing should offer the opportunity to analyze interpersonal communication. Unfortunately, the existing audio-visual recording platforms currently used are not designed to do so. As a consequence, instructors explore with learners, then discuss and summarize the factors related to interpersonal communication with a very "subjective touch." Furthermore, a number of observations are difficult to be related to communicative and behavioral evidence, mainly because evidence is from subjective observation and becomes highly disputable.

So far, in medical simulation, we are used to analyzing these aspects with our competence and our sensibility, but the risk of misinterpretation is high, and sometimes we feel like human attention is not enough to intercept all the complexity of a communication.

To solve this issue, we began thinking: what about having the possibility to record and analyze in real time both the verbal and non-verbal aspect of speech? So, three years ago, we began working on the speech analysis tool, and today we have a reliable instrument that helps us during debriefing with an objective tracking of the speech and voice.

2 The Main Features of the Voice and Speech Analysis App

2.1 Verbal Communication Analysis

In Fig. 2, we see a representative report that is usually delivered as a final output of a session.

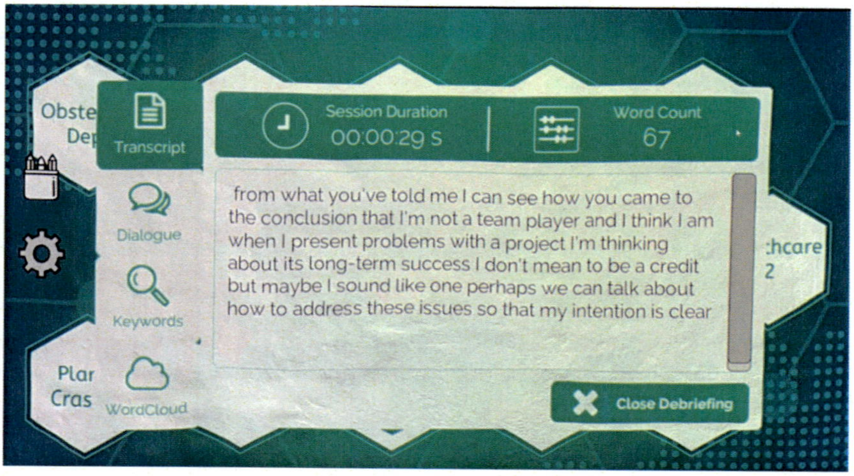

Fig. 2. Representative output from a simulation session.

On the top, we have the indication of the session duration and the total number of spoken words.

On the left side, we have four different commands:

- *Transcript* that shows the entire verbal communication (Fig. 3).

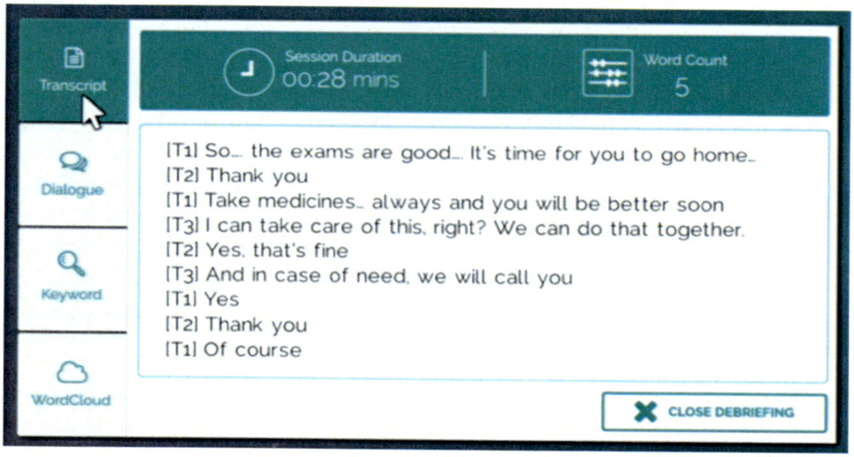

Fig. 3. Representative dialogue.

- *Dialogue* that shows in a dialogic way the spoken words.
- *Word counting and keywords spotting* that shows how many times a specific word has been pronounced during the simulation.
- *Word cloud* that shows the words used more often, providing a visual representation of the main topics shared (Fig. 4).

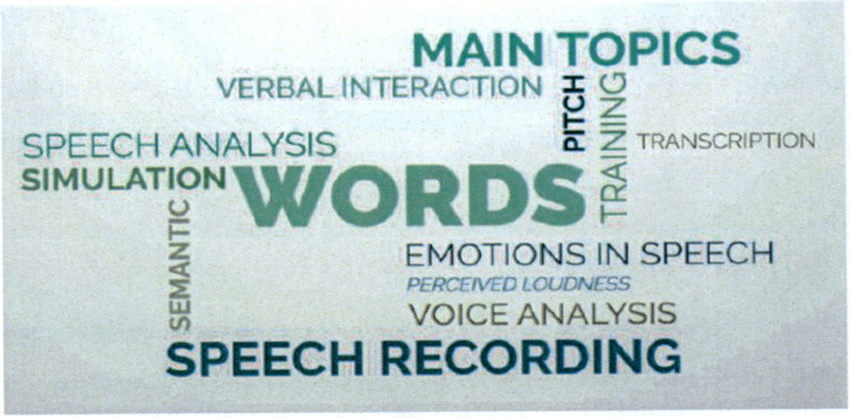

Fig. 4. Representative word cloud.

2.2 Non-verbal Communication and Emotion Analysis

In Fig. 5, the graphic visualizations show key non-verbal aspects regarding the communicative analysis.

The visualization of the non-verbal aspects of the communication is made phrase by phrase. Starting from the top, there are the Speech Waveform, then a blue line that represents the Perceived Loudness and, in black, the pitch.

Each speaker has his speaking style that gives a typical graphic representation. What is interesting is analyzing perturbations of this style during the speech. This is the field of *emotion analysis* that is about vocal behavior as a marker of affect (e.g., emotions, moods, and stress), with a focus on the non-verbal aspects of speech.

The basic assumption of the speech analysis app is that there is a set of objectively measurable voice parameters that reflect the affective state a person is currently experiencing. This assumption appears reasonable, given that most affective states involve physiological reactions that modify different aspects of the voice production process.

The speech expresses emotional meaning not only through semantics, but also by varying vocal attributes of the voice such as pitch, loudness, vocal tone, and average speaking rate (expressed in words per minute: the number on the upper right side of the screen).

As such, being able to analyze the specific words and voice-related KPIs during a simulation could be strategic to improve communicative skills. The speech analysis

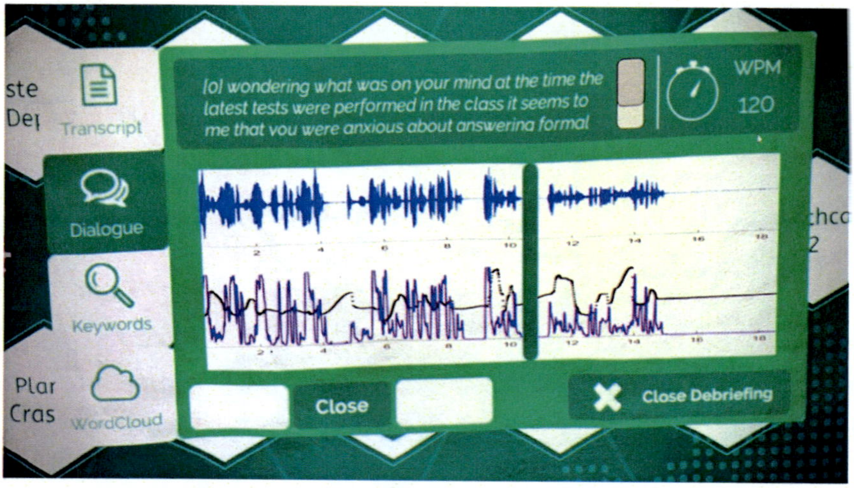

Fig. 5. Representative diagrams related to non-verbal communication.

tool aims to help simulationists on this task and also opens a window on the emotional tension, analyzing not only the spoken words, but also the physical properties of the recorded voices such as pitch, tone, and speed.

3 Emotional Intelligence and Teamwork Effectiveness as Debriefing's Key Element

3.1 Emotion Analysis and Interpersonal Communication

According to the US National Center for Voice and Speech, the average conversation rate for English speakers in the United States is about 150 words per minute, but several factors could affect the overall speaking rate, such as mental fatigue and emotional tension [5].

Hence, deduction of human emotions through voice and speech analysis has a practical plausibility and could potentially be beneficial for improving human conversational skills. With different emotions and moods, not only does the tonal quality vary, but the associated speech patterns change, too. For instance, people may tend to talk in loud voices when angry and use high-pitched voices when in a scared or panicked emotional state.

By the speech analysis app, vocal behavior can be analyzed as a marker of emotions and stress. There is increasing interest in the role that emotional intelligence has in medical training, but it is still a nascent field. Although there are many unanswered questions, an effective debriefing may offer many benefits to the development of emotional intelligence.

Emotional intelligence can be defined as an individual's alertness on emotions, together with an alertness of the emotions in other people and the capacity to manage them and act appropriately. We can say that it is the processing of emotional information and that it reflects the ability to 1. monitor one's own and others' emotions,

2. discriminate among them, and 3. use this information to guide one's thinking and actions [6]. Higher emotional intelligence contributes to interpersonal communication and relationships, increased empathy, stress management, teamworking, and leadership, among other concepts and relationships. For example, talking to each other is crucial for utilizing the collective expertise of a team.

According to the Institute of Medicine, approximately 98,000 Americans die annually due to health care errors. One of the most cited factors that lead to medical errors is communication: for example, nearly 45% of all operative adverse events stem from poor communication [7].

Simulation can be particularly effective for empowering emotional intelligence and for training non-technical teamworking skills, such as communication, decision-making, leadership, task management, and monitoring — all skills which are known to be critical to effective teamwork and vital to the prevention of errors and adverse events in healthcare. In particular, speech analysis reveals naturally occurring communication strategies that trigger actions relevant for safe practice and, thus, provides supplemental insights into what comprises "good" team communication, mainly in medical emergencies.

3.2 Teamwork and Interpersonal Communication

To optimize treatment and to coordinate team activities, communication amongst team members is crucial for utilizing the collective expertise during team interactions. Communication skills are highly emphasized both in emergency team training and in the assessment of team performance. Recommendations for standardized communication, including closed loop communication, have been obtained mainly from work in the defense and aviation communities. However, the extent to which such communication strategies are implemented in medical practice remains unclear. An additional concern is that the functions of medical emergency team talk — that is, the relationship between what is said and what is done — have remained more or less unexplored [8].

Transcription is a decisive element in communicative analysis, as well as the other features highlighted above regarding both verbal and non-verbal communication. Transcription, in particular, offers a way to translate the content and structure of an interaction into a written format that helps the simulation instructors notice details that are not readily apparent through observation. Transcription is, thus, an important tool for capturing interactional dynamics and for identifying patterns. Word cloud and counting, as well as keywords spotting, jointly with speech waveform and perceived loudness, are other essential tools to understanding more about the interconnections between team talk and actions.

We developed the e-REAL voice and speech analysis tool because to be effective, the debriefing phase has to be based on evidence from analyzing the interconnections between interpersonal communications, emotions, and actions — and not on subjective feelings.

4 Current State of Use and Future Development

So far, the Speech Analysis App is being used mainly for healthcare simulation into the field of crisis resource management. Enhancing awareness about how to use verbal communication allows for an improved performance. Moreover, it is very useful for the learners to understand when their behaviors were mainly impacted (sometimes affected) by the emotions.

The appreciation for this tool is very high: on a 0–10 Likert scale, the average output is 9 and the principal reason is that the speech analysis app allows for a comparison among the different communication's patterns and styles performed during the simulation.

By September 2021, a dry run will be performed online, into the e-REAL virtual platform: learners are expected to be provided with an analytical report about verbal and paraverbal key-aspects of their communication. Because of the high volume of remote working, a focus not only on the verbal but also on the paraverbal dimensions of the communicative performance is of the highest relevance.

By November 2021, the Speech Analysis Tool will be tested also for the debriefing training in a very peculiar way, with the learners enabled to perform in self-training with one or more digital humans able to be in a direct dialogue. As a main output, the learner is expected to be delivered with an automated feedback about: 1. The ability to perform the 3 phases of a debriefing as stated by the Center for Medical Simulation model (Reactions, Understanding and Summary, as shown below in Fig. 6); 2. key-questions asked during the debriefing; and 3. number of spoken words (in general and per minute) by both the learner and the digital humans.

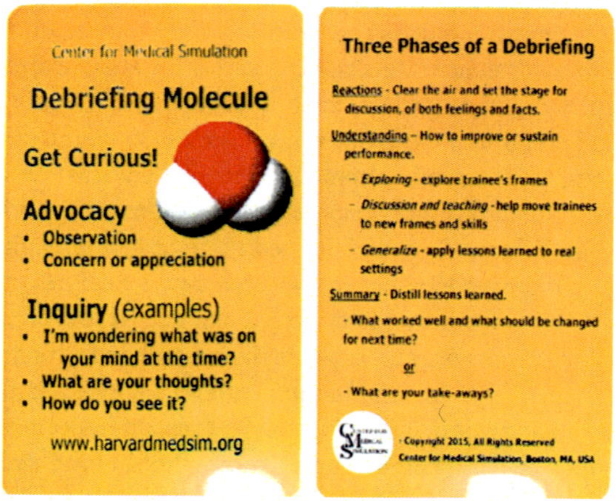

Fig. 6. Debriefing structure: Courtesy of the Center for Medical Simulation (all rights reserved), Boston, MA, USA.

References

1. Palaganas, J., Ulrich, B., Mancini, B. (eds.) Mastering Simulation. A Handbook for Success. 2nd Edition. Sigma Theta Tau, Indianapolis (2020)
2. Healthcare Simulation Dictionary: AHRQ (2016)
3. Gaba, D.: The future vision of simulation in healthcare. Quality and Safety in Healthcare. **13**(Suppl 1)

4. Reed, S. Debriefing Experience Scale. Clin. Simul. Nurs. **8**(6) (2012)
5. www.ncvs.org
6. McKinley, S., Phitayakorn, R.: Emotional intelligence and simulation. Surg. Clin. North Am. **95**(4), 855–867 (2015)
7. Lambert, J., Vanderbilt, A., Papadimos, T.: Improved emotional intelligence in perioperative care through simulation-based medical education during anesthesiology residency training: a call for implementation. Adv. Med. Educ. Pract. **10**, 39–42 (2019)
8. Gundrosen, S., Andenæs, E., Aadahl, P., Thomassen, G.: Team talk and team activity in simulated medical emergencies: a discourse analytical approach. Scand. J. Trauma Resusc. Emerg. Med. **24**, 135 (2016)

Teaching Data Science in a Synchronous Online Introductory Course at a Business School – A Case Study

Marcus Birkenkrahe(✉)

Lyon College, 2300 Highland Rd, Batesville, AR 72501, US
birkenkrahe@lyon.edu

Abstract. Data literacy can be improved by teaching data science methods and tools. Both method-kit and tool-kit are eclectic and interdisciplinary, posing potential threats for students' understanding and motivation, especially at a business school where the focus is not on IT. We found that a broad spectrum of methods and tools can be helpful, rather than a hindrance, when teaching data science to absolute beginners. This case study investigates how teaching and learning workflows can be improved when systematically deploying a wide array of digital tools in an introductory online data science class at a large German business school, taught in the winter term of 2020, during the COVID-19 pandemic lockdown. The class was offered to all students of the school as part of the extracurricular program. Several such tools were selected, introduced, and used, with a special emphasis on tools to enhance teaching and learning productivity in and between live sessions. Tools included: the GitHub global code repository, the Emacs editor with ESS and Org-Mode markdown, Kahoot! Games, screencasting, video tutorials and podcasts, external assignment tools like DataCamp and Celonis, and a Moodle platform to pull all of these together. Participation and explicit student feedback were very positive and have already initiated curricular changes at the school. In this paper, the course rationale, goals, content, tools, and preliminary findings are presented and discussed, and an outlook for further research is given.

Keywords: Data science · Tools · Business school

1 Introduction

Data science is an interdisciplinary field of applied study that investigates the use of algorithms to extract insights from structured and unstructured data. It is considered essential for developing digital business and for transforming traditional business into digitally enabled enterprises [1]. Due to the nature of typical data science problems and the tools required to tackle these problems, the skills for successful data science work are highly interdisciplinary, technical, and dependent on domain knowledge [2]. The corresponding skill stack is therefore challenging for students, especially beginners. This also means that additional tools to enhance learning performance and productivity could be important to a successful data science learning experience. Due to COVID and

D. Guralnick et al. (Eds.): TLIC 2021, LNNS 349, pp. 28–39, 2022.
https://doi.org/10.1007/978-3-030-90677-1_3

lockdown policies, teaching and learning conditions were particularly challenging. For this case study, I focus on a new introductory data science class in the extracurricular program of a large German business school. The course was taught synchronously online between October 2020 and February 2021.

The methodology is a descriptive single case study [3]. The main data for the case study consist of teaching materials and activities spread out over several online platforms, learning analytics data obtained within the Moodle learning management system (e.g., completion progress, access protocols), the web conferencing system Big Blue Button (BBB; e.g., saved chat conversations and live session recordings), and anonymous feedback solicited from the students throughout. All data are accessible upon request. The case study describes the phenomenon in question in a comprehensive manner in order to generate a compelling, dense narrative [4]. Methodologically, this seems justified, especially because we are unlikely to meet the same, or even a similar, set of teaching conditions as at the end of 2020 again any time soon. This investigation can also be classified as systemic action research in participant-observer mode [5], with the author acting as designer, lecturer, and investigator of a course.

2 Course Rationale

Data science courses are by now common electives in Master programs at business schools. However, outside of dedicated programs, the subject is often seen as too difficult, both technically and conceptually, to be an undergraduate offering. Data science skills are a potpourri of different skills rarely gathered into one classroom course. Courses in Extracurricular Studies (in German: "Studium Generale") at the Berlin School of Economics and Law are open to students at any level from all departments. Therefore, a course taught in this program can serve as a piloting platform before introducing it into the standard curriculum. However, "hard science" courses that involve maths or coding are rare in this program because the courses are optional and do not carry many credits. Those students who attend them are there because they are really interested in a subject that is not offered anywhere else. Arguably, "data science," even at an introductory level, is such a course: it is both "hard" (involves maths and coding), and it is of great interest to many students, because many jobs in the market for business administration alumni require data analytics skills and a higher than average data literacy [6]. An additional reason for offering such a course was provided by the global pandemic, which led to an increased interest in predictive methods, meaningful visualizations, and honest communication [7].

3 Course Goals

The target audience of an extracurricular course can be expected to be non-uniform (which indeed it turned out to be). Therefore, the course was announced with the subtitle "Introduction to data science: grow your data skills stack," and the course description emphasized that this course was meant for absolute beginners, with "no prior knowledge required." The only requirements were technical: an Internet connection to participate in live sessions, and a computer with an operating system capable of running the statistical programming language R [8].

The explicit course objectives were to:

1. Organize data visually
2. Find and use real-world data
3. Easily format data into graphs
4. Understand statistical data
5. See modern productivity tools in action
6. Complete an Exploratory Data Analysis (EDA) project.

These objectives were chosen because of my experience in class and during undergraduate and graduate research and thesis projects: many business students were blissfully unaware of the existence and the possibilities of state-of-the-art, open source (i.e., freely available) tools. As a result, their presentations and theses were often bland, boring, and imprecise. The tools I was offering could dramatically improve students' ability to express their thoughts, get and summarize available data, and use data to "tell a story" [9]. The secondary agenda of the course was linked to this observation. I wanted to improve students' computational and critical thinking skills. The course content should have high practical value for them and be immediately transferable to their other work. These goals are important but a little lofty, too; they are hard to measure and verify, but there is agreement in the data science literature and in industry-related publications that these goals constitute data literacy: "*the ability to consume for knowledge, produce coherently, and think critically about data.*" [10].

4 Course Content

There were four different topical parts of the course: (1) preparation, (2) programming, (3) visualization, and (4) use cases/productivity tools. Table 1 gives an overview of the content in each phase and the number of live sessions needed.

Table 1. Course phases with content and number of sessions per phase.

Phase	Content	Sessions
Preparation	Course organization, an overview of data science, and installation and first steps with R	2
Programming	Arithmetic with R and basic data structures to learn EDA (vectors, factors, data frames)	4
Visualization	Basic plotting with R and plotting with ggplot2	3
Productivity	Literate Programming and R notebooks, and process mining techniques	2

As mentioned above, **R** is a popular statistical programming language well suited for data science beginners. ggplot2 is a popular graphics R graphics package enabling users to quickly generate complex plots using the "Grammar of Graphics" (GG) framework [11]. This framework is used to map tabular data to visual aesthetic categories (e.g., dot and line types, plot types, colors).

Literate Programming (LP) is a programming paradigm created by Knuth [12], the creator of the TeX and Metafont typesetting systems. A program written in this paradigm favors documentation (to be read by humans) and also contains code (to be executed by machines). LP is also the main idea behind EDA notebooks, which are widely used in data science to collaborate on coding projects [13].

Process Mining (PM) is a method to analyse processes using event logs, simple data sets consisting of a time stamp, an activity, and an ID. PM is used for process discovery, optimization, and to check process conformance against existing models. It is an innovative technology that is quickly becoming operationally very important [14].

5 Course Process and Tools

This section provides a short account of the course processes and of the tools used. All of the listed materials and activities were made accessible via the Moodle learning platform.

Live Sessions. Optional, weekly live sessions, 90 min long, took place in BBB. The sessions began with a ritual, the "simultaneous sip": all session participants would drink one sip at the same time. The bulk of the time was spent on lectures and practice. There was always space for discussion. These sessions were kept interactive and lively through frequent polls, group exercises, collaborative Kanban boards, or small challenges. The interaction took place mostly through the chat, rarely via audio/video, though student webcams were allowed throughout. During most sessions, only my webcam was open. The sessions were attended by more than half of the participants on average, even though they were optional. After the session, a session recording was available for viewing. My presentations for the live sessions were usually created on the fly from Org-mode files in Emacs.

Tests. In between sessions, students could play an optional weekly Kahoot quiz to test their understanding of the course content, and to remind themselves of the last session. These quizzes could be played anonymously without time restriction. I offered detailed feedback via video after the quiz deadline.

Assignments. Every week, in lieu of attendance, participants had to complete one assignment at DataCamp, the well-known data science course provider. The assignment consisted in one chapter from one of DataCamp's more than 300 courses. These were available to us as part of an academic alliance deal, free of charge. Completing 50% or more of these assignments was equivalent to a passing grade. As a nice bonus, completing all assignments would give the students three DataCamp certificates.

Lectures. Lecture scripts were made available as a GitHub repository. Each lecture was pre-recorded and served in the form of a playlist consisting of an introductory video followed by screencasts. These lectures contained a wealth of additional material such as links to literature and additional exercises with solutions. The GitHub site also included a wiki where I covered off-topic issues irregularly, and a file with over 1,000 curated commented data science bookmarks.

Messages: Regular communication included one message per week at a minimum sent as a Moodle message, and reaching the students via email (19 messages were sent this way in total). An FAQ forum and file were available, partly so that I would not be overwhelmed with private emails.

Podcasts. A weekly podcast was featured in Moodle. Each podcast related to the topic covered in class that week and offered an opportunity to link the students up with practitioners of data science in industry and academia.

Further Reading. The Moodle platform featured feeds from several prominent data science blogs and discussion sites, like R-Bloggers, TowardsDataScience.com, Analytics Vidhya, and Reddit. Though technically often too demanding for beginners, these sites gave the participants something to aim at as well as many different ideas for their own projects and applications.

Table 2 lists all tools used in the course together with a short explanation (in alphabetical order). I have also indicated if an application is part of our school's infrastructure (HWR), and if it is part of an academic alliance program (aa).

Table 2. Tools and applications used in the course.

Tool/application	Description	Free
Big Blue Button	Web conferencing software	Yes (HWR)
DataCamp	Data science online course provider	No (aa)
Celonis	Process mining software	No (aa)
Emacs + ESS + Org-mode	Extensible text editor	Yes
Github (scripts, wiki)	Code & document repository	Yes
Ideaboardz	Cloud-based Kanban board	Yes
Kahoot	Cloud-based quiz platform	No
Moodle (email, forum)	Learning management system	Yes (HWR)
RStudio-cloud	R programming environment	Yes
Videos	Screencasts/presentation software	Yes/No
YouTube/Vimp	Lecture video hosting platform	Yes (HWR)
R	Statistical computing language	Yes

6 Summary Statistics

On the Moodle, Kahoot, and DataCamp platforms, personalized data were collected. I am presenting only those data here that seem to be meaningful in the chosen context.

Overall Stats. In Moodle, there were 100 materials and activities available for viewing. On average, only 9% of these online resources were viewed by the students.[1] The course had a total of 85 participants, of which 65 were enrolled for credit (and passed), while the remaining 20 were acquired during the term, as the news about the course spread in the school. Of the participants, 55% were female, 45% were male. Furthermore, 66% were enrolled in Bachelor programs and 34% in Master programs. Of the Bachelor students, 53% were undergraduates in their first year of study. In total, 34% of the remaining graduate students were business informatics majors.

Kahoot Tests. Before the course began, participants were asked to complete an entry quiz of 20 questions that covered most of the course topics. They answered 41% of the questions correctly. A total of 159 players registered. I created 13 Kahoot quizzes with 208 questions altogether, with the number of questions per quiz varying between 10

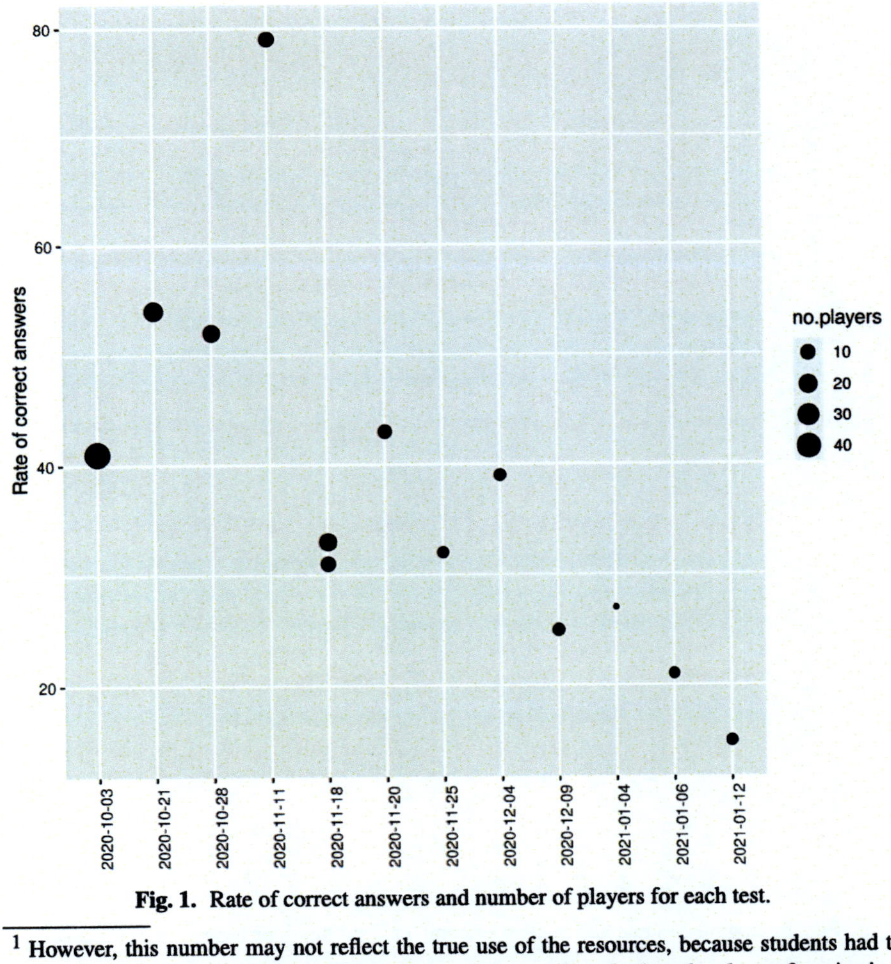

Fig. 1. Rate of correct answers and number of players for each test.

[1] However, this number may not reflect the true use of the resources, because students had to check off most of these activities themselves and may not have bothered to do so after viewing.

and 25. On average, the students who participated answered 37.85% of the questions correctly. Figure 1 shows the rate of correct quiz answers over time (for each test), alongside the number of students who played a quiz (indicated by point size). The outlier quiz (Nov. 11) was concerned with installing the R language and customizing the working environment.

DataCamp Assignments. There were 16 assignments in total, for every week of the term except during the Christmas holidays. All assignments were of comparable length, consisting of 2–4 videos and exercises in between. I estimated the time for completion of one assignment to be 30 min, or 8 practice hours in total during the term. The assignments were taken from four different DataCamp courses corresponding to the four phases of the course. Two of these courses did not involve coding ("Data science for everyone", "Visualization for everyone"), while the other two developed R coding skills

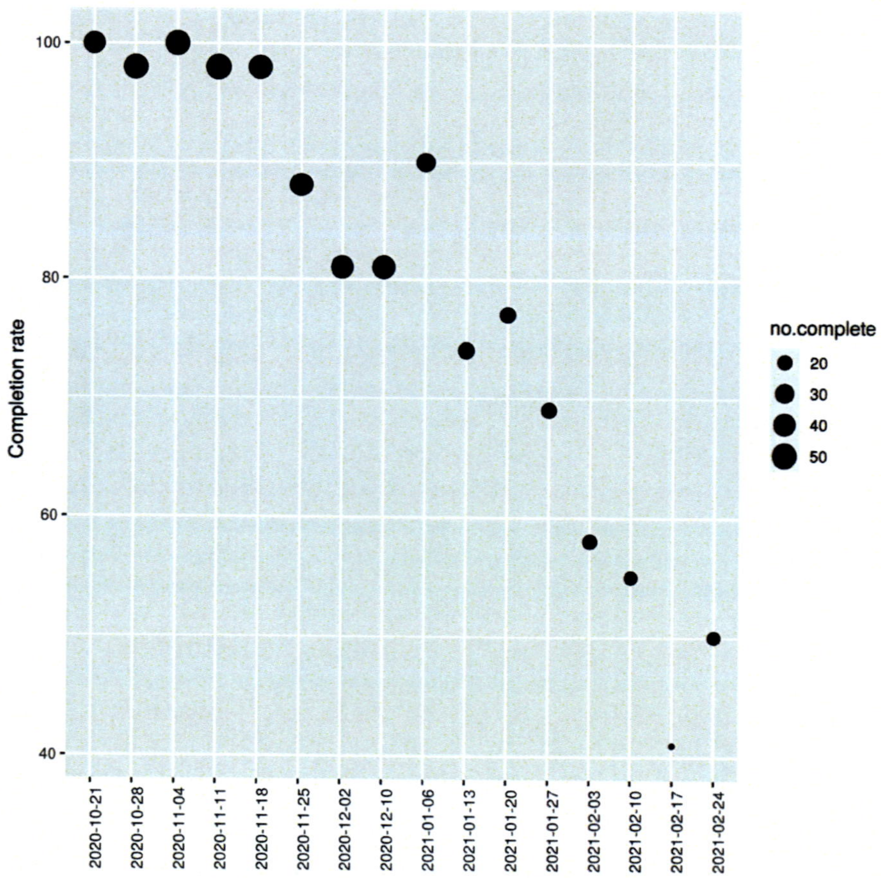

Fig. 2. Completion rate and number of completed submissions for each assignment.

("Introduction to R", "Introduction to the Tidyverse"). Figure 2 shows the rate of completed assignments over time (for each assignment), alongside the number of students who completed a specific assignment (indicated by point size).

7 Course Feedback and Findings

The students were asked formally at the start and at the end of the course to give anonymous feedback. This section contains a selection of positive and negative comments and a summary of lessons learned.

At the beginning of the course, the students were asked a single question anonymously: "*What did you like and what did you not like in this first block of the course? Please be specific!*" Twelve students (18%) answered in great detail — unusual for our students. Table 3 shows all replies that have general value and are not referring to specific course details or program specifics.

Near the end of the course, the end of the course, the students were asked to answer a large set of standard questions. This survey is always given to students of a new course. The return is typically very low, rarely higher than 10% — partly because the students are already in the exam phase and also because most of them dislike lengthy questionnaires, which they are asked to fill in for several courses in the same time period. The survey return for this course, however, was high, with 23.2% of the participants. For the summary evaluation, the students were asked to place the course on a scale from 1 (completely agree/very good) to 5 (don't agree at all/very bad). The participants gave a 1 (completely agree/very good) in all available categories:

- I really like the format of this course (block or weekly)
- The lecturer has clearly defined teaching and learning goals for this course
- The course is well structured
- The lecturer has got the technical competence to teach this course
- The lecturer is committed and enthusiastic regarding the course subject
- The lecturer makes the lessons interesting and involving for me
- The lecturer motivated me to actively contribute to the class
- Overall the course was very good

The participants were also given the opportunity to make free-style comments, listed in Table 4.

Summary of Findings

Public Goals. All but one of the original course goals could be reached. The only goal that almost completely failed was the creation of small data science projects. Here, the students had been given abundant examples, but only one team turned in a project. Its members were graduate students of business informatics, who had already done data science work in Python. They submitted a Kaggle notebook with an EDA.

Private Goals. I am very happy with the results on a professional and on a personal level. I could create a course that is both new to the school and to me. By piloting the course in the extracurricular study program, valuable data could be gathered from a wide student population.

Table 3. Feedback shortly after the start of the course.

»I liked the appropriate balance of the course structure: giving a compensated "direct" input during the wednesday live session (which was also very entertaining though! – so I got even more curious to proceed), and having the extras / self study items in the moodle course so we can increase our knowledge the way each one prefers. For me the first class that adopted its didactic concept in a good way to the online learning format. Thanks a lot!«
»You are amazing. I think this class would be my best online course experience. Thank you so much for putting so much effort into this course. I am looking forward to the rest of the sessions. What I really appreciate and like: 1. the way you clarify and explain everything 2. your really good mood 3.recording the course 4. the data camp idea and certificates 5. all the text you provided for each session 6. Kahoot quizzes 8. Clear, organized, and amazingly flexible structure of the course I am sure there are more than these.«
»I really like how structured this course is and especially the moodle page. It makes it really easy to find everything and everything you need is there. First, I was a bit overwhelmed by the amount of resources and material we have. However, as I looked at all of it started to make sense and I really appreciate all the extras and the mandatory uploads. Especially the data camp assignments seem like a fun way of learning more. The lecture itself was also really interesting. Even though it was late in the evening I was able to follow everything you said and did not zone out. I think 1,5 h of online lecture is the perfect time frame. I feel like you are really motivated to do this course and you enjoy doing it which makes it even easier to listen to you.«
»Positive Feedback: I really liked your energy and enthusiasm for the topic, it is contagious! The way you conducted the class was unlike my other online classes, it was interactive and its always nice when the Professor/ Lecturer is tech-savvy and use technology to aid the class, instead of it being as an obstacle! Access to Datacamp and getting the homework there is for me, as a total beginner really helpful, because the explanation is simple and I can re-watch the videos. Negative Feedback (but not really): Some participants have very advanced Data Science knowledge. When they ask questions about Python or R, I feel overwhelmed. It then feels like I cannot participate in the discussions because I haven't got any idea what is going on!«
»I really like how the moodle page is set up. Even though I missed the beginning of the class, I easily oriented myself thanks to the detailed path and information given on Moodle. I also feel that my interest for the course grew because of this.«
»I really liked your introduction to the topic: The kanban board made the beginning of the session a bit interactive and your hand-drawn "emotional change over time" diagrams gave me a first impression on the topic instead of jumping right into it, which is good in my opinion. Also, I have to say that it's great that we can enjoy the privilege of having access to datacamp for free since their content is really well made.«
»What did I like? – Clear structure of the course syllabus – Interaction with students – Course is very well prepared (e.g. Datacamp access, FAQ section, Github Access, Presentation of first lecture, Kahoot, Summary of previous lecture) – It feels like that you are very proficient in what you are doing. – English – Quick responses to students' questions in the moodle forum What did I not like? – So far, nothing.:)«

Table 4. Feedback near the end of the course.

»Covering different topics and demonstrating how each of them can be used and be helpful in professional life. The professor is amazing and makes the course more interesting to follow and he records the course so we have a chance to watch them later. The professor is so organized.«
»The Moodle page is very well structured, easy to understand and has an attractive layout. Much information was offered and one could continue learning outside of class. It was obvious that [the lecturer] enjoy teaching this subject and puts a lot of effort into it. I like that there are many short videos instead of one video of 90 min length. I also really liked DataCamp.«
»[The lecturer] is very committed and motivated to pass on as much as possible to his students, but sometimes the amount of material was a little overwhelming.«
»I think the course is too time-consuming for only one credit point. That's not a critique of the effort. It takes time to learn this subject but one should get at least 3 or 4 credit points for it.«

Participation. As the course statistics show, a large group of students stayed with the course to the end and completed most of the assignments. The visible trend (lowered activity towards the end of the course, visible from the test and assignment participation) is typical. The number of participants is extraordinarily high for an extracurricular (voluntary, low credit) course; the average number of participants for courses in this program is below 10. In fact, Master students from executive programs, who wanted to participate, were not allowed in — otherwise, the number would have been even larger.

Return on Investment. My personal effort for this course was unusually high — at a level that I will not easily be able to repeat any time soon. This is mostly the result of offering the students so many different, parallel avenues for entering the data science field. The uptake of these additional offerings was not pursuant to the effort. Though the feedback shows that the students clearly appreciated the effort, only few of them used all or even many of the offerings.

It would be interesting to compare the results and the feedback with a course on the same topic that offers much less — more on the level of a MOOC. However, it is well known that the success rate of MOOCs is rarely higher than 15%, which is in stark contrast to my findings [15].

8 Limitations of This Case Study

This study is missing a review of the literature on the experiences of other business schools with teaching data science. I am not able to estimate the influence of the local conditions of our school and student population. Since this is the first course of its type at our school, I cannot compare the findings with another course. Lastly, the action research methodology applied here to create the case study, with me as both creator and investigator of the course and observer of its students, leaves me wide open to errors of interpretation.

9 Conclusions and Outlook

The course presented an opportunity to teach data science in an interdisciplinary fashion, with an emphasis on improving data literacy of the participants and putting them on the path to data-driven storytelling. An interdisciplinary approach is generally accepted as a way of embedding computing in a non-STEM environment such as a business school, since "technical and computational abilities are essential within every discipline" [16]. But the business school environment also influences the teaching of computing by "bringing a holistic and multidisciplinary perspective to Computer Science that is sometimes lacking in programs at large institutions" [17]. Studies investigating the interdisciplinary teaching of data science are rare [18]. At this stage of the field, our best hope for business school education is for lecturers to integrate data science tools and methods into their classes, flanked by introductory courses like the one discussed here, to increase the data literacy of a large, diverse student body. An adaptive approach that provides tutoring and additional activities could improve student's learning [19]. More research is needed to identify the specific impact of the pandemic on both teachers and learners [20].

References

1. Andriole, S.J.: Skills and competencies for digital transformation. IT Professional **20**(6), 78–81 (2018)
2. Robinson, E., Nolis, J.: Building a Career in Data Science. Manning, New York (2020)
3. Yin, R.K.: Case study protocol. In: Mills, A.J., Durepos, G., Wiebe, E. (eds.) Encyclopaedia of Case Study, pp. 84–86 (2010)
4. Strumińska-Kutra, M., Koładkiewicz, I.: Case Study. In: Ciesielska, M., Jemielniak, D. (eds.) Qualitative Methodologies in Organization Studies. Palgrave Macmillan, Champagne (2018)
5. Burns, D.: Systemic action research – a strategy for whole systems change. Policy Press, Bristol (2007)
6. Halwani, M.A., Amirkiaee S.Y., Evangelopoulos, N., Prybutok, V.: Job qualifications study for data science and big data professions. Inf. Technol. People (2021)
7. Vinod, D.N., Prabaharan, S.R.S.: Data science and the role of Artificial Intelligence in achieving the fast diagnosis of Covid-19. Chaos, Solitons & Fractals **140**, 110182 (2020)
8. R Core Team: R: A language and environment for statistical computing. R Foundation for Statistical Computing, Vienna, Austria (2020)
9. Knaflic, C.N.: Storytelling with Data: A Data Visualization Guide for Business Professionals. John Wiley & Sons, New York (2015)
10. Gray, J., Chambers L., Bounegru, L.: The data journalism handbook: how journalists can use data to improve the news, O'Reilly Media, Sebastopol, **122** (2012)
11. Wilkinson, L.: The Grammar of Graphics. Springer, New York (2005)
12. Knuth, D.E.: Literate programming. Comput. J. **27**(2), 97–111 (1984)
13. Deutch, D., Gilad, A., Milo, T., Somech, A.: ExplainED: explanations for EDA notebooks. Proc. VLDB Endowment, **13**(12), 2917–2920 (2020)
14. Davenport, T.H., Spanyi, A.: What process mining is, and why companies should do it. Harvard Bus. Rev. **97**(2) (2019)
15. Khalil, H., Ebner, M.: MOOCs completion rates and possible methods to improve retention-A literature review. In: EdMedia+ innovate learning, Association for the Advancement of Computing in Education, pp. 1305–1313 (2014)

16. Fee, S.B., Holland-Minkley, A.M., Lombardi, T.E.: New Directions for Computing Education: Embedding Computing Across Disciplines. Springer, New York (2017)
17. Sonnier, D.L.: Computer science in a liberal arts school: convincing the sceptic. J. Comput. Sci. Coll. **28**(5), 115–121 (2013)
18. Asamoah, D., Doran, D., Schiller, S.: Teaching the foundations of data science: an interdisciplinary approach. arXiv preprint arXiv:1512.04456 (2015)
19. Troussas, C., Krouska, A., Sgouropoulou, C.: A novel teaching strategy through adaptive learning activities for computer programming. IEEE Trans. Educ. **64**(2), 103–109 (2021). https://doi.org/10.1109/TE.2020.3012744
20. Khanna, D., Prasad, A.: Problems faced by students and teachers during online education due to COVID-19 and how to resolve them. In: Proceedings of the 6th IEEE International Conference on Education and Technology, Indonesia, pp. 32–35 (2020). https://doi.org/10.1109/ICET51153.2020.9276625

Make It Matter: A Collaborative Student-Led Engagement and Persistence Program

Turner Bohlen[1(✉)], Lindy Elkins-Tanton[2], Carolyn Bickers[1], and James Tanton[2]

[1] Beagle Learning, Inc, Concord, NH 03301, USA
turner@beaglelearning.com
[2] Arizona State University, Tempe, AZ 85281, USA

Abstract. Students stay in school at higher rates if they understand the value of their learning and if they are intrinsically motivated by the topics. The student-centered program *Make it Matter* is designed to give students a reason to persist through school. This open inquiry-based process guides students through the process of doing research and producing outcomes (such as an event, a publication, or a work of art) and can be applied to any topic. Pilots are being run over a range of education levels on a variety of topics, including English Language Arts and undergraduate STEM research opportunities, which we discuss here. Initial results from the high school English Language Arts program showed a 30% reduction in failing grades as compared with the previous quarter.

Keywords: Intrinsic motivation · Persistence · Open inquiry · Technology-enabled

1 Introduction

In the United States, 38% of students who enter higher education don't graduate [1]. Of those who do graduate, 29% are underemployed due to a lack of critical job skills [2]. Recent work suggests that the reasons for this lack of persistence to graduation are not simple single variables, such as lack of financial aid, but complex ideas such as a sense of belonging or an understanding of the value of learning [3]. Similarly, programs that emphasize intrinsic motivation through student-led experiences have shown as much as 50% higher knowledge retention [4] and 12% increases in persistence to graduation [3]. These are some of the best results of any available educational intervention.

An inquiry approach is a strong way to bring intrinsic motivation to learning. Academic researchers and practitioners recognize the significance of student-generated questions. Asking productive questions relates to key cognitive skills, such as critical thinking and problem solving [5]. The quality of the questions that students ask also reveals how much they know and how well they learn [6]. Although question asking is generally encouraged and considered a sign of motivation, many researchers note that high-quality student-generated questions are scarce and that many students rarely ask questions at all [7]. For an introduction to inquiry techniques, see, for example, Littleton et al. [8].

D. Guralnick et al. (Eds.): TLIC 2021, LNNS 349, pp. 40–48, 2022.
https://doi.org/10.1007/978-3-030-90677-1_4

In order to help educational institutions take full advantage of these strong results, we have developed a student-centered, intrinsically motivated program structure called *Make it Matter*. *Make it Matter* is designed to give students a reason to persist through school, whether delivered online or in person. To support use of the program online, Beagle Learning, Inc., has developed a software platform to guide students through a student-led experience, including step-by-step processes, team collaboration, and optional grading by an instructor.

2 The *Make It Matter* Program

2.1 Program Overview

Make it Matter is a highly engaging persistence program that guides students through an open-inquiry, action-oriented process to identify a challenge relevant to their local community, to research the challenge, and to take meaningful action based on the research done. This program trains and makes explicit use of transferable skills like leadership, consensus-building, research skills, writing, speaking, and question-asking.

Make it Matter combines three supports necessary to launch new programs in today's educational system:

1. **A Program Structure,** skill rubrics, and facilitator's guide to lead participants through their research process.
2. **Professional development** to support staff in launching the program.

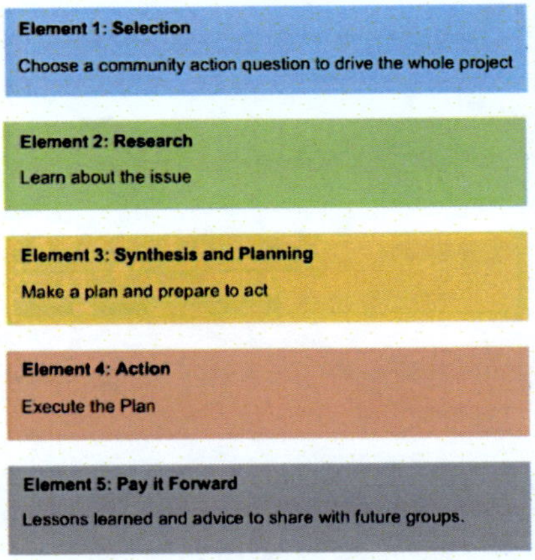

Fig. 1. *Make it Matter* is divided into five phases called *Elements,* each of which provides structure for one important step in the social action process.

3. **Beagle Learning Web App** for managing inquiry in person or remotely.

The program structure is divided into five phases called **Elements** (Fig. 1). Students are divided into small teams of 4 to 30. Every project starts with **Element 1**. After each Element, each student team reflects on progress made and chooses which Element makes the most sense to move to next. They are, thus, likely to visit the same Elements multiple times throughout a project. Each element is regarded as a tool to be used as needed throughout the program and beyond in each student's own life.

Of particular importance is **Element 2: Research**. This element takes advantage of the Beagle Inquiry Framework, a model for inquiry-based work developed by Beagle Learning in collaboration with Arizona State University (ASU). In this process, students repeatedly choose Natural Next Questions (NNQs) that they feel need to be answered to gain deeper understanding about the topic of their project. They then individually research an answer, finding sources online, in libraries, or through interviews with experts. They summarize this information and share it with their team, before repeating the cycle (Fig. 2).

Through early testing of the process, we found that teams would often allow the natural excitement about the work to pull them towards topics of research that did not directly support their goal. The Distillation cycle was added to address this. Every two to four inquiry cycles, teams review all they have learned and put together a concise graphical summary of the key ideas. This gives the team and instructor a tool for assessing progress and a way to course-correct if needed.

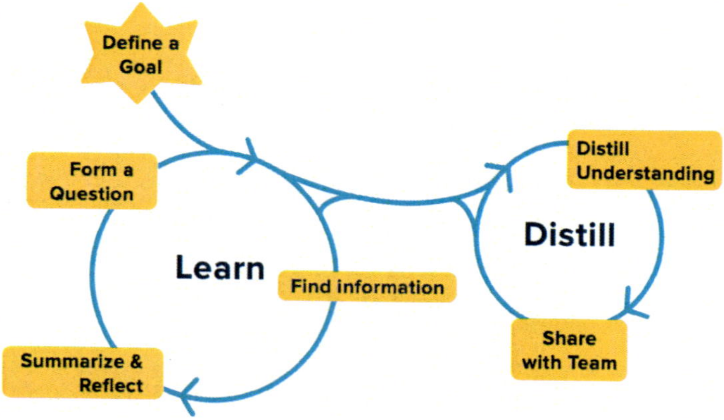

Fig. 2. The beagle inquiry framework is divided into two cycles, a "learn" cycle during which students ask questions and research answers to those questions, and a "distill" cycle, during which they summarize everything their group has learned.

Make it Matter is currently being used in 9 pilot classes across 4 locations. Each class operates as a kind of makerspace, which is — especially in technology — a place in which people can gather to work on projects while sharing ideas and equipment. In the case of *Make it Matter*, the makerspace provides a supported framework for student inquiry, research, and reflection, while maintaining student choice, decision, and action at all times.

2.2 Software Supports

To support these programs, Beagle Learning, Inc., has developed the Beagle web app. During *Make it Matter* pilot programs like those described below, three key needs were identified.

1. Consistent, step-by-step processes so students can complete work at home.
2. A collaborative space for students to share their work.
3. A competency-based grading system that allowed instructors to identify and validate learning whenever it happened in the process.

To combine assigned work with easy collaboration, the web app guides students through assignments but then saves the work done in a collaborative mind map, where students can post new ideas, comment, upvote, and share outside the context of assignments. Existing systems seem to divorce collaboration and discussion from assigned work and required learnings. The result was that discussion or team work always felt like an unnecessary burden to a large number of students. By creating a system that integrates the two, and where the outcome of required work is clearly seen in the collaborative space, our goal is to help connect team work directly to the intrinsic motivation each student has identified for themselves.

By documenting the work students do in an online system, we are able to analyze the problem-solving processes they use. To assess the value of the questions students ask as they work towards their big goal, we defined the Question Productivity Index (QPI) [9, 10]. Most methods of reviewing questions, like Bloom's Taxonomy [11] or Depth of Knowledge [12], focus on understanding and use of the content being reviewed. In the case of a project, questions have to be evaluated based on their value to achieving the student's ultimate goal. This was the goal of the QPI, which assesses based on the value to the goal, the specificity of the phrasing and the overall scale of the question.

Future work on the QPI involves the training of a neural net to automatically rate questions that students post to the Beagle platform. These ratings can be used as continuous, formative assessments of student thinking skills. In the future, the team plans to include other higher-order-thinking skill assessments in the platform to help inform teaching of a variety of vital skills.

3 Example Applications

3.1 English Language Arts Program

For the majority of at-risk students in U.S. middle and high schools who fall behind, it is nearly impossible to catch back up. Our educational partner, Riverview High School in Mesa, Arizona, is an alternative placement school that works primarily with the at-risk students who have been expelled from their district's traditional schools, been sent to juvenile detention, or have repeatedly failed classes. About 2,000 school districts across the United States have alternative placement schools that serve students with disciplinary or academic problems. These schools serve approximately 400,000 students. More than half of these schools have drop-out rates of 50% or more [13]. Class failure rates can be above 90%.

Students at these schools often miss weeks, months, or even years of school. With a traditional approach, those students are expected to catch up at home on their own time. For most at-risk students, this is all but impossible due to a combination of home environment, lack of motivation, and lack of scaffolding. When they do go back to school, the number of zero scores on missed assignments makes it feel impossible to catch up. The natural response is to disengage. Initially, students disengage emotionally, but eventually, they disengage physically and stop showing up to school. It is typically not possible to recover lost learning without attending class.

With our partners at Riverview High School, we have identified 3 key problems with the current system that push these students out:

- **Reliance on extrinsic motivation** – Highly at-risk students have often had negative experiences in school that make unmotivated content — content that is simply "part of the curriculum," *per se* — worthy of little or no attention. They need to see something more than extrinsic motivators.

  Make it Matter allows students to select their own goal question to guide their work, and anecdotally, that has been a big motivator for students to complete their work.
- **Being left behind forever** – If a student misses an assignment used to demonstrate a specific skill, they have no feasible way to catch up, making graduation feel impossible.

  Because our inquiry process follows a reliable pattern, it allows teachers to use competency-based grading. If students fail to demonstrate a skill on one assignment, they can show it on another assignment later in the semester and still get full credit. Further, once students learn the process, they can do their work even if they are home sick or, as occurred in the pilot, when suspended. In one example, a suspended student handed in none of his homework except that from our program.
- **Inconsistent classroom structures** – After returning from an absence, inconsistent structures mean students don't know what is expected of them or how to engage.

  The structure of the program is known by student and instructor alike, and expectations and assignments are understood by the whole team.

Pilot Implementation

Pilots are being run in an elementary school, two high schools, and several community colleges and four-year colleges. Only one of the high school pilots has progressed to the point that early data are available, so only that pilot will be discussed here.

Two teachers at Riverview High School used the *Make it Matter* model in their English Language Arts (ELA) and social studies (SS) programs. Mr. M taught the ELA course. In this class, Mr. M kicked things off by inviting students to brainstorm absolutely any interests or goals they had and might like to work towards. Students offered a variety of ideas, including asking how they could become professional twitch streamers. Once each student had decided on an individual goal, they came up with smaller NNQs to focus on, as dictated by the Beagle Inquiry Framework. This course used the Beagle web app to help manage the process.

Their NNQ led them through the process of identifying a source that might answer that question, annotating the source they found, and writing a short summary of what they learned from it.

After just two weeks, students had gotten into the flow and could repeat the "learn" cycle of the Beagle Inquiry Framework completely on their own, leaving Mr. M to provide feedback and individual support, rather than having to manage class. Each week as work was submitted, Mr. M reviewed the sources his students were reading, the annotations they made on these sources, and the summaries they wrote. He reported: "60% of [the skills learned were] unintentional. I didn't have to specifically focus on that." **Students were naturally encountering and demonstrating proficiency in 60% of the ELA standards without Mr. M having to say a word.**

Results

Their classes produced almost unsurpassed results in reducing failure rates. When using *Make it Matter*, failures in ELA and SS were 58% and 47%, respectively. The quarter before, when traditional teaching methods were used, the failure rates were 72% and 95%, respectively, for the same group of students. Quarter over quarter, the students' demographics were also consistent at 70% Hispanic, 10% African American, 10% Native American, and 10% White. Overall, *Make it Matter* resulted in a **30% reduction in failures**. We are awaiting data on dropout rates and hope to see our program reduce their number.

Though the data are from a small initial pilot, instructors report that the student ownership of their questions appeared to be a good motivator for doing their work, and the student ownership and knowledge of the process allowed students to do their work even when they missed class. The principal reported that the substantial improvement in SS was due to the new coherent and focused structure to the course we provided compared to the previous quarter. Assignments were no longer disconnected, and everything was aligned for the teachers and the students in how work was graded and assessed around the students' inquiry-based research.

3.2 STEM Research Program

A "leaky pipeline" is a widely acknowledged challenge in minority STEM education. Despite entering college with similar STEM aspirations as their well-represented counterparts, minority students leave the major at nearly twice the rate of white students [14]. Undergraduate research experiences (UREs) have consistently been found to be one of the most successful interventions for increasing persistence of minority students through STEM degree programs [15–17].

Based on these results, Edmonds College, ASU, and Beagle Learning have decided to implement the *Make it Matter* program as a guiding structure for student-led undergraduate research opportunities for community college students. The program offers a possible solution to a set of major barriers preventing implementation of quality undergraduate research programs for historically underserved students.

1. **Lack of Research Infrastructure and Limited Seats**: Many community colleges lack a research infrastructure and research culture [18]. Most research grants are competed and won by universities, not community colleges. Thus, implementing traditional apprenticeship-style UREs is often not a practical option. Even at schools with established research programs, the number of students who can be supported by these programs is generally small because of the constraints imposed by the traditional research structure. Apprentice-style research programs feature low student-to-faculty ratios, are expensive to administer, and place considerable burdens on faculty to supervise and mentor students [19].
2. **High Time Commitment and Outside-of-School Requirements**: Additional barriers exist to involving minority and low-income students in UREs by virtue of the unique challenges faced by these underserved groups. For instance, the volunteer-based nature of most research projects and the substantial time commitments they entail exclude many non-traditional and low-income students with significant family responsibilities [20].
3. **Stereotype Threat**: Underrepresented groups are more likely to perceive the undergraduate STEM climate as unwelcoming or even hostile and are less likely to approach faculty for research opportunities [21, 22]. In fact, first-year minority students at schools that offered structured opportunities for research were four times more likely to participate than minority students at schools without formal research opportunities [23]. Research programs that rely on students finding their own mentor are likely to dissuade underrepresented students.
4. **Prosocial Values and Science Identity**: Underrepresented students weigh altruistic or prosocial motivations more heavily than their peers when considering possible career paths [24, 25]. Unfortunately, this may be dampening the effects of undergraduate research opportunities for underrepresented students. A careful study in 2016 showed that first-generation minority students in STEM are much less likely to build a sense of science identity or persist in a STEM field if they do not perceive STEM as offering them a way to give back to their community [26]. UREs must be paired with an emphasis on the prosocial nature of STEM careers in order to offer equal benefits to all participants.

The *Make it Matter* program to address this issue and all its complexities will be launched in the fall of 2021. Students will be directed to choose a STEM-related big-goal question to drive their project. They will use the iterative Beagle Inquiry Framework to gain background knowledge on the topic, and they will then choose an action or experiment to run. Staff members at Edmonds College will work as facilitators. When students need expert input on a topic, the staff members will reach out to relevant faculty members at the institutions who will give short "topic burst" lectures on the needed topic for the students. This structure means that students will know exactly the purpose of every lecture they receive; it will, therefore, help them retain more information and stay more engaged. Faculty will have to devote far less time that they would to a traditional apprenticeship-based URE.

The Beagle web app will be used by students to document inquiry work done, letting students work remotely and removing the burden of scheduling in-person meetings, a common blocker for students who work full-time jobs or care for family members during school.

4 Conclusion

Research has demonstrated the powerful value of intrinsic motivation, interactive learning strategies, and student-led experiences. The intent of the *Make it Matter* program was to develop a clear course structure that an instructor could adopt instead of a traditional lecture-based approach. In this way, *Make it Matter* is a teaching process, just as lecturing is a teaching process, and can be applied to any content or discipline. We have seen extremely encouraging early results from our pilots and look forward to deeper research in future programs.

References

1. National Center for Education Statistics: Undergraduate Retention and Graduation Rates. The Condition of Education. https://nces.ed.gov/programs/coe/indicator_ctr.asp (April 2020)
2. Federal Reserve Bank of New York: Labor Market Outcomes of College Graduates by Major. The Labor Market for Recent College Graduates (12 February 2021)
3. Koenig, R.: To Retain College Students, Look to Academic Support and Campus Activities, New Report Finds. Edsurge. https://www.edsurge.com/news/2019-09-12-to-retain-college-students-look-to-academic-support-and-campus-activities-new-report-finds (12 Sep 2019)
4. Chi, M.T.H., Wylie, R.: The ICAP framework: linking cognitive engagement to active learning outcomes. Educ. Psychol. **49**(4), 219–243 (2014)
5. Biddulph, F., Osborne, R.: Some issues relating to children's questions and explanations. Working Paper No. 106. Distributed by ERIC Clearinghouse (1982)
6. White, R., Gunstone, R.: Probing Understanding. Routledge (2014)
7. Dillon, J.T.: The remedial status of student questioning. J. Curric. Stud. **20**(3), 197–210 (1988). https://doi.org/10.1080/0022027880200301
8. Littleton, K., Scanlon, E., Sharles, M. (eds.) Orchestrating Inquiry Learning. Routledge, Taylor & Francis Group (2011)
9. Wang, Y., Bohlen, T., Elkins-Tanton, L., Tanton, J.: What makes a question productive? – an exploratory analysis towards demystifying question productivity. Companion Proceedings 9th International Conference on Learning Analytics & Knowledge (LAK19) (2019)
10. Wang, Y., Bohlen, T., Elkins-Tanton, L., Tanton, J.: Motivating students to ask more questions. In: Isotani, S., Millán, E., Ogan, A., Hastings, P., McLaren, B., Luckin, R. (eds.) AIED 2019. LNCS (LNAI), vol. 11626, pp. 409–412. Springer, Cham (2019). https://doi.org/10.1007/978-3-030-23207-8_75
11. Bloom, B.S., Englehart, M.D., Furst, E.J., Hill, W.H., Krathwohl, D.R.: The Taxonomy of educational objectives, handbook I: The Cognitive domain. David McKay Co., Inc (1956)
12. Webb, N.: Depth-of-Knowledge levels for four content areas. Presentation to the Florida Education Research Association, 50th Annual Meeting, Miami, Florida (November 2005)
13. Vogell and Fresques: Alternative Education: Using Charter Schools to Hide Dropouts and Game the System. ProPublica/USA Today story (2017)
14. Riegle-Crumb, C., King, B., Irizarry, Y.: Does STEM stand out? Examining racial/ethnic gaps in persistence across postsecondary fields. Educ. Res. **48**(3), 133–144 (2019)

15. Lopatto, D.: Undergraduate research as a high-impact student experience. Peer Rev. **12**(2), 27 (2010)
16. Hensel, N.H., Cejda, B.D.: Embedding undergraduate research in the community college curriculum. Peer Rev. **17**(4), 27 (2015)
17. Levenson, C.W.: Enhancing undergraduate research in the arts and the humanities. Peer Rev. **12**(2), 13 (2010)
18. Hewlett, J.A.: Broadening participation in undergraduate research experiences (UREs): The expanding role of the community college. CBE Life Sci. Educ. **17**(3), es9 (2018). https://doi.org/10.1187/cbe.17-11-0238
19. Eagan, M.K., Sharkness, J., Hurtado, S., Mosqueda, C.M., Chang, M.J.: Engaging undergraduates in science research: not just about faculty willingness. Res. High. Educ. **52**(2), 151–177 (2011)
20. Bangera, G., Brownell, S.E.: Course-based undergraduate research experiences can make scientific research more inclusive. CBE Life Sci. Educ. **13**(4), 602–606 (2014). https://doi.org/10.1187/cbe.14-06-0099
21. Seymour, E., Hewitt, N.M.: Talking About Leaving, p. 134. Westview Press, Boulder, CO (1997)
22. Kim, Y.K., Sax, L.J.: Student–faculty interaction in research universities: differences by student gender, race, social class, and first-generation status. Res. High. Educ. **50**(5), 437–459 (2009)
23. Hurtado, S., Newman, C.B., Tran, M.C., Chang, M.J.: Improving the rate of success for underrepresented racial minorities in STEM fields: insights from a national project. New Dir. Inst. Res. **2010**(148), 5–15 (2010)
24. Johnson, M.K.: Social origins, adolescent experiences, and work value trajectories during the transition to adulthood. Soc. Forces **80**, 1307–1341 (2002)
25. Thoman, D.B., Brown, E.R., Mason, A.Z., Harmsen, A.G., Smith, J.L.: The role of prosocial values in motivating underrepresented minority students for biomedicine. Bioscience **65**, 183–188 (2015)
26. Jackson, M.C., Galvez, G., Landa, I., Buonora, P., Thoman, D.B.: Science that matters: the importance of a cultural connection in underrepresented students' science pursuit. CBE Life Sci. Educ. **15**(3), ar42 (2016). https://doi.org/10.1187/cbe.16-01-0067

Video Consumption with Mobile Applications in a Global Enterprise MOOC Context

Max Bothe[1](✉), Florian Schwerer[2], and Christoph Meinel[1]

[1] Hasso Plattner Institute, Potsdam, Germany
{max.bothe,christoph.meinel}@hpi.de
[2] SAP SE, Walldorf, Germany
florian.schwerer@sap.com

Abstract. With the proliferation of mobile devices in daily lives, users access Massive Open Online Courses (MOOCs) more often with devices other than stationary desktop computers. The introduction of mobile applications for MOOC platforms improved the learning experience by enabling Internet-independent learning sessions. Despite the university-like setting, MOOCs have also been utilized in a corporate context. Learners from different countries can integrate mobile applications into their learning process, which may result in varying usage patterns. In an observational study, the video consumption with mobile applications in different countries was examined. Based on over 2.9 million tracking events from 2019, the learning behavior was evaluated in the ten most active countries of an Enterprise MOOC platform. India and the United States showed unique patterns in terms of mobile application usage by having the highest and the lowest adoption rates. For other countries, no overarching multi-dimensional patterns were detected. In general, the use of cellular data correlates with the costs of mobile data in the respective countries. However, users who avoid cellular data do not automatically download video content to the mobile devices. Small to medium practical effects were proven for all dimensions.

Keywords: MOOCs · Mobile learning · Enterprise MOOCs

1 Introduction

Massive Open Online Courses (MOOCs) have been proven to be effective in providing free educational resources to people with access to the Internet. They have been deployed to a variety of fields — from computer science to health-related topics [7]. Despite the original university-like setting, most MOOC learners are professionals who want to deepen their knowledge or create new job opportunities [11]. Hence, MOOCs have been utilized in a corporate context [8]. Companies offer courses, e.g., to train employees or to provide detailed documentation of their products.

Most learning activities in MOOCs are performed in a stationary setting on desktop computers [6]. However, mobile devices present a shift in traditional learning environments [2]. With the proliferation of mobile devices in daily lives, users accessed MOOC

D. Guralnick et al. (Eds.): TLIC 2021, LNNS 349, pp. 49–60, 2022.
https://doi.org/10.1007/978-3-030-90677-1_5

platforms with those devices [13]. MOOC platforms, in turn, started to optimize their user interfaces and developed dedicated mobile applications to provide a better learning experience on mobile devices [13]. With the help of mobile devices, MOOC platforms allow users to choose their preferred device for learning activities while at the same time reaching people who don't have access to desktop computers [10]. Operating with cellular networking capabilities can be both — advantage and disadvantage — for mobile devices. On one side, it allows users to escape the traditional learning setting. On the other side, broadband Internet access, cellular coverage, and the respective usage rates are influenced by regional factors [10] and prices [21]. Triggered by this, mobile applications provide the functionality to download and store learning resources on mobile devices [4]. Thereby, learners also gain more options in creating their preferred learning process, being time-, device-, and network-independent without leaving the designed learning environment.

Especially in a global enterprise context, learners from different countries will join the courses, thus creating a diverse learning community. With such diversity, different usage patterns in Enterprise MOOCs will surface [5]. Analyzing learning behavior in a global context has become of increasing interest [15] as regional factors have to be considered and learning characteristics may vary. MOOC platform providers and course designers have to consider differences by adapting their technical setup and content delivery. Also, mobile devices are becoming more relevant in the corporate learning landscape [20]. To create insights on how mobile applications are used for video consumption in different countries, we formulated these research questions:

RQ1: How does the usage of mobile applications for video consumption in Enterprise MOOCs differ between countries?

RQ1a: How is the acceptance rate for mobile applications?

RQ1b: How is the acceptance rate of using a cellular connection?

RQ1c: How is the acceptance rate of downloading content beforehand?

2 Foundations

This section provides foundational information to better illustrate the underlying principles and rationales of this work. In this regard, the concepts of Enterprise MOOCs and mobile learning are introduced.

2.1 Enterprise MOOCs

The term Enterprise MOOC is used to describe online courses following xMOOC design principles, implemented in a broader business context. These courses are scalable to serve a larger audience, open to everyone, providing videos and tools for interaction — e.g., quizzes, exercises, surveys, or forum discussions — in a digital classroom over a fixed period with subsequent on-demand availability of content [14]. In contrast to Corporate MOOCs, the delivery does not exclusively focus on a company's employees but provides relevant business knowledge to all stakeholders of a company's ecosystem [8]. openSAP[1]

[1] Available at open.sap.com.

is a digital learning platform with a focus on delivering video-based online courses. The free-of-charge offering plays an important role in SAP's continuous transformation of its education business to share knowledge about its latest innovations and relevant business or product-related topics within and beyond its global ecosystem more openly, time and device-independent, and at scale [16]. Combining short videos with additional interactive elements, and the flexibility of content consumption via desktop, mobile web, mobile applications, online or offline, Enterprise MOOCs on openSAP are an effective format for knowledge transfer in a fast-moving business context.

2.2 Mobile Learning

Research on Mobile Learning already started with the availability of the first handheld devices. Back then, studies centered around non-stationary learning methods [17]. When mobile devices became an integral part of users' lives, the focus extended to providing a ubiquitous learning experience. Compared to learning with traditional desktop computers, mobile devices offer smaller screen sizes but allow learning in non-stationary situations. Mobile applications on mobile devices are utilized to enable network-independent learning sessions [3], thus offering a higher degree of flexibility to the users' learning process. On mobile devices, the learning behavior is influenced by various aspects — such as the device's screen size, the utilized user interface (mobile website or mobile application), or the current network connectivity state [1, 5, 12, 19]. Because of this, not all learning activities are suitable for being performed in a mobile context [9]. Learning with mobile devices should rather be considered as a complement to learning with desktop computers [18]. However, by applying Mobile Seamless Learning techniques [2], the interplay of desktop computers and mobile devices is strengthened. Here, different constraints are minimized that can influence the learning experience negatively. The ubiquitous access to learning resources is aligned this work.

3 Study Design

We conducted a long-term observational study to capture the real-world interactions of the learners with course videos on openSAP. Started video playbacks have been recorded as events using the tracking capabilities of the MOOC platform. Those tracking events are annotated with the learners' contextual environment — e.g., the used device category, operating system, and interacted content. Based on these context attributes, three different metrics have been defined and calculated — each to answer one respective research question. These metrics are discussed in more detail in Sect. 3.2. By considering only recorded tracking events, this study focused on learners who watched at least one video, while disregarding learners who didn't participate in a course (no shows). Furthermore, only the ten most active countries have been considered in this study.

3.1 Sample Data

This study is based on the users' learning activities of the year 2019 (2019–01–01 until 2019–12–31). At the end of 2019, openSAP offered 166 courses, and approximately

1 million learners have been registered to the platform. Learning activities may vary depending on the offered guided MOOC courses. We acknowledge the fact that mobile usage rates evolve constantly. However, the COVID-19 outbreak in 2020 would risk distorting the recorded data disproportionately. Therefore, no data from this period were included in this study. The effects of the COVID-19 outbreak on mobile usage rates should be examined in a separate study.

3.2 Data Processing and Metric Definition

In total, over 2.9 million events for video playbacks have been recorded in the year 2019 on openSAP. Due to this huge amount, the data processing had to be performed close to the data storage. By the means of SQL queries, each metric was first calculated for all active users. Along with that, the user's country was determined based on the IP address. If a user accessed the MOOC platform from multiple countries, only the country with the most interactions was considered. The following metrics have been defined for this study:

Adoption of Mobile Applications
Regarding RQ1a, the usage of the mobile applications had to be identified. For this, two context attributes have been utilized: *platform* and *runtime*. The *platform* attribute states the operating system of the user's device. In the case of mobile devices, this is *iOS*, *Android*, and others. This is complemented by the *runtime* attribute, which describes the accessing application. For browsers on desktop or mobile, this would be the name of the browser. Concerning the mobile applications, the value would again be *iOS* or *Android*. By these means, video playbacks with mobile applications can be distinguished from other devices or platforms. Regarding the metric, the ratio of mobile video playback to all video playbacks has been determined for each active user. A ratio of 0 indicates no usage of the mobile application, while a ratio of 1 implies that the user only consumed video with mobile applications.

Network State
As mobile devices can be connected to the Internet in different ways and offered mobile applications can operate without an Internet connection, the context of the tracking events had to be extended for these devices. The *network* attribute captures the current connectivity method when triggering new tracking events. The system can differentiate between WiFi connections, cellular connectivity, and no connection to the Internet. The connectivity method, however, does not reliably correlate to the available bandwidth. A WiFi connection does not have to be sufficiently fast, and low cellular connectivity could appear as non-existent to the user. However, due to the broad cellular coverage, real offline usage of mobile applications is rare [3]. Therefore, we focus on video playbacks with a cellular connection as part of this study (RQ1b). Similar to determining the usage rate of mobile applications, the ratio of cellular video playbacks has been determined for each active user of the mobile applications. Mobile users who avoid video playbacks

with a cellular connection show a ratio of 0. Users who always rely on cellular data will have a ratio of 1.

Download Functionality
As described before, video content can be downloaded by mobile applications to enable network independent learning sessions. To capture the respective context, the *current_source* attribute in the tracking context is utilized. This attribute is also specific to the mobile application, as other device groups (desktop web and mobile web) can only operate with an Internet connection. If the user has downloaded the video item before playback, the *current_source* attribute will hold the value *offline*, and the video is loaded from the local storage of the mobile application. Otherwise, the video content is streamed over the Internet and the attribute holds the value *online*. Users might also download video content for non-offline learning sessions [3]. This is partly due to the varying network condition described as part of the previous metric. In that way, users are also prepared for possible future offline learning activities. In this study, we thus examine video playbacks with downloaded content (RQ1c). The ratio of video playbacks with an offline source has been calculated in proportion to the overall number of the user's playbacks with the mobile application. If the user always pre-downloads video content, the metric will state a value of 1. In contrast, if all video content is streamed, the ratio will be 0.

4 Results and Discussion

As a preparation for the evaluation, the countries with the highest number of active users have been determined. This study is limited to the ten most active countries of openSAP. Those countries are (in descending order of active users): India (IN), Germany (DE), United States of America (US), Great Britain (GB), Brazil (BR), Canada (CA), Spain (ES), Australia (AU), France (FR), and the Netherlands (NL). These countries combined accounted for about 69% of all started video playbacks, whereas the majority of playback events (53%) were created for users from India, Germany, and the United States.

4.1 Video Playbacks in Mobile Applications

Table 1 displays the results of the mobile application usage metric grouped for users of these ten countries along with the number of active users per country (RQ1a). India shows the highest adoption of mobile applications (14.6% of all video playbacks), while the United States registers the lowest adoption rate (3.5%). When comparing the metric results of all countries, the Kruskal-Wallis test returned highly significant differences between the countries ($H = 3596.49$; $p \leq 0.001$). Table 2 displays the pairwise Bonferroni corrected probabilities of the adjoining post hoc test with the respective effect sizes shown in Table 7. Based on the identified high statistically significant differences, the countries have been clustered into groups. Regarding this, countries with highly significant differences have not been placed into the groups. As a result of the general usage of mobile applications, four groups were formed. First, India shows the highest adoption rate, with 10% of the users only relying on mobile applications (small to medium effect). Second, in Brazil and Australia, at least 10% of the users make use of mobile applications (small effect). Third, users from European countries (Germany, Netherlands, Great Britain, Spain, France) and Canada have a similar attitude (small effect). Fourth, the United States shows the lowest adoption rate (small effect). These groups are visualized in Fig. 1. The majority of users still rely fully on the desktop environment. However, mobile applications are still utilized by users. Surprisingly, there are always users who only rely on their mobile devices for video consumption.

Table 1. Ratio of video playbacks on mobile applications.

Country	User count	Metric		Extreme users	
		Mean	Std.Dev.	0%	100%
IN	35877	0.1460	0.3275	80%	10%
DE	25144	0.0556	0.2084	91%	3%
US	23856	0.0348	0.1662	94%	2%
GB	5839	0.0587	0.2152	91%	4%
BR	4275	0.0849	0.2542	87%	5%
CA	4122	0.0578	0.2141	91%	4%
ES	3516	0.0669	0.2249	89%	4%
AU	3233	0.0846	0.2535	87%	5%
FR	3041	0.0621	0.2234	91%	4%
NL	2976	0.0506	0.2002	92%	3%

Table 2. Video playbacks on mobile applications: pairwise Bonferroni corrected probabilities.

	IN	DE	US	GB	BR	CA	ES	AU	FR	NL
IN	–	<0.001	<0.001	<0.001	<0.001	<0.001	<0.001	<0.001	<0.001	<0.001
DE	<0.001	–	<0.001	1.000	<0.001	1.000	0.057	<0.001	1.000	1.000
US	<0.001	<0.001	–	<0.001	<0.001	<0.001	<0.001	<0.001	<0.001	0.032
GB	<0.001	1.000	<0.001	–	<0.001	1.000	0.293	<0.001	1.000	1.000
BR	<0.001	<0.001	<0.001	<0.001	–	<0.001	0.111	1.000	<0.001	<0.001
CA	<0.001	1.000	<0.001	1.000	<0.001	–	1.000	<0.001	1.000	1.000
ES	<0.001	0.057	<0.001	0.293	0.111	1.000	-	0.374	1.000	0.021
AU	<0.001	<0.001	<0.001	<0.001	1.000	<0.001	0.374	–	<0.001	<0.001
FR	<0.001	1.000	<0.001	1.000	<0.001	1.000	1.000	<0.001	–	1.000
NL	<0.001	1.000	0.032	1.000	<0.001	1.000	0.021	<0.001	1.000	–

4.2 Video Playbacks with Cellular Data

For RQ1b, only users of the mobile applications have been considered. In Table 3, the data of the cellular usage metric are presented. In India, mobile users most often rely on cellular data for video playbacks (57%), whereas mobile users from Germany and Canada are more reluctant (19%–22%). The Kruskal-Wallis test revealed highly statistical differences for the usage patterns across the studied countries ($H = 1687.71$; $p \leq 0.001$). The pairwise probabilities of the post hoc test and the respective effect sizes are shown in Table 4 and Table 8.

Table 3. Ratio of video playbacks with cellular data.

Country	User count	Metric		Extreme users	
		Mean	Std.Dev.	0%	100%
IN	7481	0.5681	0.4328	27%	39%
DE	2321	0.2186	0.3515	62%	10%
US	1429	0.2990	0.3968	54%	17%
GB	538	0.2988	0.3918	53%	16%
BR	570	0.3377	0.4011	47%	19%
CA	386	0.1872	0.3314	66%	9%
ES	410	0.3195	0.4050	52%	17%
AU	428	0.4772	0.4393	36%	30%
FR	293	0.5035	0.4503	38%	32%
NL	253	0.2867	0.3963	57%	15%

Table 4. Video playbacks with cellular data: pairwise Bonferroni corrected probabilities.

	IN	DE	US	GB	BR	CA	ES	AU	FR	NL
IN	–	<0.001	<0.001	<0.001	<0.001	<0.001	<0.001	0.001	0.087	<0.001
DE	<0.001	–	<0.001	0.003	<0.001	1.000	0.002	<0.001	<0.001	1.000
US	<0.001	<0.001	–	1.000	1.000	<0.001	1.000	<0.001	<0.001	1.000
GB	<0.001	0.003	1.000	–	1.000	0.004	1.000	<0.001	<0.001	1.000
BR	<0.001	<0.001	1.000	1.000	–	<0.001	1.000	<0.001	<0.001	1.000
CA	<0.001	1.000	<0.001	0.004	<0.001	–	0.002	<0.001	<0.001	0.362
ES	<0.001	0.002	1.000	1.000	1.000	0.002	–	<0.001	<0.001	1.000
AU	0.001	<0.001	<0.001	<0.001	<0.001	<0.001	<0.001	–	1.000	<0.001
FR	0.087	<0.001	<0.001	<0.001	<0.001	<0.001	<0.001	1.000	–	<0.001
NL	<0.001	1.000	1.000	1.000	1.000	0.362	1.000	<0.001	<0.001	–

The countries were categorized into groups based on the highly significant differences. First, mobile users from India are the most willing to use cellular data for video streaming (small to medium effect). Second, the behavior shown in France and Australia is only slightly less in favor of cellular data usage (small effect). Here, at least 30% of the mobile users are using only cellular data. Third, the United States, Great Britain, Brazil, Spain, and the Netherlands show a similar acceptance rate (small effect). Fourth, mobile users in Germany and Canada are the least in favor of using cellular data, with 62%–66% of mobile users avoiding cellular data at all (small effect). These grouping results are shown in Fig. 1. The described findings correlate with prices for cellular data in the respective countries [21]. The more affordable mobile data plans are in a country, the more users consume MOOC video content on cellular data.

4.3 Video Playbacks with Downloaded Content

Following the result of the cellular usage, the download state metric was evaluated for RQ1c (see Table 5). Also, for this metric, only users of mobile applications have been considered. In Germany, most video content was downloaded before playback (18%). In contrast, mobile users from the United States made only little use of the download functionality (9%). The applied Kruskal-Wallis test yielded highly statistical differences in usage patterns across countries ($H = 102.35$; $p \leq 0.001$). In Table 6, the probabilities of the post hoc test are displayed, with the effect sizes shown in Table 9. The grouping process of countries based on highly significant differences resulted in three groups. First, mobile users from Germany are the most active in downloading video content, with 7% of the mobile users only consuming downloaded materials (small effect). Second, mobile users from India, Great Britain, Brazil, Canada, Spain, Australia, France, and the Netherlands all show a non-distinguishable download behavior (13%–17%; small effect). Third, most users in the United States (83%) don't consume downloaded video content (small effect). Mobile users who are more reluctant in using cellular data for video playback might be more willing to download the video content beforehand and vice versa. This assumption holds up from users from Germany and India. However, the usage patterns shown by users from the United States break the assumption. Therefore, there must be another factor influencing the users' actions, which should be discussed in future studies.

Table 5. Ratio of video playbacks with downloaded content.

Country	User count	Metric		Extreme users	
		Mean	Std.Dev.	0%	100%
IN	7481	0.1273	0.2870	77%	6%
DE	2321	0.1771	0.3287	71%	7%
US	1429	0.0889	0.2464	83%	4%
GB	538	0.1445	0.3074	75%	7%
BR	570	0.1681	0.3187	71%	6%
CA	386	0.1537	0.3084	74%	6%
ES	410	0.1489	0.3074	74%	6%
AU	428	0.1149	0.2651	79%	3%
FR	293	0.1401	0.2927	74%	5%
NL	253	0.1209	0.2720	79%	2%

Table 6. Video playbacks with downloaded content: pairwise Bonferroni corrected probabilities.

	IN	DE	US	GB	BR	CA	ES	AU	FR	NL
IN	–	<0.001	<0.001	1.000	0.088	1.000	1.000	1.000	1.000	1.000
DE	<0.001	–	<0.001	0.743	1.000	1.000	1.000	0.004	1.000	0.099
US	<0.001	<0.001	–	0.005	<0.001	0.002	0.004	1.000	0.033	1.000
GB	1.000	0.743	0.005	–	1.000	1.000	1.000	1.000	1.000	1.000
BR	0.088	1.000	<0.001	1.000	–	1.000	1.000	0.168	1.000	0.693
CA	1.000	1.000	0.002	1.000	1.000	–	1.000	1.000	1.000	1.000
ES	1.000	1.000	0.004	1.000	1.000	1.000	–	1.000	1.000	1.000
AU	1.000	0.004	1.000	1.000	0.168	1.000	1.000	–	1.000	1.000
FR	1.000	1.000	0.033	1.000	1.000	1.000	1.000	1.000	–	1.000
NL	1.000	0.099	1.000	1.000	0.693	1.000	1.000	1.000	1.000	–

4.4 Overarching Usage Patterns

By combing the findings from each discussed metric, we were able to visualize multi-dimensional usage patterns for video consumption in mobile applications across the considered countries (RQ1). Figure 1 shows all groups of identified countries with similar usage patterns. Hereby, each metric was encoded differently, while at the same time, more general patterns become visible. Most notably, users from India and the United States show both relatively unique usage patterns. India from a separate group in terms of mobile usage and cellular usage, while the United States represents a unique group in terms of mobile usage and download behavior. For the remaining groups and countries, no overarching multi-dimensional patterns were revealed. Similar usage traits are shared across continents and no further grouping with multiple metrics can be detected. Therefore, customization and enhancements to course delivery methods for different regions should rather be made on a per-metric level.

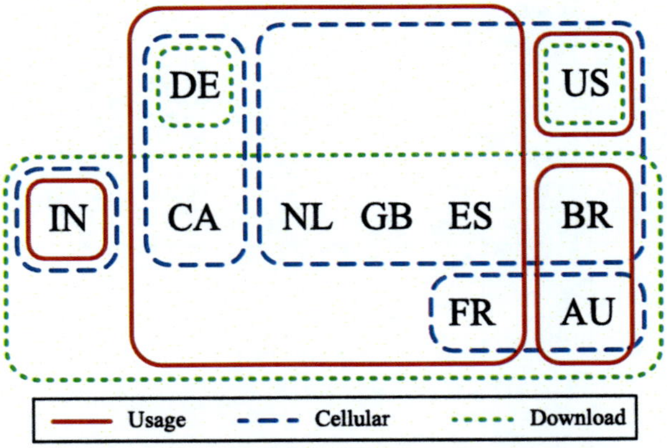

Fig. 1. Grouping of countries with similar mobile application usage patterns.

5 Conclusion

In this observational study, we evaluated the users' usage of mobile applications in a global Enterprise MOOC context. For this, three metrics about the video playback context have been defined and applied to events tracked in 2019. Users in India showed a higher adoption rate of mobile applications compared to other countries, while users in the United States are more reluctant in their usage (RQ1a). The use of cellular data for video playbacks was higher in India, whereas users in Germany and Canada avoided the usage of cellular data. The use of cellular data correlates with the costs of mobile data in the respective countries (RQ1b). Users in Germany downloaded more videos before playback. However, users who avoid cellular data do not automatically download video content to the mobile device (RQ1c). By combining the results from all three metrics, it becomes apparent that India and the United States form unique groups in terms of mobile usage. For the remaining groups and countries, no overarching multi-dimensional patterns could be revealed (RQ1). Enhancements to course delivery methods for different regions should, therefore, be made for each metric individually. For all studied dimensions, small to medium practical effects could be proven. Nevertheless, we are confident that this study further contributes to a better understanding of the different aspects of how mobile applications for MOOCs are used in a global enterprise context. In turn, this allows MOOC platform providers and course designers to improve the overall learning experience.

Appendix

Table 7. Video playbacks on mobile applications: effect sizes by Cohen's d.

	IN	DE	US	GB	BR	CA	ES	AU	FR	NL
IN	–	0.318	0.405	0.278	0.190	0.277	0.247	0.191	0.262	0.299
DE	0.318	–	0.110	0.015	0.136	0.011	0.054	0.135	0.031	0.024
US	0.405	0.110	–	0.135	0.275	0.132	0.184	0.278	0.157	0.093
GB	0.278	0.015	0.135	–	0.113	0.004	0.037	0.113	0.016	0.039
BR	0.190	0.136	0.275	0.113	–	0.115	0.075	0.001	0.094	0.147
CA	0.277	0.011	0.132	0.004	0.115	–	0.041	0.115	0.019	0.035
ES	0.247	0.054	0.184	0.037	0.075	0.041	–	0.074	0.021	0.076
AU	0.191	0.135	0.278	0.113	0.001	0.115	0.074	–	0.094	0.148
FR	0.262	0.031	0.157	0.016	0.094	0.019	0.021	0.094	–	0.054
NL	0.299	0.024	0.093	0.039	0.147	0.035	0.076	0.148	0.054	–

Table 8. Video playbacks with cellular data: effect sizes by Cohen's d.

	IN	DE	US	GB	BR	CA	ES	AU	FR	NL
IN	–	0.842	0.630	0.626	0.535	0.889	0.576	0.210	0.149	0.652
DE	0.842	–	0.218	0.223	0.329	0.090	0.280	0.706	0.783	0.191
US	0.630	0.218	–	0.001	0.097	0.291	0.051	0.438	0.503	0.031
GB	0.626	0.223	0.001	–	0.098	0.303	0.052	0.432	0.495	0.031
BR	0.535	0.329	0.097	0.098	–	0.402	0.045	0.334	0.396	0.128
CA	0.889	0.090	0.291	0.303	0.402	–	0.356	0.740	0.817	0.277
ES	0.576	0.280	0.051	0.052	0.045	0.356	–	0.373	0.434	0.082
AU	0.210	0.706	0.438	0.432	0.334	0.740	0.373	–	0.059	0.450
FR	0.149	0.783	0.503	0.495	0.396	0.817	0.434	0.059	–	0.509
NL	0.652	0.191	0.031	0.031	0.128	0.277	0.082	0.450	0.509	–

Table 9. Video playbacks with downloaded content: effect sizes by Cohen's d.

	IN	DE	US	GB	BR	CA	ES	AU	FR	NL
IN	–	0.167	0.137	0.059	0.141	0.091	0.075	0.044	0.044	0.023
DE	0.167	–	0.294	0.100	0.027	0.072	0.087	0.195	0.114	0.174
US	0.137	0.294	–	0.210	0.295	0.248	0.230	0.104	0.201	0.128
GB	0.059	0.100	0.210	–	0.075	0.030	0.014	0.102	0.015	0.080
BR	0.141	0.027	0.295	0.075	–	0.046	0.061	0.179	0.090	0.155
CA	0.091	0.072	0.248	0.030	0.046	–	0.016	0.135	0.045	0.111
ES	0.075	0.087	0.230	0.014	0.061	0.016	–	0.119	0.029	0.095
AU	0.044	0.195	0.104	0.102	0.179	0.135	0.119	–	0.091	0.022
FR	0.044	0.114	0.201	0.015	0.090	0.045	0.029	0.091	–	0.068
NL	0.023	0.174	0.128	0.080	0.155	0.111	0.095	0.022	0.068	–

References

1. Bothe, M., et al.: A quantitative study on the effects of learning with mobile devices in MOOCs. In: 2019 IEEE International Conference on Engineering, Technology and Education (TALE), pp. 1–7. IEEE, Yogyakarta, Indonesia (2019). https://doi.org/10.1109/TALE48000.2019.9226023.
2. Bothe, M., Meinel, C.: Applied mobile-assisted seamless learning techniques in MOOCs. In: Calise, M., et al. (eds.) Digital Education: At the MOOC Crossroads Where the Interests of Academia and Business Converge, pp. 21–30. Springer International Publishing, Cham (2019). https://doi.org/10.1007/978-3-030-19875-6_3
3. Bothe, M., Meinel, C.: On the potential of automated downloads for MOOC content on mobile devices. In: 2020 IEEE Learning With MOOCS (LWMOOCS), pp. 58–63. IEEE, Antigua Guatemala, Guatemala (2020). https://doi.org/10.1109/LWMOOCS50143.2020.9234338
4. Brady, K., et al.: Who downloads online content and why? In: Proceedings of the Fifth Annual ACM Conference on Learning at Scale, pp. 1–4. ACM, London United Kingdom (2018). https://doi.org/10.1145/3231644.3231699

5. Brooks, C., et al.: Who you are or what you do: comparing the predictive power of demographics vs. activity patterns in massive open online courses (MOOCs). In: Proceedings of the Second ACM Conference on Learning @ Scale, pp. 245–248. ACM, Vancouver BC Canada (2015). https://doi.org/10.1145/2724660.2728668
6. Dalipi, F., Imran, A.S., Idrizi, F., Aliu, H.: An analysis of learner experience with MOOCs in mobile and desktop learning environment. In: Kantola, J.I., Barath, T., Nazir, S., Andre, T. (eds.) Advances in Human Factors, Business Management, Training and Education. AISC, vol. 498, pp. 393–402. Springer, Cham (2017). https://doi.org/10.1007/978-3-319-42070-7_36
7. Dhawal Shah, D.: By the numbers: MOOCs in 2019. https://www.classcentral.com/report/mooc-stats-2019/. Last Accessed 10 Mar 2021
8. Egloffstein, M., Schwerer, F.: Participation and achievement in enterprise MOOCs for professional development: initial findings from the openSAP university. In: Sampson, D., Spector, J.M., Ifenthaler, D., Isaías, P., Sergis, S. (eds.) Learning Technologies for Transforming Large-Scale Teaching, Learning, and Assessment, pp. 91–103. Springer, Cham (2019). https://doi.org/10.1007/978-3-030-15130-0_6
9. Gamage, D., et al.: Exploring mOOC user behaviors beyond platforms. Int. J. Emerg. Technol. Learn. **15**(8), 161–179 (2020)
10. International Telecommunication Union: ICT facts and figures 2019 (2019)
11. Kizilcec, R.F., Schneider, E.: Motivation as a lens to understand online learners: toward data-driven design with the OLEI scale. ACM Trans. Comput. Hum. Interact. **22**(2), 1–24 (2015). https://doi.org/10.1145/2699735
12. Marco, F.A., et al.: What happens when students go offline in mobile devices? In: Proceedings of the 17th International Conference on Human-Computer Interaction with Mobile Devices and Services Adjunct, pp. 1199–1206. ACM, Copenhagen Denmark (2015). https://doi.org/10.1145/2786567.2801609
13. Renz, J., et al.: MOOC to go. ERIC (2014)
14. Rohloff, T., et al.: openSAP: learner behavior and activity in self-paced enterprise MOOCs. Int. J. Adv. Corp. Learn. **13**(2), 30 (2020). https://doi.org/10.3991/ijac.v13i2.16531
15. Ruipérez-Valiente, J.A., et al.: Macro MOOC learning analytics: exploring trends across global and regional providers. In: Proceedings of the Tenth International Conference on Learning Analytics & Knowledge, pp. 518–523. ACM, Frankfurt Germany (2020). https://doi.org/10.1145/3375462.3375482
16. SAP SE: SAP Integrated Report 2020
17. Sharples, M., Spikol, D.: Mobile Learning. In: Duval, E., Sharples, M., Sutherland, R. (eds.) Technology Enhanced Learning, pp. 89–96. Springer, Cham (2017). https://doi.org/10.1007/978-3-319-02600-8_8
18. Sher, V., et al.: On multi-device use: using technological modality profiles to explain differences in students' learning. In: Proceedings of the 9th International Conference on Learning Analytics and Knowledge, pp. 1–10. ACM, Tempe, AZ, USA (2019). https://doi.org/10.1145/3303772.3303790
19. van der Sluis, F., et al.: Mobile screen size limits multimodal synergy. In: Proceedings of the 36th European Conference on Cognitive Ergonomics, pp. 1–4. ACM, Utrecht, Netherlands (2018). https://doi.org/10.1145/3232078.3232101
20. Van den Berg, J., Van der Lingen, E.: An empirical study of the factors affecting the adoption of mobile enterprise applications. South Afr. J. Ind. Eng. **30**, 1 (2019). https://doi.org/10.7166/30-1-1992
21. Worldwide Mobile Data Pricing: The cost of 1GB of mobile data in 228 countries. https://www.cable.co.uk/mobiles/worldwide-data-pricing/. Last Accessed 01 Mar 2021

"All Zoomed Out": Strategies for Addressing Zoom Fatigue in the Age of COVID-19

Angela N. Bullock[1](✉), Alex D. Colvin[2], and M. Sebrena Jackson[3]

[1] University of the District of Columbia, Washington, DC 20008, USA
angela.bullock@udc.edu

[2] Department of Social Work, Texas Woman's University of Denton, Denton, TX 76204, USA
acolvin3@twu.edu

[3] School of Social Work, The University of Alabama, Tuscaloosa, AL 35487, USA
msebrena@ua.edu

Abstract. The COVID-19 pandemic has impacted institutions of higher education, and as a result, many educators are now experiencing Zoom fatigue. Institutions of higher education have moved to online or remote learning and are integrating web conferencing tools such as Zoom, WebEx, and Adobe Connect for course delivery. Many educators have reported an overutilization of technology, which has resulted in technostress. Technostress has been shown to affect physical and mental health. As more institutions of higher education move to remote learning, it will require a collective response from educators and from colleges and universities. The purpose of this article is to explore how the technostress model can be used as the framework to provide strategies to recognize and address Zoom fatigue.

Keywords: COVID-19 · Remote learning · Technology · Technostress · Online education · Zoom fatigue

1 Introduction

COVID-19 has impacted millions of people worldwide [1], and many people are now adopting new ways to interact, learn, and work as a result of the coronavirus pandemic. Specifically, COVID-19 has significantly impacted how institutions of higher education (IHEs) operate, many of which have implemented school/university closures [2]. As a continuity plan and to avoid massive disruptions of instruction due to these closures, many IHEs are now offering courses online [3]. Consequently, IHEs are utilizing technologies similar to those used for remote work [2]. These technologies include video conferencing tools such as Zoom, WebEx, Blackboard Collaborate, Skype, Adobe Connect, GoToMeeting, and Microsoft Teams for online synchronous course delivery [4].

Video conferencing tools operate as methods of communication that allow connected users to share video and audio in real time by allowing users to share files, slides, static images, and text through the platform being used [5]. One video conferencing platform

D. Guralnick et al. (Eds.): TLIC 2021, LNNS 349, pp. 61–68, 2022.
https://doi.org/10.1007/978-3-030-90677-1_6

that has grown in popularity since the COVID-19 pandemic is Zoom. Approximately 90,000 schools in 20 countries use Zoom [6]. Video conferencing tools such as Zoom have increased the social presence of the learning environment, provided instructors with the ability to observe students working in real-life learning experiences, and increased the effective use of academic time [7]. Along with the new adoptions of video-based platforms, the number and length of video-based meetings and classes for instructors and students have significantly increased, which has resulted in individuals experiencing what many describe as Zoom fatigue [8].

Zoom fatigue or video conferencing fatigue arises when individuals spend too much time looking at computer/phone screens, and it manifests as emotional, psychological, and physical exhaustion. This is sometimes further compounded by the stress caused by COVID-19 and can impact individuals' health. Zoom fatigue can have notable biological, psychological, and social impacts and can be recognized by a number of indicators. Video conferencing requires individuals to communicate in different ways with prolonged eye contact, a reduction in body gestures, and an increase in facial expressions [9]. Adding to this, individuals often remain in the same spot, which can result in users experiencing headaches or migraines, eye irritation and pain, blurred and double vision, and excessive tearing and blinking. Other recognizable markers may include a lack of focus or finding it difficult to stay on track during discussions [9]. Although the factors that contribute to Zoom fatigue are complicated and multifaceted, a closer examination can lead to strategies to deal with Zoom fatigue [10]. This article will examine how the technostress model can be used as the framework to provide strategies to recognize and address Zoom fatigue.

2 Conceptual Framework

Modern information and communication technologies (ICTs) have changed the work environment and culture with the increased use of email, electronic scheduling, and video conferencing [11]. Although the growth in ICTs has had many positive outcomes, this same growth has also led to misuse and overuse, resulting in over 25% of the general working population complaining of lack of time and energy due to digital technology [12]. Due to this technology overuse, the resulting lack of energy and distress has been referred to as technostress [13]. Technostress has been shown to affect physical health, with symptoms that include headaches, stiff shoulders, backaches, anxiety, and fatigue [14, 15].

The definition of technostress has seen many iterations. Clinical psychologist Craig Brod [16] first defined technostress as a "disease of adaptation caused by an inability to cope with the new computer technologies in a healthy manner" (p. 16). Weil and Rosen [17] expanded the definition of technostress to include "any negative impact on attitudes, thoughts, behaviors, or body psychology caused directly or indirectly by technology" (p. 5). Ragu-Nathan et al. [11] defined technostress as an "individual's attempts to deal with constantly evolving ICTs and the changing physical, social, and cognitive responses demanded by their use" (p. 418). Tarafdar et al. [18] extended the definition and defined technostress as a stress experienced by users as a result of emerging applications, multitasking, constant connectivity, information overload, frequent system upgrades, constant uncertainty, and continual relearning, as well as technical problems associated with the organizational use of ICT. Other researchers have taken a closer look at technostress and revealed main aspects of this phenomenon, such as conditions that create technostress, adverse effects of technostress on work life, and antecedents of technostress [19]. La Torre et al. [20] noted that the definition of technostress changed from the 1980s — when it was defined as the inability to cope with new ICTs in a healthy manner — to the negative impact of physical or mental attributes caused directly or indirectly by technology.

Conditions that create technostress include techno-overload, techno-invasion, techno-complexity, techno-insecurity, and techno-uncertainty [18, 19]. Techno-overload describes situations in which use of information systems (IS) forces professionals to work more and to work faster. Professionals are forced to change their work habits to adapt to new technologies. Techno-invasion describes situations in which professionals can potentially be reached anywhere and anytime and feel the need to be constantly connected. Professionals feel that their personal lives are being invaded by the IS they use at work. Techno-complexity describes situations in which the complexity associated with IS forces professionals to spend time and effort in learning and understanding how to use new applications. The professionals feel that they cannot find enough time to study and upgrade their technology skills. Techno-insecurity emerges in situations in which users feel threatened about losing their jobs to other people who have a better understanding of new IS. Users believe they have to constantly update their skills to avoid being replaced. Techno-uncertainty refers to contexts in which continuing changes and upgrades to IS do not give professionals a chance to develop a base of experience for a particular application or system [18].

Tarafdar et al. [18] found seven consequences of technostress conditions: (1) technostress conditions exacerbate role overload, (2) technostress conditions are associated with increased role conflict, (3) technostress conditions are linked to reduced job satisfaction, (4) technostress conditions decrease professionals' innovation in their tasks while using IS, (5) technostress conditions reduce professionals' productivity while using IS in their work, (6) technostress conditions causes professionals to be dissatisfied with the IS they use, and (7) technostress conditions are associated with reduced commitment of professionals to their current organizations' goals and values. The researchers also found that gender and computer confidence have major influences on technostress. Interestingly, men experienced more technostress than women. Professionals with greater computer confidence experienced less technostress because they likely had more faith in their ability to handle any disruptions arising from technostress-creating conditions [18].

3 Challenges Experienced

The rapid pivot to remote learning necessitated the use of video conferencing tools for real-time meetings and synchronous class sessions with students. This rapid pivot created multiple challenges for students and faculty alike. First, many students and faculty were not familiar with the video conferencing tools that universities rolled out. With little time for training, faculty and students essentially moved from being casual users to expert users of video conferencing tools. Many had to learn the tool as they used it. During an already stressful time due to multiple changes as a result of the COVID-19 pandemic, this lack of preparedness increased stress levels. Second, as many students chose to return to their homes or remain home to continue the semester, faculty and students had to navigate the challenges of various time zones for synchronous meetings [21]. Requiring students to actively engage in class discussions during a live session at 11:00 p.m. or 5:00 a.m., or to take exams during these times, certainly is unrealistic and creates stress and fatigue.

Another challenge experienced by faculty and students, particularly located in rural areas, is the demand on internet bandwidth required when using webcams. Although the Federal Communications Commission (FCC) set a minimum data transmission speed for broadband service that would be deemed adequate to stream video and participate in other high-traffic online activities, those minimum speeds still are not readily available in many rural areas [22]. Students in rural areas may experience difficulty connecting to video conferencing tools or streaming video for a full 90-min class session. According to Bullock and Colvin [23], lack of access to the minimally accepted broadband speed can create a digital divide, as students try to access course content and participate in live video conferencing activities. This lack of access places students at a disadvantage, and instructors need to be mindful of this as assignments are created.

Finally, the increased usage of video conferencing tools like Zoom has made it challenging for faculty and students to concentrate during the live sessions. According to researchers, video calls require users to focus more intently on conversations in order to absorb information. Users attempt to show that they are paying attention by looking at the camera, which can make one feel uncomfortable and tired [24].

4 Strategies for Addressing Zoom Fatigue

Although technostress is not a new phenomenon, COVID-19 has led to the widespread impact of what is now being called Zoom fatigue. Practical strategies exist that can help educators control physical space, set boundaries, and ultimately prevent and/or neutralize Zoom fatigue. Recommendations for educators to control physical space include keeping the laptop or desktop at a comfortable height, avoiding cell phone usage during Zoom sessions, and setting up an external webcam instead of using the laptop/computer camera so the webcam can be moved into a comfortable position. Furthermore, making specific adjustments to the workspace, such as propping the screen up with a couple of books to create a straight line from one's face to the people on the screen can help one to see micro-expressions and feel more connected [25]. Additionally, incorporating built-in breaks for educators and students, along with setting up and adhering to a specific start time and end time, can also help to control physical space and avoid feelings of sluggishness [26]. Fossilen and West Duffy [24] also propose taking mini breaks during long video calls by minimizing the window, moving it behind an application, or occasionally looking away from the computer completely for a few seconds.

McWhirter [26] suggests that educators should create boundaries: set virtual office hours that are separate from teaching hours, physically leave the teaching/office space once tasks are complete for the day, continue to dress professionally when teaching and meeting with students, and change clothes once daily tasks are complete. Leazenby

[27] further supports these strategies by recommending the following: (1) schedule time to vent to friends or family, (2) change scenery between meetings, and (3) put away computers when finished with work or school to complete that part of the day. McWhirter [26] reminds educators that, although a strong desire to assist and be available to students exists, educators must recognize the importance of personal time, family time, and creative time, along with the need for rejuvenation.

5 Recommendations

Colleges and universities have the potential to support faculty through the challenges of long-term engagement with video conferencing tools as a result of the pandemic. Studies show that university support can help reduce a faculty member's stress levels caused by technology use [28–30]. Stress reduction can occur by providing faculty with end-user training for video conferencing tools. This can help to increase technology literacy and confidence in technology use, which have been shown to reduce technostress. Stress-reduction tools also include creating spaces for faculty to engage in discussions and an exchange of ideas that focus on self-care and wellness. Since faculty may experience social and psychological effects of isolation, creating an initiative to ensure that faculty are aware of how to access psychological and social support services, and resources offered through the institution, can be helpful. This initiative can include ensuring that faculty understand how to access college or university Employee Assistance Program (EAP) resources virtually, in the event that faculty are not able to return to the institution. For instance, colleges and universities can promote their EAP online resources and provide instructions on how to access them. Specifically, institutions can make faculty aware of an After-hours Crisis Line offered through EAP, through which faculty can access and speak directly with a crisis counselor. EAP may also provide additional services such as the LiveHealth Online Psychology program. LiveHealth Online Psychology is a program that involves virtual visits with a licensed counselor, therapist, or psychiatrist at no cost. Through the LiveHealth Online Psychology program, typical in-person counseling sessions can be accessed by video via Zoom, Microsoft Teams, or other telehealth platforms. Through this program, college or university employees can make a virtual counseling appointment [31]. For additional appointments, the EAP counselor can give more contact information for a therapist who accepts the insurance plans of the employee.

Furthermore, institutions can work to implement health and well-being initiatives. Notably, since physical activity is associated with reduced risk of fatigue [32], creating a health initiative where faculty can share their mindfulness/meditation or exercise routines can offer ideas for reducing stress and fatigue and for increasing wellness. An innovative approach can include developing a Health and Well-being Canvas course. The course can be broken up into individual modules with specific theme areas, such as Eat Well or Mind Well, Introduction to Mindfulness, Building a Meditation Practice, Nutrition and Stress, and Anxiety Awareness and Management. Different modules can be offered each week. For instance, the Introduction of Mindfulness module can offer short, 20-min interactive activities to learn about the science and practical application of mindfulness, which can include strategies on how to reduce high blood pressure, improve sleep quality, and improve immune function. Within the module addressing Building a Meditation Practice, a webinar can be introduced to explore strategies to build a daily meditation practice [33]. Additionally, within this module, on-demand meditation videos can be offered, which can provide instructions on how to reboot the mind in just 10 min a day to help build sustainable habits. The Stress and Nutrition module can offer recorded webinars and resources that focus on stress and emotional eating, along with its impact on physical well-being. The module can also introduce readings or healthy eating options and provide strategies to address cravings in times of stress. Lastly, Anxiety Awareness includes a podcast that can be integrated into the course; it explores the stigma and misinformation of anxiety and offers a variety of coping strategies for managing the experience of anxiety due to stress [33]. Additional supplemental materials such as readings, videos, podcasts, and apps can be included within the Canvas modules to augment the interactive activities. When health and wellbeing are addressed, faculty are more productive and can more easily find satisfaction in life.

6 Conclusion

As IHE continue to shift to remote learning in response to the COVID-19 pandemic, it will take a collective response from educators and from colleges and universities as a whole. For this reason, both educators and institutional administrators must work together to integrate strategies to aid in alleviating Zoom fatigue for both themselves and their students [26]. If these measures are integrated and successfully accomplished, the future of planning and executing an educational strategy for IHEs during an emergency situation would be less arduous for all.

References

1. Centers for Disease Control and Prevention: Coronavirus disease 2019 (COVID-19). https://www.cdc.gov/coronavirus/2019-ncov/your-health/need-to-know.html 11 December (2020)
2. Xiao, Y., Fan, Z.: From 10 technology trends to watch in the COVID-19 pandemic. World Economic Forum. https://www.weforum.org/agenda/2020/04/10-technology-trends-coronavirus-covid19-pandemic-robotics-telehealth/ 27 April (2020)
3. LeBlanc, P.: COVID-19 has thrust universities into online learning-how should they adapt? Brookings. https://www.brookings.edu/blog/education-plus-development/2020/03/30/covid-19-has-thrust-universities-into-online-learning%E2%81%A0-how-should-they-adapt/ 30 March (2020)
4. Sidpra, J., Gaier, C., Reddy, N., Kumar, N., Mirsky, D., Mankad, K.: Sustaining education in the age of COVID-19: a survey of synchronous web-based platforms. Quant. Imaging Med. Surg. **10**(7), 1422–1427 (2020). https://doi.org/10.21037/qims-20-714
5. Al-Samarraie, H.: A scoping review of videoconferencing systems in higher education. Int. Rev. Res. Open Distrib. Learn. **20**(3), 121–140 (2019). https://doi.org/10.19173/irrodl.v20i4.4037
6. Miller, M.: Zoom CEO says company reached 200 million daily users in March. The Hill. https://thehill.com/policy/cybersecurity/490794-zoom-ceo-says-company-reached-200-million-daily-users-in-march 2 April (2020)
7. Mader, C., Ming, K.: Videoconferencing: a new opportunity to facilitate learning. Clearing House **88**(4), 109–116 (2015). https://doi.org/10.1080/00098655.2015.1043974

8. Venugopal, A.: Psychology professor and students weigh in on Zoom fatigue. https://www.newsbreak.com/news/1580292443322/psychology-professor-and-students-weigh-in-on-zoom-fatigue 8 June (2020)
9. Dodge, M.: 4 signs you have "zoom fatigue" (and what you can do about it). https://www.jobillico.com/blog/en/4-signs-you-have-zoom-fatigue-and-what-you-can-do-about-it/ 22 May (2020)
10. Lee, J.: A neuropsychological exploration of zoom fatigue. Psychiatric Times. https://www.psychiatrictimes.com/view/psychological-exploration-zoom-fatigue 17 November (2020)
11. Ragu-Nathan, T.S., Tarafdar, M., Ragu-Nathan, B.S., Tu, Q.: The consequences of technostress for end users in organizations: conceptual development and empirical validation. Inf. Syst. Res. **19**(4), 417–433 (2008). https://doi.org/10.1287/isre.1070.0165
12. Korunovska, J., Spiekermann, S.: The effects of digitalization on human energy and fatigue: A review. https://arxiv.org/pdf/1910.01970.pdf (2019)
13. Brivio, E., et al.: Preventing technostress through positive technology. Front. Psychol. **9**, 1–5 (2018). https://doi.org/10.3389/fpsyg.2018.02569
14. Okonoda, K.M.: Prevalence and correlates of technostress among academic staff at the University of Jos, Nigeria. J. Med. Sci. Clin. Res. **05**(03), 18616–18624 (2017). https://doi.org/10.18535/jmscr/v5i3.57
15. Marchiori, D.M., Mainardes, E.W., Rodrigues, R.G.: Do individual characteristics influence the types of technostress reported by workers? Int. J. Hum. Comput. Interact. **35**, 218–230 (2019). https://doi.org/10.1080/10447318.2018.1449713
16. Brod, C.: Techno Stress: The Human Cost of the Computer Revolution. Addison-Wesley (1984)
17. Weil, M.M., Rosen, L.D.: Technostress: Coping with Technology @ Work @ Home @ Play. John Wiley & Sons (1997)
18. Tarafdar, M., Tu, Q., Ragu-Nathan, T.S., Ragu-Nathan, B.S.: Crossing to the dark side: examining antecedents and consequences of technostress. Commun. ACM **54**(9), 113–120 (2011). https://doi.org/10.1145/1995376.1995403
19. D'Arcy, J., Gupta, A., Tarafdar, M., Turel, O.: Reflecting on the "dark side" of information technology use. Commun. Assoc. Inf. Syst. **35**(5), 109–118 (2014). https://doi.org/10.17705/1CAIS.03505
20. La Torre, G., Esposito, A., Sciarra, I., Chiappetta, M.: Definition, symptoms and risk of techno-stress: a systematic review. Int. Arch. Occup. Environ. Health **92**(1), 13–35 (2018). https://doi.org/10.1007/s00420-018-1352-1
21. Zheng, K.: Students and faculty discuss time zone difficulties in adjustment to online classes. The Michigan Daily. https://www.michigandaily.com/section/academics/students-and-faculty-discuss-time-zone-difficulties-adjustment-online-classes (2020)
22. Strover, S.: Reaching rural America with broadband internet service. https://theconversation.com/reaching-rural-america-with-broadband-internet-service-82488 16 January (2018)
23. Bullock, A., Colvin, A.: Rural access. In: Danver, S. (ed.) The Sage Encyclopedia of Online Education. Mesa Verde Publishing, Washougal, WA (2016). https://doi.org/10.4135/9781483318332.n309
24. Fossilen, L., West Duffy, M.: How to combat Zoom fatigue. Harvard Business Review. https://hbr.org/2020/04/how-to-combat-zoom-fatigue 29 April (2020)
25. Walker, T.: How Zoom fatigue impacts communication with students. NEA News. https://www.nea.org/advocating-for-change/new-from-nea/how-zoom-fatigue-impacts-communication-students 16 October (2020)
26. McWhirter, J.L.: Are you zoomed out? Dealing with zoom fatigue in the virtual classroom. Choral J. **61**(2), 41–43 (2020)

27. Leazenby, L.: Zoom fatigue, the downside of constant video conferences. The Columbia Chronicle. https://columbiachronicle.com/zoom-fatigue-the-downside-of-constant-video-conferences 6 May (2020)
28. Dong, Y., Xu, C., Chai, C.S., Zhai, X.: Exploring the structural relationship among teachers' technostress, technological pedagogical content knowledge (TPACK), computer self-efficacy and school support. Asia Pac. Educ. Res. **29**(2), 147–157 (2019). https://doi.org/10.1007/s40299-019-00461-5
29. Fuglseth, A., Sørebø, Ø.: The effects of technostress within the context of employee use of ICT. Comput. Hum. Behav. **40**, 161–170 (2014). https://doi.org/10.1016/j.chb.2014.07.040
30. Zorn, T.E.: The emotionality of information and communication technology implementation. J. Commun. Manage. **7**(2), 160–171 (2003)
31. Blue Cross/Blue Shield: LiveHealth online psychology supports members with stress, anxiety, depression. https://www.bcbs.com/news/press-releases/livehealth-online-psychology-supports-members-stress-anxiety-depression (2020)
32. Puetz, T.W.: Physical activity and feelings of energy and fatigue: epidemiological evidence. Sports Med. **36**(9), 767–780 (2006)
33. Employee Retirement Systems of Texas: Wellness webinars. https://ers.texas.gov/event-calendars/wellness-events (2021)

Online Learning Always Happens Somewhere: Where and When Will Office Workers Learn Post-pandemic?

Imogen Casebourne(✉)

Oxford University, Oxford, UK
imogen.casebourne@education.ox.ac.uk

Abstract. While advances in learning technologies offer potential for new approaches to creating effective and impactful workplace learning, they are unlikely to reach their full potential if consideration is not given to *where* learners will be when they engage with the technology and *when* they are expected to do so. The pandemic has highlighted the various roles of place in supporting learning for work, even for those people who don't have to go to a specific location in order to do their work. It has also brought about considerable changes related to where and when people work, some of which may not be reversed even if vaccine programs make it safer for people to mix and travel. By giving thought to issues of place and time, organizations may be more likely to be successful in introducing new technologies.

Keywords: Pandemic · Work · Place · Time · Online learning

1 Introduction

In 2020, the LinkedIn organization surveyed 1,260 learning and development (L&D) professionals and 814 learners across multiple countries, as well as 3,080 English-speaking people managers, asking them a series of questions about workplace learning and reporting the findings in their 2021 WorkPlace Learning Report. This makes for interesting reading. A number of the organizations apparently indicated that they intended to remain fully remote after the pandemic, while others intended to offer a hybrid model in the future, with some office working and some remote working. LinkedIn also found, perhaps as a result of this, that 73% of L&D professionals expected to spend less on ILT (instructor-led training) and 79% expected to spend more on online learning [25].

If it is the case that some organizations are switching away from face-to-face instructor-led learning and investing more in online learning *because* they anticipate that fewer of their employees will be co-located in an office in the future, this raises a question: What are the experiences of undertaking work-related learning both at and away from the office?

D. Guralnick et al. (Eds.): TLIC 2021, LNNS 349, pp. 69–79, 2022.
https://doi.org/10.1007/978-3-030-90677-1_7

2 How Did the Pandemic Change Work-Related Learning Practice?

Work-related learning includes a broad spectrum of activities with a variety of aims. For example, individuals may be focused on learning related to their current role, or on career development. Learning may involve formal professional accreditation required for continuing professional practice; it may also involve formalized employer-provided ILT, or eLearning. It will also likely include a component of what has been called *informal* work-related learning, perhaps online in-the-moment research initiated by the learner, or learning occurring as a result of 'just being there': immersed in the physical workplace environment. These various activities involve practices that have different relations with time and place and are likely to have been differently impacted by the changes brought by the pandemic.

When people moved to working entirely from home, this move raised questions about how their approach to what had been ad hoc unplanned place-related learning changed, how scheduled ILT was adapted by Learning and Development departments for the new dispersed audience, and whether people working from home were more or less able to find or make time to undertake asynchronous online learning (such as eLearning) during working hours.

2.1 Learning and Development Strategy During the Pandemic

In 2020 and 2021, the US-based Learning Guild surveyed L&D professionals to find out how the pandemic had changed current activities and plans. Unsurprisingly, one immediate response involved a shift of instructor-led classroom training to online learning[1] especially virtual classrooms [16]. For individual learning practitioners, UK-based CIPD (Chartered Institute of Professional Development) research found that this pandemic-related shift could be challenging for classroom instructors, who suddenly found themselves delivering training online—something that they may not have been used to or trained to do [6]. When UK-based Emerald Works surveyed learners, they found they were nearly twice as likely to report learning online in 2020 than in pre-COVID years [7].

US-based SHRM surveyed 2,278 HR professionals and found that 54% of employers surveyed were offering additional pandemic-related education for employees, and 24% were considering doing so [23]. When UK-based FOSWAY group surveyed directors, managers, and learning professionals (largely in Europe) in May 2020, they found that demand for digital learning had increased both from senior stakeholders and from learners themselves. However, curated content and videos, rather than interactive eLearning courses, were reportedly viewed as more *successful* by respondents[2]. This may have been because of the relatively long lead times typically required to create new interactive SCORM-based eLearning modules in comparison to the time needed for recording video or curating existing content.

---

[1] Nuts and Bolts: How Will You Meet COVID-19 Challenges? | Learning Solutions Magazine.

[2] COVID-19 L&D Research | Fosway Group.

When UK-based Emerald Works surveyed 260 managers and leaders globally, they found that learning departments were viewed somewhat negatively by managers, as disorganized, isolated from the rest of the business, and reactive and transactional rather than strategic [8]. Perhaps as a result of such perceptions, Learning Guild researchers found that training budgets might be cut during the pandemic[3], with L&D professionals losing work either temporarily or permanently and budgets for technology reduced, leading to difficulties in providing online learning (although there was an expectation that budgets would be increased by 2022). Another reported difficulty was a slowness of HR teams to formulate strategies and policy for working and learning from home [16], with SHRM finding that may businesses did not have emergency preparedness plans that covered pandemics or diseases [23]. When the Learning Guild surveyed L&D professionals in 2021, they were likely to report being further ahead or on plan with virtual classrooms, eLearning, and video but behind on more technologically complex longer-term projects such as augmented or virtual reality, chatbots, and artificial intelligence-based approaches [24].

2.2 Learning as a Result of Just Being There

While some work-related learning takes the form of online courses designed and delivered by an educational professional, other work-related learning is dependent on the location and situation. As Billett [4] points out, people learn in workplaces by watching and listening to others around them, by asking questions to clarify understanding, and by trying things out with others nearby to help and give feedback—all things that contribute to learning by *just being there*. Such learning is often described as informal (although there may sometimes by planned and formal aspects to making this happen, such as in internships, induction programs, and apprenticeships).

In his account of mimetic learning, Billett [4] draws on Barsalou's [3] concept of *grounded cognition* and the importance of bodily presence within a workplace, and suggests that the apparent value of *just being there* (being physically or bodily present at a specific place and time within a workplace context) may be associated with how representations in memory are organized using aural, visual, and haptic data—a hypothesis supported by Baddesley's [1] experiment where divers were asked to memorize random lists while under water and later did better at recalling them if once again under water but less well when on dry land. Billet goes on to point out that, if situational and circumstantial factors (related to place) shape learning and cognition, as he argues they do, it may not be straightforward to adapt knowledge from one setting to another, *unless* there are relevant situational cues in both settings. When everyone was separated into their own homes, many physical cues and prompts associated with the office were presumably lost.

Additionally, it was no longer possible to overhear useful discussions or ask questions of colleagues in quite the same way as before. Ad hoc questions of colleagues can be recreated via digital chat, which can also create a sense of immediacy, although there are reports of people feeling overwhelmed by *digital noise* coming from a multiplicity

[3] https://home.kpmg/in/en/home/insights/2020/08/covid-19-and-the-renaissance-of-organisational-learning.html.

of online communication platforms in cases where these have been deployed without clear guidance on use, so it is important for organizations to offer guidance on etiquette[4]. Attempts to virtually recreate more informal social situations, such as virtual cocktail hours, have also been reported to have had mixed success, perhaps because people are tired of looking at screens after a day of online meetings[5].

Global induction strategies deployed by multi-nationals, where everyone was not in the same office (even pre-pandemic when most people were in *an* office), can be helpful in plugging the lack of ad hoc water cooler moments and colleague interactions. Such approaches often draw on libraries of online video stories. Stories, if told well by colleagues who are plausible, may be closer to the way people learn informally in conversation with colleagues [19]. Another important strategy reported by successful departments was to ensure that regular virtual mentoring and feedback sessions were scheduled to make up for the lack of the unscheduled ad hoc conversations that can more easily occur when people are co-located.

2.3 Instructor-led Training Moving Online

Perhaps the most significant change reported by L&D departments involved a widespread and sudden move from physically located training session workshops to synchronous virtual classrooms. Studies have found that virtual classrooms can be as effective as physical classrooms in terms of knowledge retained; however, students may expect virtual sessions to be shorter [17]. Since some moves to online learning could apparently involve virtual classroom sessions lasting seven hours [16], it is likely that many learners will have struggled to maintain concentration. Reports from designers suggest, perhaps not surprisingly, that active group-based sessions have been more successful than sessions involving long periods of listening to a speaker. Splitting large groups into virtual breakout sessions, using virtual seating features so that everyone appears in the same location on screen for all colleagues, and using the Zoom spotlight feature to highlight a group to present are all recommendations[6]. Another group described successfully transforming a multi-hour face-to-face session involving presentations and discussions into a 90 min small group discussion session[7].

A complication that might arise regardless of pedagogic format, depending on the platform, is something that has been called "Zoom fatigue" [21]. It has been suggested that this experience may be related to being forced to stay motionless to remain in frame, the impact of maintaining close eye contact in a way that wouldn't happen in a physical environment (eye gaze), the additional cognitive load involved in creating intentional non-verbal cues, and the effect of seeing oneself online, as though in mirror, on platforms such as Zoom or Teams [2]. It has also been suggested that the ill effects of self-monitoring may impact women more than men [9]. McKinsey recommends instructors

[4] Lars Hyland: Overcoming the lack of HR alignment between learning, employee engagement and performance management – HRreview.

[5] The Business of Learning, Special Episode: Virtual Learning Engagement (trainingindustry.com).

[6] https://kineo.com/resources/the-story-of-the-virtual-classroom-part-two.

[7] 9 Important Things We've Learned About Distance Learning in Lockdown | LEO Learning.

keep video options *on* to create a community feel[8], but virtual classrooms where self-video is not present, or can be switched off, might be less tiring for participants.

Additionally, as Malpas argues, people act in ways that are shaped by the characteristics of the places, spaces, and times within which they are embodied and located [14]. Historically, ILT for office workers took learners *away* from their desks, to a separate location dedicated to learning. Such learning happened at a set *time*, but importantly was *also* associated with rituals and conventions associated with *place*—a classroom where learners were typically encouraged to switch off phones and only check email during breaks, where everyone was concentrating on the same thing, where there was no background noise, and where no-one interrupted. With a move to live synchronous virtual sessions, the set time point remained, but the conventions and affordances of a dedicated place fell away, with learners joining from their (possibly makeshift) desk using the computer they used for other work tasks. Where people don't have access to places dedicated to work and study, they may struggle to deal with interruptions. Not only might the urge to check email or try to undertake other work tasks be more difficult to combat in such circumstances, but participants at home might also be subject to interruptions from household members.

As Gratton points out, commentating on *working* from home, boundaries matter[9]. While the occasional intrusion of children into a workplace context can be charming, as with the BBC news commentator who was suddenly joined by his children while live on air, the reality is that trying to undertake a full-time job while also caring for children is challenging. The experience of being interrupted might be more frequent for some than for others, depending on the size of the home and the arrangements within it. A study found that, while 41% of male respondents had a dedicated room for work (and work-related learning) in their home during the pandemic, only a quarter of women were in the same position [13]. Those without a separate space dedicated to work may be more often interrupted. As Selwyn [21] puts it:

> "The lockdowns have starkly illustrated the limitations of presuming digital education to be an immaterial process. Digital education always takes place somewhere."
>
> "Soon after settling into our lockdown working-from-home routines, we rediscovered some oft-neglected and politically side-lined aspects of online education. Those working from nice large houses have it much better than those working from tiny apartments. Childless couples and couples with grown-up children have it much better than those who home-school their children on top of working full-time jobs."

More flexible asynchronous learning appears suited to those with caring responsibilities in a pandemic, as it can be undertaken at a time of the learner's choosing, rather than at a set time that may clash with competing domestic responsibilities. Rather than longer eLearning courses, this can involve short courses that learners can fit into pockets

---

8 https://www.mckinsey.com/business-functions/mckinsey-accelerate/our-insights/adapting-workplace-learning-in-the-time-of-coronavirus.

9 How to Help Employees Work From Home With Kids (mit.edu).

of dead time. As Sharples found, such designs may help busy learners: *"computer-based learning must fit into the gaps in their busy schedule — in the hospital, at home, when travelling — which means a personal and portable system"* [22]. However, if the scheduling of asynchronous online learning is left to individuals, they may experience difficulties in deciding *when* to engage, as is discussed below.

2.4 Asynchronous Online Learning

Asynchronous online learning has historically been considered to be detached from the specifics of time and place. In fact, it has been described as *anytime, anywhere* [27]. It might include eLearning, virtual simulations (enabling practice in a virtual environment), or libraries of reference material or videos, as well as blended programs involving asynchronous study in addition to virtual or face-to-face workshops. It has been suggested that learning *anytime, anywhere* may be freeing for individuals, as it allows them to choose when and where to learn and to fit learning into their personal schedule [12, 22].

However, my research found that some people struggle to find time for asynchronous learning when in the office during office hours. Offices could be high-tempo environments where learners found it hard to switch to the lower tempo they found necessary in order to concentrate on learning while in the office. As one interviewee said, *"I am full pelt []I don't have that slightly slacker time, so I guess for me it's a choice of do I do the learning or not and I'd rather do it flexibly in the times I can manage."*

Additionally, as interviewees explained, if online learning can be done at any time, rather than being scheduled to happen at a set date in the calendar (as with a workshop), it is all too easy to postpone in favor of something that seems more urgent, and most other work-related tasks were perceived as more urgent, as another interviewee stated: *"Well, you always plan it. And then let's see how that works out, really, depending on what else jumps in the way. You might plan half a day doing some eLearning, or even just two, three hours. And, you know, sometimes that'll work out really well and sometimes loads of things will just pop up."*

Others reported similar difficulties: *"People ask me questions and things need attending to quickly, so… It's hard to just say 'OK, I won't be contactable for this afternoon or whatever'."* A German study similarly found people and organizations failing to schedule time for learning during working hours: *"Although training is an essential part of the corporate competitive ability, the planning of training in terms of time does not comply with other economic/temporal flows. This is a contradiction. Transferring training to the weekend has become commonplace."* [10]. Research undertaken over several years by UK-based Emerald Works consistently found that a perception of lack of time was cited by learners as one of the top *barriers* to completing learning [7]. Romero and Gentil suggest that as time scarcity is acknowledged to be a major factor in online learning *failure*, eLearning should be considered to be "spaceless but not timeless" [18]. However as discussed, places often map to specific practices and associated temporal routines, so there is a question as to how far it can be considered to be spaceless.

Difficulties with setting aside time at work for asynchronous learning appeared related to a perception that online work-related learning was a low-priority activity *"if you say, 'No, I can't do that because I'm doing [eLearning]...' I think that would be a little bit, 'Oh, can you not* **moveit***? Can you not do it another time?'"*.

Relatedly, the fact that eLearning could be undertaken at *any time* made it all too easy to postpone, as this interviewee made clear: *"You can always deprioritize training. Unless you have to do it, you can always deprioritize it."*

Furthermore, not everyone found it easy to *concentrate* in their office environment because of noise: *"I think the open-plan office environment is not conducive to learning. It is, it can be very conducive if you're doing a group-learning thing, but if you're on your own and everyone else is working; it's just constant, you know, phones ringing, people having conversations."*

Before the pandemic, difficulties with concentration in busy offices, and with scheduling resulted in eLearning being undertaken on the commute (by train or bus) or at home during evenings and weekends, when other more urgent tasks were not being undertaken. I examined an anonymized data set from a Learning Management System (LMS) that indicated the times and days (weekdays or weekends) that online eLearning courses were accessed by UK government workers and found that, while most courses were accessed, as one might expect, during working hours, they were also sometimes accessed at evenings and weekends. I went on to interview 24 individuals in a variety of professions and at various levels of seniority. The majority of interviewees described at least occasionally learning for work during their commutes (by train or bus) and also learning for work during the evening or at weekends. During the pandemic, commuting fell away for most, potentially giving some learners more time at home. However, for those spending that time at home with children rather than alone on a train, it may have been less easy to devote the time to study.

At home and outside of working hours, individuals surrounded by family or housemates might struggle to concentrate, and learning could be postponed to marginal times and places, when the learners in question might be tired and less able to concentrate, as these two interviewees explained. One stated, *"they're kind of around, so I often just move to a place where they're not there. Yeah, so that's why. I mean, I don't really like studying in my bedroom, but if it's the only place, I'll go there,"* while another said, *"I tend to do it in the evening after the children have gone to bed."*

Jamieson et al. make clear that *locations* have always been associated with a possible requirement to undertake *duties* and the time needed to undertake such duties may dictate how much time is available for online learning: *"a participant's physical location and their relationship to it (e.g., a domestic site where household duties impose) may dictate the amount of access they have to an online site, thereby restricting their capacity to contribute to shared activities or to access essential course materials"* [11].

The availability of discretionary time is related to household traditions; not everyone experiences the same level of demand to undertake household duties. Prior to the pandemic, Wacjman [26] found that working mothers were most likely to feel time pressured and be constantly multitasking, whereas married fathers were significantly less likely to feel "always rushed," which is perhaps not surprising, given that she also found that the combined *work* hours of mothers exceeded those of men by five hours and those of non-working mothers by 15 h. During the pandemic, Kott [13] found that 43% of women and 26% of men reported spending more than four hours of additional time caring for children. There appears to be anecdotal evidence that, during the pandemic, publication rates decreased for female academics[10], and Selwyn suggests that we need, in the future, to *"locate our understandings of digital education within the gendered politics of domestic labour"* [21].

Interestingly, the Emerald Works learner survey found reports of time as a barrier to learning actually *dropped* during the pandemic. However, this should be treated with caution: considerably fewer people completed the survey during the pandemic in comparison with previous years, and it is possible that people who found themselves with more time on their hands as a result of not having to commute may have been more likely to complete the survey than those whose discretionary time decreased.

My research into mobile work-related learning found some interviewees listening to work podcasts while cooking or ironing. This enabled them to fulfil urgent location-related household duties while also making time for work-related learning. The manual and automated nature of activities such as ironing made the combination possible. Podcasts can be an engaging way of delivering stories, or sharing conversations where colleagues discuss relevant issues or techniques. However, work-related podcasts intended to be listened to at home should be introduced with care, as not everyone will want to do this — as early as 2006, researchers expressed concerns about potential negatives, including stress, that might be involved in feeling expected or required to undertake formal learning while also engaged in informal activities at home:

> "As the boundaries between formal and informal environments for education and learning become ever more permeable [], we may experience an educational version of the negative effects that are emerging in the workplace when formal and informal work time is blurred. These include added stresses due to the invasion of externally controlled work activities into all of one's life-spheres" [5].

In summary, making time for learning is something that workers can find difficult [15] and the fact that asynchronous online learning can be done at *any time*, and in *many places* can make it hard for individuals to decide that it should be their most urgent priority any *particular time*. In considering issues related to the scheduling of learning time, Holm says: "The increasing flexibility of working time and free time in today's society has given rise to the development of a new type of requirement, to be able to manage our time in and amongst many areas of our lives" [10]. This requirement intensified during the pandemic, when the *areas* of work and family no longer mapped to distinct spatial locations separated by physical boundaries and it was also more difficult for some than for others.

2.5 Access to Technology

There are other potential difficulties associated with learning from home: unequal access to devices and unequal quality of connection may introduce problems. A study into responses to home working during the pandemic found that a quarter of respondents had

[10] https://www.thelily.com/women-academics-seem-to-be-submitting-fewer-papers-during-coronavirus-never-seen-anything-like-it-says-one-editor/?.

not received technological support from their organizations or had not received financial support for increased internet connection or software costs; again, such support appeared to be distributed unevenly, with men more likely to be the recipients of this type of employer assistance than women, for example [13].

Differences in support levels leading to variations in quality of connectivity might present a variety of issues, ranging from inability to participate at all, to negative perceptions from colleagues. In 2004, Schoenenberg et al. found that individuals whose poor bandwidth had the effect of slowing their speech as it was heard by other online participants might be perceived as mentally slower by those other participants [20]. Perhaps with these types of issue in mind, some commentators on equality have added the term 'situational impairment'[11] to their lexicon. As organizations move to hybrid ways of working, they should review the support they offer to those employees working some or all of the time from home and ensure that financial support for Wi-Fi or other access to technology is distributed according to a considered and consistent policy.

3 The New Normal

If as the LinkedIn survey predicts, employers continue with remote working or move to hybrid working, they will need to consider issues of time and place. As discussed in Sect. 2, people often found it difficult to engage in online learning in offices — *"it's not quiet in the office and there's always other stuff to do in the office"* — and this is unlikely to change without a deliberate strategy. Online learning could be scheduled for work-from-home days, assuming homework offers a slower tempo where it is easier to reflect and concentrate, as several of my interviewees said: *"at home on my working at home days, that's when I tend to get the time to do that kind of stuff."*

However, before the pandemic, home workers often had their home to themselves during the day, or had a dedicated study. If more people are asked to work from home in the future, this is less likely to be the case. Working from home is unlikely to be conducive to concentration in small and crowded households. Gratton considers this issue[12] and, like Selwyn [21], points to the importance of differences in people's home circumstances, arguing that people with larger houses may be keener to work from home in the future, whereas those with smaller, more crowded living conditions may prefer to return to the office[13].

In moving to hybrid working, organizations could take action to make it easier to concentrate at work by designating quiet areas (especially for those for whom it is not easy to concentrate at home). Organizations could also do more to address uncertainty around priorities and use of time. As mentioned earlier, it may be possible to design online learning to better fit into gaps between other activities. Senior executives could also model and champion online learning during working hours. People attending face-to-face courses were not previously expected to be available for other tasks; encouraging

[11] https://www.usabilityfirst.com/glossary/situational-impairment/index.html.

[12] How to Do Hybrid Right (hbr.org).

[13] Neighborhood cafés were used by some interviewees as an occasional refuge from noisy homes and might be part of a hybrid solution as cafés open up. However, cafés do not have an obligation to provide ergonomic desks and chairs.

a culture where it is understood that people learning online are also not available for other work tasks might help employees feel more confident in engaging in online study at work. Some interviewees proposed exactly this:

> "Would be more likely to do it if my manager said I really think you should take a day a month for learning, if they formalised it in that way and gave me the permission to take time out in work time and understood it would inevitably impact on other stuff"

Even if specific times cannot easily be scheduled by the organization, it can be made clear that learning is considered an organizational *priority* and not to be undertaken only when all other more urgent work tasks are complete, as was the understanding of this interviewee: *"so my main priority is getting the work delivered basically... So, learning, kind of, ends up being the last thing in the pecking order of stuff to do."*

In summary, while advances in AI and other approaches to learning via technology may have exciting potential for work-related learning, this potential is unlikely to be fully realized for workplace learning if careful organizational consideration is not given to where learners will be studying and when they will be doing so.

Acknowledgments. I would like to acknowledge the Learning Technologies Group for their financial support of my doctoral study, as well as my supervisors, Professor Niall Winters and Dr. Susan James Relly for their invaluable guidance and support.

References

1. Baddesley, A., Eyesenck, M.: Memory (2nd ed.). Taylor & Francis (2014)
2. Bailenson, J.N.: Nonverbal overload: a theoretical argument for the causes of zoom fatigue. Technol. Mind Behav. **2**(1) (2021). https://tmb.apaopen.org/pub/nonverbal-overload/release/1
3. Barsalou, L.W.: Grounded cognition. Annu. Rev. Psychol. **59**(1), 617–645 (2008). https://doi.org/10.1146/annurev.psych.59.103006.093639
4. Billett, S.: Mimetic learning in circumstances of professional practice. In: Technology-Enhanced Professional Learning: Processes, Practices, and Tools, pp. 85–96 (2013)
5. Chan, T., et al.: One-to-one technology-enhanced learning: an opportunity for global research collaboration. Res. Pract. Technol. Enhanced Learn. **01**(1), 3–29 (2006)
6. CIPD: Impact of Covid19 on the L&D Profession – Perspectives from independent learning practioners. March, 3–21 (2021)
7. Emerald Group: Learner Intelligence Report (2021a)
8. Emerald Group: Same team, different sides ? Business leaders' thoughts and perceptions of L&D (Issue March) (2021b)
9. Fauville, G., Luo, M., Queiroz, A.C.M., Bailenson, J.N.: Nonverbal mechanisms predict zoom fatigue and explain why women experience higher levels than men (2021)
10. Holm, U.: Lifelong learning and "time competence." Int. J. Lifelong Educ. **26**(1), 25–43 (2007). https://doi.org/10.1080/02601370601151364
11. Jamieson, P., Fisher, K., Gilding, T., Taylor, P.G., Trevitt, A.C.F.: Place and space in the design of new learning environments. High. Educ. Res. Dev. **19**(2), 223–235 (2000)
12. Jones, A., Issroff, K., Scanlon, E.: Affective factors in learning with mobile devices. In: Big issues in Mobile Learning: Report of a Workshop by the Kaleidoscope Network of Excellence Mobile Learning Initiative, pp. 17–22 (2007)

13. Kott, S.: Work in times of COVID-19: what is new and what is not. A Western European perspective. Corona and Work Around the Globe, vol. 7, pp. 225–230 (2020). https://doi.org/10.1515/9783110718249-026
14. Malpas, J.: Place and Experience: A philosphical topograhy (Kindle), Routledge (2018)
15. McNeill, B.: Time and the working online learner. In: Barbera, E., Reimann, P. (eds.) Assessment and Evaluation of Time Factors in Online Learning and Teaching, pp. 24–62. IGI Global (2014)
16. Megan, T., Bozarth, J., Jackson, J.: Covid-19 and L & D: Present and Future. https://www.learningguild.com/insights/252/covid-19-and-ld-present-and-future/?from=home (2020)
17. Morice, A., Jablon, E., Delevaque, C., Khonsari, R.H., Picard, A., Kadlub, N.: Virtual versus traditional classroom on facial traumatology learning: evaluation of medical student's knowledge acquisition and satisfaction. J. Stomatol. Oral Maxillofac. Surg. **121**(6), 642–645 (2020). https://doi.org/10.1016/j.jormas.2020.03.001
18. Romero, M., Gentil, C.: Characterising online learners' time regulation. In: Barbara, E., Reimann, P. (eds.) Assessment and Evaluation of Time Factors in Online Teaching and Learning, pp. 91–110. Information Science Reference (an imprint of IGI Global) (2014)
19. Schank, R.C.: Lessons in Learning, E-learning, and Training: Perspectives and Guidance for the Enlightened Trainer. Pfeiffer, Print (2005)
20. Schoenenberg, K., Raake, A., Koeppe, J.: Why are you so slow? – Misattribution of transmission delay to attributes of the conversation partner at the far-end. Int.J. Hum. Comput. Stud. 477–487 (2014)
21. Selwyn, N., Jandric, P.: Postdigital living in the age of Covid-19: unsettling what we see as possible. Postdigit. Sci. Educ. **2**, 989–1005 (2020)
22. Sharples, M., Sanchez, I.A., Milrad, M., Vavoula, G.: Mobile learning: small devices, big issues. In: Balacheff, N., Ludvigsen, S., Jong, Tde, Barnes, S. (eds.) Technology-Enhanced Learning: Principles and Products, pp. 233–249. Springer, Heidelberg, Germany (2009)
23. SHRM.: COVID-19 RESEARCH HOW THE PANDEMIC IS CHALLENGING AND CHANGING EMPLOYERS.pdf (p. 25). https://shrm.org/hr-today/trends-and-forecasting/research-and-surveys/Documents/SHRM_CV19_Research_Presentation_Release_2.pdf(2020)
24. Torrance, M.: Effects of the Pandemic on Learning Technology Roadmaps (2021)
25. Van Nuys, A.: Workplace Learning Report 2021. Linked In (2021)
26. Wacjman, J.: Pressed for Time the Acceleration of Life in Digital Capitalism. University of Chicago Press (2016)
27. Wong, L.-H.: A brief history of mobile seamless learning. In: Wong, L.-H., Milrad, M., Specht, M. (eds.) Seamless Learning in the Age of Mobile Connectivity, pp. 3–40. Springer Singapore, Singapore (2015). https://doi.org/10.1007/978-981-287-113-8_1

Best Practices for Assessing Digital Literacy and Strengthening Online Teaching Pedagogy of Digitally Immigrant Stakeholders in Higher Education

Elisabeth Counselman-Carpenter(✉) and Jemel Aguilar

Southern Connecticut State University, New Haven, CT 06515, USA
{counselmane1,aguilarj2}@southernct.edu

Abstract. Here, we add to the growing body of e-learning and teaching literature by examining the digital literacy of full-time and adjunct faculty at a public regional secondary education institution, examining the relationship between digital literacy and technology self-efficacy and understanding the role that digital immigration status plays in faculty's technology-based choices. How faculty's digital literacy translates into hybrid and online teaching environments and how faculty categorize their technologically based self-efficacy is also explored, particularly the role of user-based decision making through the lens of the Unified Theory of Acceptance and Use of Technology (UTAUT). Findings suggest that, to engage digitally immigrant and digitally native faculty teaching with technology, training should be scaffolded, with a particular focus on screening and assessing potential participants, with a focus on rigorous training that moves faculty from one stage or behavior change to another. Other suggestions and implications for strengthening online teaching pedagogy are discussed.

Keywords: Digital literacy · Higher education · Online teaching pedagogy

1 Introduction

For a considerable amount of time, faculty members at many higher education institutions across the United States had some level of control over whether to engage in online and hybrid teaching as institutions moved towards increased online and hybrid educational programming. However, e-learning and e-teaching has taken a prominent position in higher education given the COVID-19 pandemic and the 2020 emergency move to online learning with the ongoing public health need for e-learning to slow the transmission of COVID-19 in the United States and around the world. Technology was already rapidly exploding across classrooms and higher education even prior to the pandemic, but the necessary rapid mid-semester transition from face-to-face to e-learning and teaching that followed was unprecedented. The growing focus on teaching online and the forced mandatory technological changes challenged many faculty members to identify opportunities through which they could integrate technology into their teaching

D. Guralnick et al. (Eds.): TLIC 2021, LNNS 349, pp. 80–88, 2022.
https://doi.org/10.1007/978-3-030-90677-1_8

to facilitate student learning. Here, we add to the growing body of e-learning and teaching literature by examining the digital literacy of full-time and adjunct faculty at a public regional secondary education institution, examining the relationship between digital literacy and technology self-efficacy, and understanding the role that digital immigration status plays in faculty's technology-based choices. More specifically, we also explore how faculty's digital literacy translates into hybrid and online teaching environments and how faculty categorize their technologically based self-efficacy.

2 Literature Review

2.1 Digital Immigration

While teaching with technology is an opportunity to enhance teaching and student experiences, as well as course delivery, it can also be overwhelming when faculty are not familiar with instructional design and other education related technologies and yet attempt to integrate them into their teaching practices [9]. For faculty who did not grow up with technology in every-day life, regularly integrating technology into course development can provide unique challenges. Referred by some as digital immigrants, these faculty were exposed to digital technologies as adolescents or adults and are found to have an overall different attitude towards technology compared to digital natives [21] (Kesharwani, 2020). Digital immigrants, in short, categorize faculty as those born before 1980 who did not have access to computers or the internet during their formative years [5, 15]. Digital immigration, as a concept, acknowledges that exposure to and interaction with technology during these formative years or 'apprenticeship socialization period' shapes personal beliefs and attitudes towards technology [4]. Research on digital immigration and stakeholder involvement with technology demonstrate that there are specific factors that inhibit growth within higher education institutions including personal motivation, the amount of time to dedicate to improving digital literacy, and a lack of institutional support [5, 15].

The apprenticeship socialization period for faculty born before 1980 involved typewriters or word processors, so they are more likely to incorporate research and other paper-based assignments because they identify with these modes of instruction. Hence, attitudes towards technology and the apprenticeship socialization periods can interact and, if positive, result in an increase likelihood of integrating technology into one's teaching [4]. Institutions governed by a majority of digitally immigrant faculty members are more likely to experience resistance to or discomfort with technological integration in teaching. Controversies around the concept of digital immigration and related research often call out the 'us versus them' mentality where the focus is exclusively on the age-based comparation of two groups (Kesharwani, 2020). References [21, 22], and Tufts (2010) call for a spectrum view when comparing digital natives, digital immigrants, and exploring technological preferences.

2.2 Digital Literacy

Digital literacy is defined as "…the technological critical thinking skills needed for advancement as new types of digital formats evolve in society" (Littlejohn, Beetham, &

McGill, 2012; p. 547). In general, faculty are commonly identified as an institution's most valuable asset, yet research indicates that even prior to the COVID-19 pandemic, faculty continue to report that they feel the least equipped to teach in technologically based environments when entering higher education (Gardner, Waters & McLaughlin, 2017). Present-day life requires digital skills and knowledge, yet if faculty have limited technological skills, or are not yet digitally literate, then how can they prepare students to work in increasingly technologically based professional sectors [2]? While there is growing research on the role of the flipped classroom and hybrid learning, there continues to be a paucity of knowledge in overall educator self-efficacy in the area of digital literacy of those who have struggled to progress to teaching models based exclusively within the Information Age [3, 5].

Digital literacy scholarship calls upon researchers to demonstrate the level of importance of digital and information literacy and encourages institutions to continually assess faculty member's understanding and implementation of technology in teaching [14]. Social work as a profession must now consider the new context of human development, including technology and human-technological interactions and how this knowledge bears down upon social work education, practice, and research [7]. Despite this mandate to the profession, intervention research is scarce in social work, and more focus is on descriptive aspects of technological integration [5, 23]. Is the scarcity of intervention research driven by digital literacy or digital immigration status? Answers to these questions might be informed by digital literacy's relationship to faculty member's experiences with technology because past experiences with technology influences subsequent comfort levels with available technology [13]. Digital immigrants in contrast to digital natives are more likely to encounter seemingly insurmountable barriers and have greater difficulty identifying how and when to use technology in teaching and learning. Theoretical frameworks integrated into digital literacy studies can create connections that frame the questions posed in the previous paragraphs.

3 Theoretical Frameworks

It is critically important to understand the faculty's usage levels and behavioral determinants influencing technology choices prior to assessing digital. Reference [20] developed the Unified Theory of Acceptance and Use of Technology (UTAUT) to explore technology intentions, technology use, and technological behaviors. Four constructs serve as the foundation for UTAUT: (1) performance expectancy, (2) effort expectancy, (3) social influence, and (4) facilitating conditions. Additionally, gender, age, experience, and voluntariness of use moderate UTAUT relationships. During the pandemic, voluntariness of use became a moderating factor for digitally immigrant faculty teaching as they were required to move to online teaching. In contrast to voluntariness, performance expectancy relates to users' perception that technology improves performance while effort expectancy focuses on users' perceived ease of use. Performance and effort expectancies, therefore, can theoretically relate to digital migrancy though faculty's background [13].

Social influence is users' perception that valued or important stakeholders believe instructors should use technology. Finally, facilitating conditions examines if faculty members believe infrastructure and support for teaching with technology exists. UTAUT explains the relationship between faculty skills, familiarity with devices, and perception [11]; however, UTAUT is rarely used to frame university faculty members' technology adoption. Contrastingly, the Lazy User Model (LUM) posits that faculty members choose solutions to problems based on effort [19]. Collan (2012) adds that faculty opt for solutions they do not need. These theories provide the impetus for assessing digital literacy.

4 Best Practices

4.1 Assessing Digital Literacy

The concept of being 'digitally native' is under considerable debate, including its validity, learning styles, multi- versus mono-tasking, distractibility, and applicability related to student and faculty diversity (Evans & Robertson, 2020). The debates about the concept seem to be less about the veracity of the concept and instead focus on digital literacy's application or measurement. In essence, digital nativity exists on a four-by-four table influenced by the apprenticeship period, facilitating conditions, interest, and risk taking. The four-by-four table includes on the x-axis those born before and after 1980 and technological interest on the y-axis. While reference [18] focuses on mostly technological use, adding measures of technological training, facilitating conditions, and technological risk-taking can improve the classification of being digitally native and strengthen technology integration.

4.2 Strengthening Online Teaching

Technological use and choices are typically overlooked when strengthening online teaching because studies focus on building community and fostering engagement for students [1, 12]. While community and engagement are important, examining how faculty members integrate technology into curriculum design can strengthen online teaching. Consequently, the definition of online teaching to include community building, engagement, curriculum design, technological choices, content delivery, context, and interactions matches the earlier discussion about assessing digital literacy [16, 17].

5 Pilot Study

Personal motivation, time dedicated to digital literacy, and levels of institutional support can inhibit technological adaptation [5]. We, therefore, conducted a mixed method exploration of the relationship between digital immigration, digital literacy, and technology self-efficacy using the Digital Natives Assessment Scale (DNAS) [18], the Technology Self-Efficacy scale [10], and a semi-structured interview. Adjunct and full-time faculty at a New England university were sampled before the emergency transition to online learning. Data were collected sequentially, and survey respondents were invited to a focus group about training, teaching, and technology use. Of the 11 who completed the surveys, four agreed to participate in the 60-min focus group. Qualitative data were analyzed in three rounds. The first round was completed before the qualitative data was analyzed and the second round was completed as the data was mined and analyzed through SPSS, and then finally a final time, after both results were written up. Concurrent triangulation was used to compare results and generate a comprehensive understanding of the research questions [6].

5.1 Findings

A sample of full- and part-time faculty (n = 11) completed the survey. The sample is primarily white (n = 9), female (n = 6), and heterosexual (n = 10) and included those between 45 and 54 years old (n = 4) and those over 65 years of age (n = 4); so, more than 72% of the respondents are digital immigrants. Most respondents in the sample taught mostly face-to-face classes (36.4%) at the Master of Social Work level (54%) for more than 16 years. Despite being digital immigrants, graduate students are entering highly technological practice contexts and practicing with technology immersed clients.

Digital immigrant status based on age is confirmed in the sample mean scores for growing up with technology (M = 31.09), comfort with multitasking (M = 23.73), reliant on graphics for communication (M = 17.36) and thriving on instant gratification and rewards (M = 14.90) are low. The sample's technological self-efficacy as in selecting media to support teaching and learning (M = 3.45), evaluating software for teaching and learning (M = 2.18), integrating technology (M = 3.36), determining when and how to use technology in the curriculum (M = 3.18), selecting assistive technologies (M = 2.91), and using technology to enhance teaching and learning (M = 3.45) is also low.

Using Pearson's R, statistically significant bivariate correlations were identified between: (1) digital immigration and evaluating software for teaching and learning, (2) digital immigration and technology self-efficacy, (3) level of technological self-efficacy and evaluating software for teaching and learning, (4) level of technological self-efficacy and growing up with technology (M = 31.09), (5) technological self-efficacy and comfort with multitasking, (6) growing up with technology and integrating technology, (7) growing up with technology and evaluating software, (8) comfort with multitasking and selecting media to support teaching and learning, (9) reliance on graphics for communication and evaluating software and using technology to enhance teaching and learning, and (10) comfort with multitasking and selecting media to support teaching and learning. These bivariate relationships were supported in the qualitative data through the following five themes: institutional, pedagogical, personal beliefs about technology, personal behaviors related to teaching with technology, and beliefs about faculty development and/or technologically based training (Tables 1, 2 and 3).

Table 1. Selected sample demographics.

	N	%
Male	5	45.5
Female	6	54.5
Transgender	0	0
White	9	81.8
African American	2	18.2
Latino	0	0
Heterosexual	11	100
Gay, lesbian, bisexual	0	0
MSW	6	54
BSW	4	36.4
DSW	1	9.1

Table 2. Digital natives scale scores.

	Mean	SD
Growing up with technology	31.09	4.04
Comfort with multitasking	23.73	6.37
Reliant on graphics for communication	17.36	8.21
Instant gratification and rewards	14.91	4.72
Digital literacy	114.91	27.26

Table 3. Technological self-efficacy.

	Mean	SD
Media to support teaching and learning	3.45	0.82
Determining when and how to use technology in the curriculum	3.18	1.17
Integrating technology	3.36	0.92
Selecting assistive technology	2.91	0.87
Using technology to enhance teaching and learning	3.45	1.04
Evaluating software	2.18	0.82

5.2 Discussion

Embedding best practices for digitally immigrant faculty can include (1) skill level scaffolding, (2) incentives, (3) diversity in training offered to faculty, and (4) best practices supported by theory and research. Organizing training around participant experience and levels of digital literacy can occur through pre-training assessment instruments. Next, scaffolded training programs can build participants' skills and comfort level while exposing them to technological options offered in various lengths of training times, interactivity, and prior knowledge. Shorter one- or two-hour soundbite trainings, or half- and full-day workshops in synchronous and asynchronous formats, permits faculty to choose how and when to strengthen technological teaching skills. Topics such as (1) general comfort building with overall technology through setting alerts, effectively use of email, and online calendar systems; (2) tasks to infuse technology into teaching; (3) decision making in for the science of teaching and learning with technology; and (4) code switching language for technologically based teaching and learning might be interesting to faculty members.

Assessing and addressing negative perspectives of faculty related to technology is critically important in training digitally immigrant faculty. Instructing faculty on the benefits of teaching with technology and the pedagogical reasoning for technology can identify course aspects to consider in the planning and execution phase. Similar to our findings, reference [8] suggested a 'model course' for new faculty that illustrates the learning management system mechanisms, demonstrates ideal instructional design, and formats course modules along with webinars that can orient faculty to the possibilities while also orienting them to the expectations.

Offering incentives for faculty to engage to build technological skills and then demonstrate teaching with technology might increase faculty skill building and engagement. Incorporating student expectations for learning with technology, uniform technology expectations, learning management systems, and instructor availability in syllabi and learning management systems might address some of the concerns faculty have about student facets of teaching with technology. Additionally, concurrent student training in technological skills is also key and reinforces instructors' use of technology in teaching. While digitally immigrant faculty members might experience a steeper learning curve, students also come into the classroom with a wide range of knowledge and skills and need support in diverse ways.

This was conducted as a pilot study, and the sample size remained small; thus, the findings cannot be generalized. The original study design was also meant to include the student perspective; however, due to the COVID-19 pandemic, both faculty and students became overwhelmed by the forced transition to online learning and other pandemic related factors. Faculty were clear that they needed to focus on increased teaching responsibilities rather than donating their time for in-depth follow-up qualitative interviews. In future iterations of this research, the study would be replicated across the entire campus and throughout the consortium of schools through which the research team is affiliated in order to significantly expand the number of participants.

6 Conclusion

Faculty typically conducted face-to-face sessions until widespread quarantines and lockdowns changed the face of online education. Theories of digital immigration in general, and the technological apprenticeship period in particular, shed light on why faculty choose face-to-face or online teaching, as well as choices about technology integration. To engage digitally immigrant and digitally native faculty members teaching with technology, training should be scaffolded after screening potential participants, and participants should then experience rigorous training that moves faculty from one stage or behavior change to another. Given the new 'COVID normal' of online or hybrid teaching, screening can also be conducted at the point of hire with a specific integration technology in mind.

Acknowledgments. The authors gratefully acknowledge the CSU-AAUP Research Grant for funding this study.

References

1. Adams, C., Rose, E.: Will I ever connect with the students? Online teaching and the pedagogy of care. Phenomenol. Pract. **8**(1), 5–16 (2014)
2. Amstutz, D., Whitson, D.: University faculty and information literacy: who teaches the students? Res. Strateg. **15**, 18–25 (1997)
3. Aslan, S., Reigeluth, C.: Educational technologists: leading change for a new paradigm of education. Techtrends Linking Res. Pract. Improve Learn. **57**, 18–24 (2013). https://doi.org/10.1007/s11528-013-0687-4
4. Baek, J.H., Jones, E., Bulger, S., Taliaferro, A.: Physical education teacher perceptions of technology-related learning experiences: a qualitative investigation. J. Teach. Phys. Educ. **37**(2), 175–185 (2018)
5. Belluomini, E. Digitally Immigrant Social Work Faculty: Technology Self-Efficacy and Practice (Doctoral thesis). Retrieved from ProQuest Dissertations and Theses database (UMI No. 10250293) (2016)
6. Creswell, J.W., Plano Clark, V.L.: Designing and Conducting Mixed Methods Research, 2nd edn. Sage Publications, Thousand Oaks CA (2011)
7. CSWE Commission on Educational Policy: Final 2015 Educational Policy (EP). http://www.cswe.org/File/aspx?id=79793 (2015)
8. Gomez, C.: Use existing resources to better support adjunct faculty teaching online: recruiting and retaining adult learners. J. Educators Online **17**, 1–5 (2015)
9. Henderson, M., Selwyn, N., Aston, R.: What works and why? Student perceptions of "useful" digital technology in university teaching and learning. Stud. High. Educ. **42**(8), 1567–1579 (2015). https://doi.org/10.1080/03075079.2015.1007946
10. Kent, A.M., Giles, R.M.: Preservice teachers' technology self-efficacy. SRATE J. **26**(1), 9–20 (2017)
11. Koivumaki, T., Ristola, A., Kesti, M.: The perceptions towards mobile services: a empirical analysis of the role of use facilitators. Pers. Ubiquitous Comput. **12**, 67–75 (2008)
12. Meyers, S.: Using transformative pedagogy when teaching online. Coll. Teach. **56**, 219–224 (2008)
13. Reid, P.: Supporting instructors in overcoming self-efficacy and background barriers to adoption. Educ. Inf. Technol. **22**(1), 369–382 (2015)

14. Pinto, M.: Assessing disciplinary differences in faculty perceptions of information literacy competencies. Aslib J. Inf. Manage. **68A**, 227–247 (2015)
15. Presky, M.: Digital natives, digital immigrants. On the Horiz. **9**, 1–6 (2001)
16. Serdyukov, P.: Does online education need a special pedagogy? J. Comput. Inf. Technol. **23**(1), 61–74 (2015). https://doi.org/10.2498/cit.1002511
17. Steele, J., Holdbeck, R., Mandernach, J.: Defining Effective Online Pedagogy. J. Instruction. Res. **8**, 5–8 (2019)
18. Teo, T.: An initial development and validation of a digital native's assessment scale (DNAS). Comput. Educ. **67**, 51–57 (2013)
19. Tetard, M., Collan, M.: Lazy user theory: a dynamic model to understand user selection of products and services. In: 42nd Hawaii International Conference on System Sciences, pp. 1–9 (2009). https://doi.org/10.1109/HICSS.2009.287
20. Venkatesh, V., Morris, M., Davis, G., Davis, F.: User acceptance of information technology: toward a unified view. MIS Q. **27**, 425–478 (2003)
21. Vodanovich, S., Sundaram, D., Myers, M.: Digital native and ubiquitous information systems. Inf. Syst. Res. **21**, 711–723 (2011)
22. Wang, Q., Myers, M., Sundaram, D.: Digital natives and digital immigrants. Bus. Inf. Syst. Eng. **5**, 409–419 (2013). https://doi.org/10.1007/s12599-013-0296-y
23. Watling, S.: Digital exclusion: potential implications for social work education. Soc. Work. Educ. **31**, 125–130 (2012). https://doi.org/10.1080/02615479.2010.539605

Analyzing User Behavior in a Self-regulated Learning Environment

Sarah Frank[1](✉), Alexander Nussbaumer[2], and Christian Gütl[2]

[1] CAMPUS 02 University of Applied Sciences, Körblergasse 126, 8010 Graz, Austria
sarah.frank@campus02.at
[2] Graz University of Technology, Inffeldgasse 16c/1, 8010 Graz, Austria

Abstract. E-learning systems present an opportunity for both students and the field of education that has been steadily increasing in importance, especially over the past few years. This paper examines an intra-university e-learning course built on the principle of self-regulated learning, with the goal of finding behavioral patterns in the way learners navigated the content. The limitations associated with the amount of data generated by the only 31 students who took part in the course presented a challenge to the popular machine learning algorithms, which ultimately led to a combination of sequence extraction methods and heat maps being used in the evaluation. By classifying users both according to their learning behavior and their course performance, it was possible to put organised and unorganised learning behavior into relation with course performance. The findings show a higher likelihood for students who displayed unstructured learning behavior to repeat assessments than for students demonstrating structured pathways through the learning content, who tended to attain full points on their first attempt. Additionally, students who repeated assessments often reviewed more previously studied content and had shorter sequences where they only viewed content that was new to them. Overall, the specific setting of the course and the data limitations mean that it would be beneficial to reconstruct the research with larger data sets, which would also allow for the usage of machine learning algorithms in the analysis.

Keywords: Self-regulated learning · E-learning · Sequence mining

1 Introduction

Over the past year, use of distance learning tools in traditional education sectors surged, which continues to drive the search for ways to make online courses serve their students better. Often, students have to display a large amount of self-efficacy in distance learning settings, necessitating the look into technological support structures already built into the course. While there is much focus on MOOCs in recent research, this paper focuses on analysing a self-regulated learning (SRL) system that was originally part of a mostly in-person class at an Austrian university partway through the semester.

Zimmerman [1] describes SRL as the meta-cognitive, motivational, and behavioural activities that enable learners to take over control of their own learning process. Instead

D. Guralnick et al. (Eds.): TLIC 2021, LNNS 349, pp. 89–98, 2022.
https://doi.org/10.1007/978-3-030-90677-1_9

of following external instructions of teachers, learners decide autonomously on their own learning behaviour. Dabbagh and Kitsantas [2] describe six key processes that are essential for performing SRL, namely goal setting, help seeking, time management, self-monitoring, task strategies, and self-evaluation. Hence, self-regulated learners define their own learning goals, manage the timing of their learning, ask for help in case of difficulties, select and apply appropriate learning strategies, monitor their knowledge acquisition, and evaluate their learning outcome in light of the self-defined goals. In general, applying self-regulatory activities in the learning process has a moderate positive effect on academic achievement, but it can also be not beneficial if learners choose to apply inefficient techniques [3]. However, support or training of SRL activities can improve the achievements of learners [3]. Consequently, it's a key question how such support for SRL can be designed and provided to learners.

Supporting SRL in the right way is a crucial factor. While some learners are capable of applying meta-cognitive activities and learn self-regulated, others face difficulties in carrying out meta-cognitive activities and perform less successfully [4]. Therefore, concrete support techniques can particularly help increase SRL capabilities. A variety of methods are presented in [5] that provide specific support based on the SRL skills. For example, explanatory videos introduce the basic concept, training courses help apply the key concepts, an activity recommender gives hints regarding what to do next, and a tool recommender supports the compilation of the learning environment. In particular the creation and use of a learning environment tailored to one's own capabilities is a powerful way of increasing SRL skills [6].

As outlined above, tailored support for particular meta-cognitive activities is a key aspect to help learners develop and increase their SRL skills. This paper provides a contribution on how to provide such support by identifying individual weaknesses. For the sake of an experimental setting, a simplified learning environment has been created that provides access to a short computer science course covering a selected topic in Information Search and Retrieval. This online activity was created to serve as a substitute for a two-hour lecture unit as part of an on-site lecture. The course consists of a set of web pages organised in three sections, whereby each section provides an overview of the learning goal, five content pages, and an assessment. Furthermore, the instruction page gives general guidelines to the learning environment, the index page at the beginning gives an overview and direct access to each unit, and the progress page presents an overview of the achieved progress and learning history. Beside this structure, the learning environment does not provide support for the learning behaviour, but the learner can freely navigate through the course. Navigation is conducted by either clicking on the Next-Button (continuing with the next page) or by using the index page (selecting any page). The only requirement consists in conducting the three assessments by providing correct answers to all items (multiple attempts are allowed).

This paper analyzes the learner behaviour in detail by using the recorded interactions with the learning environment. Such analysis provides a valuable foundation for designing learning systems, as it discovers support methods for weaker students, and indicators that can be used to determine learning difficulties automatically in advance. The next section describes the methods used to analyse the interaction data. Section 3 presents the results from the analysis, and Sect. 4 discusses these results.

2 Study Design and Methods

This paper addresses the question of how students move through a course that supports SRL, as well as which patterns appear to be the more efficient ones in regards to the learning results. This knowledge allows us to support students in their learning processes and give them the tools to choose efficient learning pathways in learning environments that require a high degree of self-efficacy. It also gives us the means to prepare courses in a way that best serves the students in their learning.

Most machine learning algorithms require large data sets to create reliable models, which presents a problem in cases where this is not possible. In the case of this paper, the user group in the analysed course consisted of only 31 students and 3 assessments per student. Thus, the idea was to statistically analyse the available data, as well as to apply a sequence mining approach as described in [7]. To prepare the data, it was first cleaned of any test and invalid users, after which the recorded actions were used to create separate user profiles that listed each user's actions sorted by time. Actions that took place within 30 min of each other were assigned to the same session [10]; otherwise a login/logout pair was added between the two actions if not already the case, as described by [8]. This visualises the session break in the later user behavior heat map (see Fig. 2).

Table 1. Action codes with short descriptions [10].

Code	Description	Code	Description
IN	Login	U3	Section 3 content
OUT	Logout	G1	Goal page of Sect. 1
IND	Index page	G2	Goal page of Sect. 2
INS	Instruction page	G3	Goal page of Sect. 3
PRO	Progress page	A1	Assessment of Sect. 1
U1	Section 1 content	A2	Assessment of Sect. 2
U2	Section 2 content	A3	Assessment of Sect. 3

By defining codes for all actions/action types (see Table 1), each student's path through the course could be visualised as a more simplified list of codes. Any accesses that involved a content chapter of a particular section were coded with that section — e.g. an access of an item belonging to Sect. 2 would simply be referred to as "U2" [10]. Accesses to the overview page for the section's exam were dropped, as they function only as information before the assessment.

After dividing the students into the two classes — R, for students who repeated one or more assessments, and N, for students who did not repeat assessments — a heat map was created to visualise the typical paths through the course pages for each of the two classes (see Fig. 2). After this, it was possible to further simplify the data by removing the information of how students moved through the course, as this was already visualised in the heat maps and does not give information on assessment performance and the type of content interaction. This was done by extracting only content accesses and assessment attempts with their results [7]. This way, it was possible to determine whether a student was viewing a content chapter for the first time or reviewing previously viewed content. Assessments were simplified into attempts where the students achieved full points and attempts where they reached less than 100%. Once again, codes were specified for each case (see Table 2). Due to the small data set, this work also decided not to make the distinction between retries of the same assessment as the immediately previous one and those of different assessments to the previous ones, as was the case in [7].

Table 2. Codes used in sequence mining, adapted from [7, 10].

Code	Description
v	The user viewed content for the first time
rv	The user reviewed previously viewed content
a-p	The assessment was passed with 100% of the points
a-f	The assessment was passed with less than 100% of the points

This change also mitigated some of the effects of the small number of assessment attempts when compared with actions concerning content chapters, which caused significantly lower support scores for sequences that included assignments, as opposed to those with only view and review actions.

Referring once more to [7], sequence extraction with lags from 1 and 9 then took place using the entire data set, as well as the class data of N and R, separately. At each lag iteration, the support values were calculated for the extracted sequences, with sequences that did not attain a 10% minimum support value being dropped from the result set.

3 Results

The course involved 31 students, each with an average of 65.53 actions. The content was divided into three sections, each with its own graded assessment at the end. To pass the course, each of the assessments needed to be passed — in 44 of 93 cases, the activity was completed with full points on the first attempt, with 16 users repeating assessment 1 at least once, 12 users repeating assessment 2 at least once, and 21 users repeating assessment 3 at least once. In all other cases, the students repeated the assessments until they did reach 100%, which took 4 attempts at the most and 1.74 attempts on average.

Overall, students did not tend to leave the content chapters during active sessions, which — for the purpose of this work — was specified to start at the first interaction with a section page and the first attempt of the section's assessment. While the instruction page was viewed 55 times in total, only 15 of those took place during an active session. This was even more pronounced when it came to the progress page, where only 11 of 107 page views took place during an active session. Fifteen students switched between sections during active sessions. The number of students who viewed the content chapters of a section in order varied by section (see Fig. 1). Of those who did not, 5 students skipped learning units or passed them over entirely.

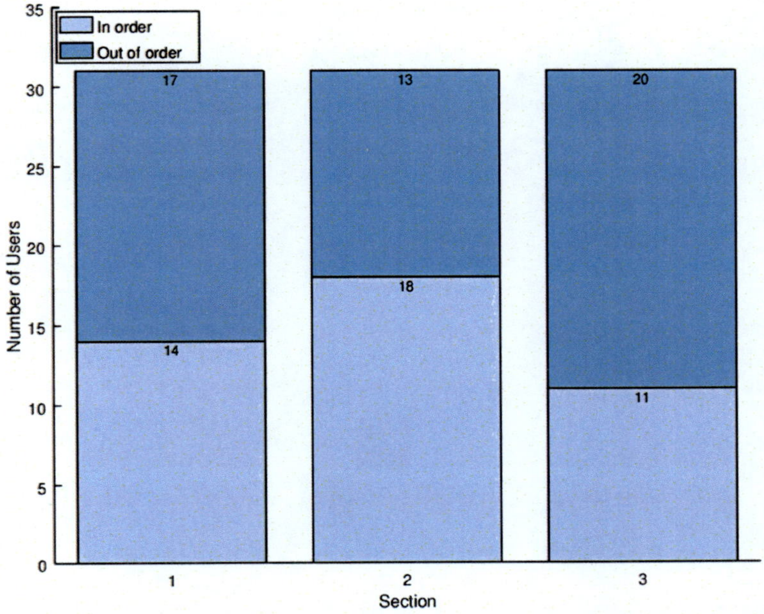

Fig. 1. The number of students who viewed the content of the section in question in order/out of order [10].

Differences in behavior patterns became more visible when the paths through the course were visualised in a heat map using transition probabilities between pairs of states (see Fig. 2). These transition probabilities were calculated separately for each of the classes N and R, and in the following step, they were used to determine the difference between those two values. Negative values denote transitions that scored higher for repeating students (R), while positive values were more important for students who did not repeat assessments (N). This allows a direct comparison of where the paths through the course differ.

As Fig. 2 shows, transitions along the suggested path through the course were often more likely for members of class N, such as those from each section's course content to the same section's assessment. In contrast, members of class R tended to utilise the index more heavily, with an overall higher emphasis on the transitions both coming from section contents, goal pages and assessments to the index page, and the opposite way around. In a further distinction, class N showed a higher likelihood to view the instructions, with most transitions taking place from the index, which represents the entrypoint into the course; even more pronounced is the strong tendency to review their progress after finishing the final section's assessment that members of this class displayed.

Overall, the heat maps suggest stronger adherence to the suggested path through the course from students who did not repeat assessments, whereas students who repeated were more likely to use the index to view the course content in an order of their choice.

Following this analysis, the specification of ordered/unordered behavior as additional classes allowed the investigation of the interaction with the two previously defined classes

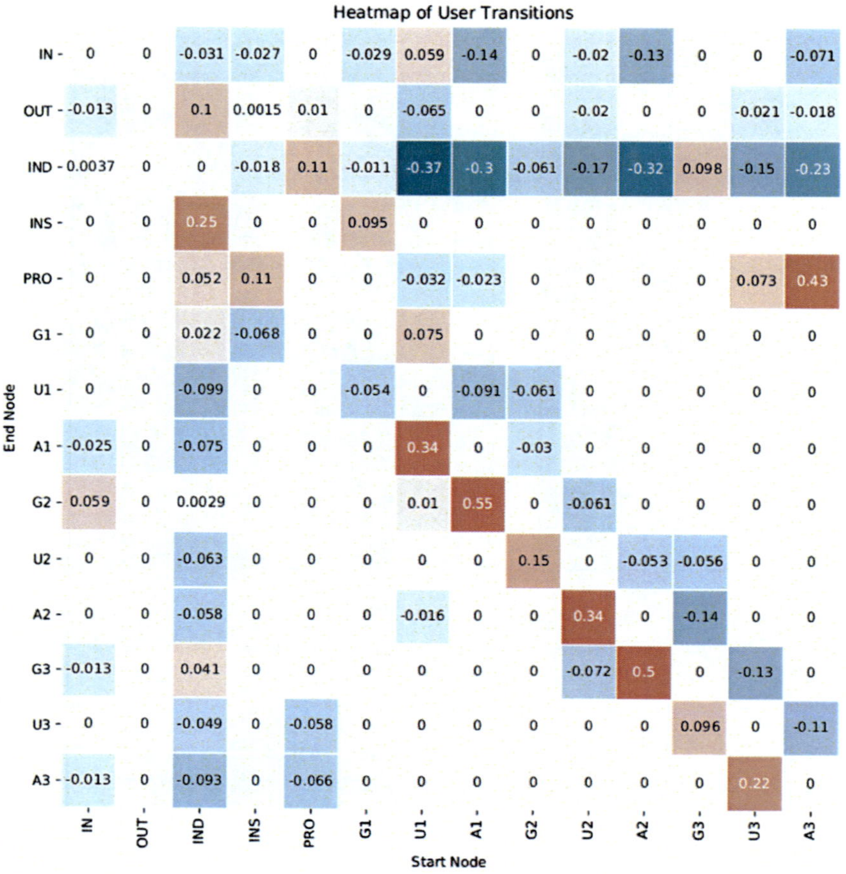

Fig. 2. The difference in transition probabilities between students in class N and those in class R, calculated by subtracting transition probabilities for R from N [10].

N and R, which refer to the learning outcome. While class affiliations for N and R are section-specific for this, ordered and unordered behavior classes (O and U, respectively) depended on the learning behavior displayed throughout the whole course. Thus, it was possible to draw connections between the two different pairs of classes (see Table 3).

Table 3. Percentage of users who repeated (R) and did not repeat assessments (N), for ordered (O) and unordered (U) learners [10].

	U			O		
	A1	A2	A3	A1	A2	A3
N	29.41%	46.15%	25.00%	71.43%	72.22%	45.45%
R	70.59%	53.85%	75.00%	28.57%	27.78%	54.55%

Finally, we used the approach from [7] to examine the most common user action sequences in the course. The 20% minimum support used in [7], where support is calculated by dividing the number of occurrences of a sequence by the total number of extracted sequences [10], proved too high for this data set, with no valid sequences starting from lag 6 when considering all students. This was exacerbated when only considering the class N, where no valid results existed starting from lag 4, while R failed to return results from lag 7. With a lowered minimum support of 10%, both the combined-class data set and R returned results for all nine lags, though N did not return sequences for lags larger than 4 once more (see Fig. 3).

To visualise the results, the sequences were visualised as sequence diagrams, as in [7]. In these, arrows signify a possible transition from state to state, with the numbers in the square brackets above the arrows denoting the number of repetitions to be found among the set of extracted sequences — e.g., "[1…3]" being the repetition of the state any of 1 to 3 times [10].

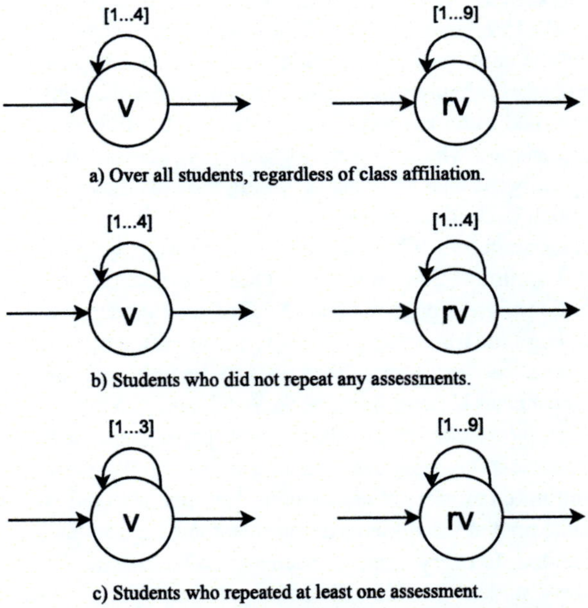

Fig. 3. State transitions over all students, class N, and class R [10].

4 Discussion

Whether students displayed organised learning behavior differed by section, with 45.16% of students accessing the chapters of section one in order, 58.06% for section two, and 35.48% for section three (see Fig. 1). These numbers appear to negatively correlate with assessment attempts, as assessment 1 was repeated only 1.48 times per student on average, but assessment 2 had an average of 2.06 attempts per student [10]. This possible correlation is also shown in Table 3, where learners who demonstrated disorganised learning behavior show higher probabilities for repeating an assessment and students who viewed learning content in order more often finished the assessment with full points on the first try. This reflects the results from previous research [9], where higher-performing students typically showed more structural learning behavior than lower-performing students.

Conversely, the lower number of consecutive view actions also reflects the more disjointed behavior by students who repeated at least one assessment (see Fig. 3) and also shows the tendency to review a larger number of already-viewed content chapters than is the case for class N. This mode-changing behavior is also visible in the transition probabilities visualised in the heat maps for the two classes. For students who did not repeat assessments, there is a clear preference in the path through the course, with very few outliers (see Fig. 2). They appear to put higher emphasis on checking the instructions, especially after accessing the index, which is the start page after logging into the system, with 40% likelihood of the transition. On the other hand, students from the class R made this move with only 15% probability and displayed a broader range of state transitions. This seems to be the case because the students in this class appear to use the index to jump to different content chapters in the course more often. Additionally, users from class R more often returned to an assessment's course content after an assessment attempt, serving as an explanation why this behavior was more present in the class' sequences. This was not typically the case for class N, where the only exception was the transition from section content back to the section's goal.

High importance can be attributed to the progress page, with each student checking it an average of 3.45 times. For members of class N especially, they transitioned from assessment 3 to the progress page in 88% of the cases. Repeaters showed this behavior less markedly, though the transition probability for the same move was still 45%. After viewing the progress case, most transitions were to the index or logout action, with only 5% of transitions for class N accessing assessment 3, and 12% for class R. This suggests that students generally checked and evaluated their performance after an initial full pass through the course and its assessments, and returned to improve their results if necessary.

Due to the unlimited number of assessment attempts, with only the best result being counted, the course invited the students to improve their performance, and this proved to be very well accepted. In every case, assessment results improved with each additional attempt, with every student finishing the course with full points at the end. Due to the fact that the online course was part of a longer, in-person class and took place partway through the semester where other assignments had already been finished, the students were likely already motivated to perform well due to previous investment of their time and work [10]. This mindset is visible in the number of sessions it took the students to complete the online course — for 61.30% of students, this was done in only a single session, with only a slightly higher average of 1.55 sessions overall.

However, the specific setting of the course presents a limitation in the results by itself, as any behaviors found may be due to the way the online course was structured as part of an in-person class. Furthermore, due to the fact that all students completed the course, there were no data specific to dropouts, and their behavior is thus not accounted for in the results.

Overall, it turned out that learners with organised learning behaviour performed better than learners with disorganised learning behaviour. This finding is also in line with Winne [3], who states that SRL is not always beneficial, especially if learners decide to choose non-efficient or self-handicapping techniques, such as procrastination, which is also a kind of SRL because learners can regulate their engagement in and disengagement from learning. So beneficial SRL also needs some kind of structure. Consequently, supporting or training learners to apply beneficial SRL-related actions can improve achievement [3]. In this view, both organised and disorganised learning behaviour can be seen as SRL, whereby organised learning is beneficial and disorganised learning is less beneficial. This emphasizes that at least unorganised learners should be supported to apply proper SRL techniques, as stated in [3]. In our case, such support would consist of scaffolds for organised learning, such as recommendations or visual cues for meaningful learning paths and indicators or hints if learners deviate from organised learning. Consequently, the findings of this paper suggest that future work should focus on developing concepts for scaffolding organised learning.

5 Summary and Future Work

This paper investigated user behavior patterns in a system based on SRL. Using heat maps and a sequence mining approach, it examined and visualised user learning behavior patterns. The consequent findings revealed that students who displayed organised learning behaviors generally repeated fewer assessments than those who did not. Furthermore, students who repeated assessments had longer sequences of review actions, as well as shorter sequences of view actions than those who did not repeat.

With the findings being specific to a course that was part of a longer, in-person class, changing this setting opens up possibilities for further research. By examining the results for a SRL course set at the beginning of an in-person class rather than partway through, it would be possible to examine the change in motivation on the students' parts, in regards to their drive to reach full points or even finish the course. Furthermore, the SRL course on its own, without an in-person portion of the class, would lead to further results on the same topic.

Both of these changed settings would also show the effects that social connectedness (due to the in-person part of the class) has on student retention and motivation. It would be possible to investigate whether the student's pre-existing connection to the student group when the course takes place partway through the in-person course has positive effects that are less or not present in the other two settings.

Finally, analysing larger courses would open the possibility of using machine learning algorithms on the data, gaining even further insights into the user behavior.

References

1. Zimmerman, B.J.: Becoming a self-regulated learner: an overview. Theory Pract. **41**(2), 64–70 (2002). https://doi.org/10.1207/s15430421tip4102_2

2. Dabbagh, N., Kitsantas, A.: Supporting self-regulation in student-centered web-based learning environments. Int. J. e-Learn. **3**(1), 40–47 (2004)
3. Winne, P.H: Self-regulated learning. In: International Encyclopedia of the Social & Behavioral Sciences, 2nd edn., pp. 535–540. Elsevier (2015). https://doi.org/10.1016/B978-0-08-097086-8.25091-5
4. Bannert, M.: Effects of reflection prompts when learning with hypermedia. J. Educ. Comput. Res. **35**(4), 359–375 (2006)
5. Nussbaumer, A., Dahn, I., Kroop, S., Mikroyannidis, A., Albert, D.: Supporting self-regulated learning. In: Kroop, S., Mikroyannidis, A., Wolpers, M. (eds.) Responsive Open Learning Environments, pp. 17–48. Springer, Cham (2015). https://doi.org/10.1007/978-3-319-02399-1_2
6. Nussbaumer, A., Dahrendorf, D., Schmitz, H.C., Kravcík, M., Berthold, M., Albert, D.: Recommender and guidance strategies. Comput. Sci. Inf. Syst. **11**(1), 321–342 (2014). https://doi.org/10.2298/CSIS121210011N
7. Malekian, D., Bailey, J., Kennedy G.: Prediction of students' assessment readiness in online learning environments: the sequence matters. In: Proceedings of the Tenth International Conference on Learning Analytics & Knowledge, pp. 382–391. Association for Computing Machinery, New York, NY, USA (2020). https://doi.org/10.1145/3375462.3375468
8. Saint, J., Gašević, D., Matcha, W., Uzir, N. A., Pardo, A.: Combining analytic methods to unlock sequential and temporal patterns of self-regulated learning. In: Proceedings of the Tenth International Conference on Learning Analytics & Knowledge, pp. 402–411. Association for Computing Machinery, New York, NY, USA (2020)
9. Mukala, P., Buijs, J., Van Der Aalst, W.M.P.: Exploring students' learning behaviour in moocs using process mining techniques. BPM Rep. **1510** (2015)
10. Frank, S.: Data Analysis and Learning Behavior in Two Different Learning Settings. Master Thesis. Graz University of Technology (2020)

ICBCI: An Integrated Model of Group and Individual Development for Learning Facilitators

Josh Friedman(✉)

Teachers College, Columbia University, New York, NY 10027, USA
jsf2167@columbia.edu

Abstract. This paper outlines a conceptual proposal that integrates theories of group development, individual learning, and therapeutic growth to diagram how individuals, and the groups they are a part of, support or hinder learning. The model has implications for group facilitators' roles and actions to maintain inter- and intra-personal balance throughout the learning process. Balance is defined as the context where productive and sustainable action is possible. The model considers five 'modes of being': Introduction, Conflict, Balance, Creation, and Identity (ICBCI), and four 'balance tools' that facilitate balanced change from one mode of being to another: Purpose, Norms, Goals, and Meaning. Further, the model identifies two modes of imbalanced expressions (rigidity and chaos) that provide behavioral hints for facilitators to utilize the balance tools to maintain sustainable learning and engagement. Finally, there is a discussion of the impacts of power and leadership within the group context on this learning process. ICBCI works in the classroom, the living room, the boardroom, and the Zoom room, and knowing its dynamic relationships empowers facilitators to maintain sustainable growth across contexts and characters.

Keywords: Group dynamics · Leadership · Facilitation

1 Purpose of the ICBCI Model

1.1 A Synthesis of Giant Shoulders

Learning never occurs in a vacuum, yet far too often, the theories we work with imply that it does [1]. The following model synthesizes seminal work from the history of social, clinical, and learning psychology to overcome this implication and make clear that an individual's clinical self, their learning self, and their interpersonal self are always concurrently present in the individual, as well as in their expressions in a group. Specifically, Tuckman's model of group development [2] (updated by Wheelan [3]), Kolb's model of experiential learning [4], and Bowlby's model of attachment [5] (interpreted by Siegel [6]) are synthesized into a single framework. This strategy provides a metacognitive model to guide learning facilitators across contexts to maintain sustainable growth for groups and individuals with which they work.

D. Guralnick et al. (Eds.): TLIC 2021, LNNS 349, pp. 99–110, 2022.
https://doi.org/10.1007/978-3-030-90677-1_10

The present purpose is to describe the model itself, rather than the seminal work from which it is derived. Thus, focus will be maintained, not on fine points of experimental results that have provided good faith for each of these theories alone, but on how integrating these well-established models aids the work and action tendencies of learning facilitators. In other words, a reader of this paper should reach the end with an understanding of *how* their *actions* can be theoretically informed, as opposed to simply their perceptions of *what* is going on in the group.

1.2 A Model for How, More than What

Far too often, researchers and practitioners play a reluctant, ambivalent, and ignorant game of cat-and-mouse carousel around the individuals, groups, and contexts that need their help the most and who frequently get only one or the other's help at any one time [7]. By focusing this model on balancing actions, necessarily embedded in a dynamic context, practitioners will find easily implementable actions to solve encountered problems, and researchers will see those same actions as theoretically grounded valuable data points that illuminate deeper intra- and interpersonal processes.

Below is an outline of the models synthesized into Introduction, Conflict, Balance, Creation, and Identity (ICBCI). Next, I present the model in the third section, how it integrates individual and group actions in the fourth, and — in the final section — how the model guides facilitators' actions to balance intrapersonal emotion regulation and interpersonal social interaction with the same set of tools.

2 Theoretical Background of the ICBCI Model

2.1 Tuckman's Model of Group Development

Bruce Tuckman described a theory of group development that includes four stages groups experience from formation to termination: *forming*, *storming*, *norming*, and *performing* [2]. Essentially, groups come together, fight over their differences, resolve those differences, and then produce their outcomes. While various theorists have added updates or nuance to the theory since then [3], it has largely remained well accepted and untouched as a stage-based descriptive theory. Despite its applications today, the idea of 4–5 stages of groups begs the question of transition between and maintenance of the very processes described [8]. ICBCI was born out of this expressed need: how to deepen the theory from being descriptive *of* groups to being prescriptive *for* their learning, development, and success.

ICBCI recasts Tuckman's stages by 1) integrating their progression with Kolb's experiential learning theory [9], 2) defining the *norming* stage in Tuckman's linear model instead as *balance*, in a non-linear dynamic model (Fig. 1), and 3) by making concrete and explicit a mode of being absent from Tuckman's theorizing that regularly occurs in both groups and individuals, and goes beyond *adjourning*; what ICBCI refers to as moments of Identity [22]. The effect of these changes is two-fold: 1) they allow learning facilitators to see the concrete and necessary norming processes within other modes of being, and 2) they highlight the central importance of balancing in the group context relative to the other actions of orienting, negotiating, producing, and appraising (captured by Introduction, Conflict, Creation, and Identity, respectively), which can each occur in a balanced or unbalanced fashion. These ideas are revisited below after first reviewing the other two shoulders.

2.2 Kolb's Model of Experiential Learning

David Kolb's experiential learning cycle claims that deep and retainable learning experiences move through 4 stages of knowledge formation and integration: Active Experimentation (AE), Concrete Experience (CE), Reflective Observation (RO), and Abstract Conceptualization (AC), also described as Try (AE), Do (CE), Reflect (RO), and Plan (AC) [4]. For example, learning to swing a hammer requires taking a first swing (AE), 'hammering' a nail (CE), noticing finger pain and a bent nail (RO), and adjusting grip and gaze direction (AC) before taking the next swing (AE), ad infinitum. Relying on research in 'embodied,' 'active,' and 'experiential' learning, Kolb sought to centralize the *learner's* experience of the to-be-learned content as opposed to the *teacher's*, as is the norm in pedagogical models [9]. ICBCI takes a similar stance, while taking care to delineate the individual *and* group experiences.

2.3 Siegel's Model of Integrative Development

Commonly referred to as 'attachment theory,' Bowlby proposed that individual differences could be predicted based on whether as a child the individual developed a *secure, avoidant*, or *anxious* attachment style [5]. Since Bowlby, researchers, and practitioners have used this theory to understand interpersonal relationships across the lifespan. Siegel proposed clinical state corollaries to each of these attachment styles that explain swaths of neurobiological behavior: secure individuals experience 'integration,' avoidant individuals 'rigidity,' and anxious individuals, 'chaos' [6]. Essentially, anxious individuals change so often there is little stability, avoidant individuals maintain stability by resisting change, and secure individuals 'go with the flow.'

Taking a dynamic systems perspective, Siegel suggests that all individuals experience moments of each state, where integration is a sustainable and balanced experience, rigidity is one without differentiation, or openness to change, and chaos is one without linkage or adoption of structure [10]. For individuals and groups, how to achieve integration and balance depends on whether one is currently experiencing more chaos or more rigidity. This basic idea, that balance and sustainable growth result from responding adaptively to needs for differentiation and linkage in individual and group experience, serves as the lynchpin to connect Kolb and Tuckman. An individual expressing rigidity will be resistant to change, or not changing as the group does, while an individual expressing chaos will be changing too often to produce direction, or changing too fast for the group to adapt to. On a group level, rigidity is stagnation, or a lack of action, while chaos is indeterminacy, or a lack of direction.

3 The ICBCI Model

3.1 Introduction

ICBCI follows the last half century of work in psychology using a nonlinear dynamic systems approach [11, 12]. Further, by refraining from describing exact modes and instead focusing on transition or change processes *between* modes, ICBCI seeks to equip facilitators with clear ideas of what they can *do*, over and above simply what they are *observing*. By loosely defining 5 modal spheres, while centralizing balance and explicitly defining actions at the group and individual levels that 'scaffold' sustainable development from one sphere to the next, learning facilitators can use ICBCI to both observe *and* act in the presence of the groups and individuals they work with. Examine Fig. 1 for the overall model, described in more detail below.

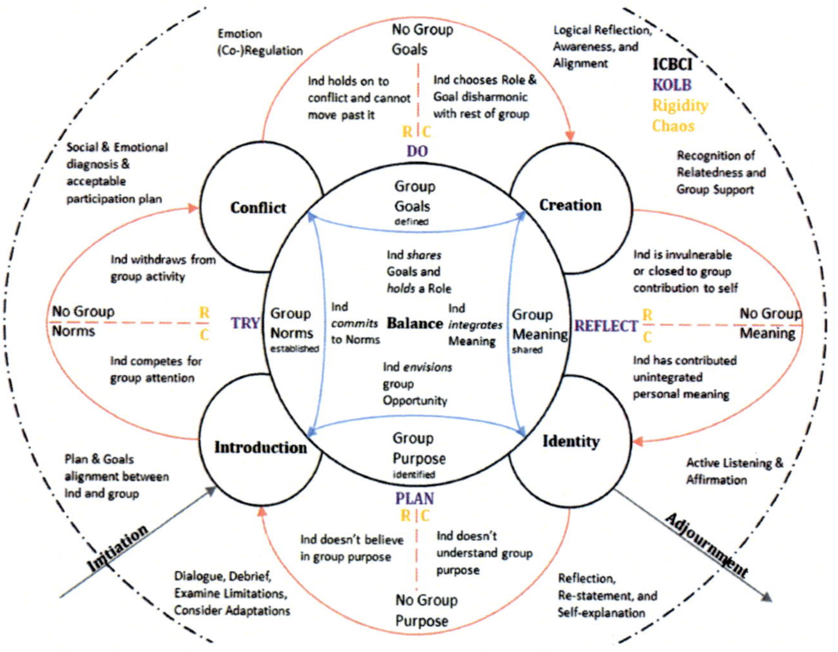

Fig. 1. The introduction, conflict, balance, creation, and identity (ICBCI) group-individual integration model of learning. Spatial logic is intentional, e.g., there are balanced and unbalanced forms of all outer 'state spheres.' Balance tools described along cardinal directions. Outer ring highlights facilitator actions. Note: not all possible transition processes (arrows) are depicted (Color figure online).

3.2 The Four Outer Spheres

The four outer spheres are defined as Introduction, or more explicitly, external perception of challenge (orienting); Conflict, or internal action against challenge (negotiating);

Creation, or external action towards actualization (producing); and Identity, or internal perception of actualization (appraising). Each of these 'states of being' are considered possible for moments of any duration within both groups and individuals and are not necessarily always aligned between the two levels (i.e., a group may be in conflict while an individual is still in introduction). The detailed descriptions are derived from replicable mechanisms in psychology [13–15] and align largely with Tuckman's characterizations, with the exception of the Identity stage — derived from the transition between RO and AC — and recent work in educational psychology highlighting the value of meaning, and the reflective activities that create it [4, 16].

3.3 The Balance Sphere

Of central importance to ICBCI is the concept of transition between these outer 'mode' spheres. Transition can occur in a balanced (blue arrows in Fig. 1) or unbalanced (red arrows in Fig. 1) way. Whereas Tuckman conceived of *norming* as a stage that groups were in only between conflict and creation (or "storming" and "performing"), ICBCI further details the nuance of group dynamics and individuals' integration within them by highlighting that balance in a dynamic system is something that is constantly sought, both within other states — hence the overlap of the outer spheres with the balance sphere — and between them (transition arrows and balance tools). Balance can occur for the sake of balance — as often happens with storytelling, singing, dancing, and rituals — as well for scaffolding or enabling other states of group activity — as happens with defining the purpose, norms, goals, and meaning of an activity for individuals and/or the groups they exist in. These 'balance tools' are described next.

3.4 Balance Tools

The contribution that balance pays to sustainable group and individual engagement and learning is the central advance of ICBCI, and it is hard to overstate [16, 21–23]. The four 'balance tools' described here, derived by squaring Kolb learning cycle actions, are the first of two main theoretical advances proposed for the field by ICBCI.

Norms as Trying^{n}. When an individual actively experiments (AE), they are 'trying,' or attempting to establish a contextual foundation of experience for their learning. For example, when learning to swing a hammer, one may begin by simply feeling the weight of it and one's strength to move that weight flexibly. Similarly, when groups establish a contextual foundation of experience for their learning, a necessary component for sustainable success are norms of interaction, or in other words, *how individuals will try to act in the presence of others.* This is the first balance tool. When groups establish explicit or implicit norms that the group agrees to adhere to — for conflicts, creations, reflections, and perceptions (or basic assumptions) — the group can learn and develop sustainably (without engaging in unbalanced conflict or cyclical re-orientation to a stifled environment). Examples of these norms are often found in group agreements, values contracts, and classroom rules (e.g., 'this group agrees to disagree with ideas, and not people,' 'don't yuck their yum,' etc.). Similarly, individuals must each *commit* to these norms so that they have a balanced learning experience in the group context. Norms are alterable and never complete and are thus descriptions of individuals *trying* to balance their actions to create group structure.

Goals as Doing^{n}. When an individual engages with their CE, they are *doing with presence* in the service of a stated or unstated goal [17, 18]. When many individuals are simultaneously doing in a group, that stated or unstated goal demands roles for each individual to play. If a group goal for the actions and/or a role for each individual is not established, progress in pursuit of goals will take longer, encounter disruption, and result in individual strife [17, 18]. Thus, the second balance tool suggests that goals must be made explicit at the group level, *and* roles must be established in pursuit of those goals for each individual. By seeing the doings of the group across the individual and group levels, the necessary establishment of goals *and roles* integrated with those goals emerges.

Meaning as Reflecting^{n}. The missing piece from a biopsychosocial perspective in the process of group development as outlined by Tuckman and those that follow in his tradition is the naturally emergent process of consolidation and identity formation that occurs at both the individual and group level [19]. Every individual scaffolds their identity with reference groups, and every group in one way or another defines and appraises its activity as it happens (e.g., 'that was a great conference we pulled off' vs. 'we have a lot to learn…'). Individuals reflect on experience to create meaning for themselves, and when they do this in groups, they contribute to and shape the other individuals' appraisals of events and how they identify the group, themselves within in it, and themselves in reference to it [20]. When learning facilitators make this an explicit action in their process, by group reflection, assignment 'wrappers' and other metacognitive activities, learning, and transfer consistently improve on both group and individual levels [21]. This effect is larger the more the appraisal or meaning is shared among the group and the more each individual integrates that shared meaning within their own identity formation process [22].

Purpose as Planning^{n}. When an individual engages AC, or the building of a concept, it is both necessarily purposeful and embedded in preparation for future action through the predictive capacity of concepts in the brain [23]. When this conceptualization occurs between individuals in a group around the group's identity, it is necessarily an ontological claim about the purpose of the group or its function in a larger context, much like a plan is the future intended function of any individual in a larger context. With this balance tool, learning facilitators advance the likelihood of sustainable group development by identifying the group's 'unique selling point' for every individual, defined as the problem spaces that the individuals will look to the group for support in solving. This link completes the reinforced learning loop by instigating opportunities for continued group learning and individual growth in the context of the ever-increasing capacity and capability of the group [23].

With these four balance tools (norms, goals, meaning, and purpose), learning facilitators can: 1) specify theoretically grounded, trackable actions that can be measured empirically within the sociocultural context of the group and the socioemotional context of the individual (e.g. 'did the group agree on, and the individual commit to, the stated/unstated norms, goals, meaning, and/or purpose?'); 2) highlight the importance of 'Why' through the doubled-down focus on goals (concrete task targets) and purpose (abstract developmental targets); and 3) readily hypothesize sources of imbalance in the group (assumed, ambitious, inappropriate, or missing norms, goals, meaning, or purpose). The next section examines observable signs of imbalance in the individuals who make up that group.

4 How ICBCI Integrates Individual and Group Action

4.1 Complex Dynamic Systems Overview

In his original work, as in the work of countless other psychologists [11, 12, 19], Dan Siegel described the brain as a massively complex dynamic system that benefits from the integration — or clear differentiation *and* linkage — between its many intertwined networks and that struggles in the context of any lack of integration or unresolved, disorganized states [6]. A developing brain in this context is conceptualized as a series of increasingly complex states that either increase, decrease, or maintain an overall level of integration among its component parts. Each increase in complexity requires a rewiring process through differentiating the new complexity into component parts and linking them together into an integrated, coherent, flexible, and adaptive whole. Any lack in achieving this integrated whole can arise from rigidity, or a failure to differentiate, and/or chaos, or a failure to link. Each imbalance's contribution to ICBCI is generally described below and further detailed in Fig. 1.

Group action is necessarily the collection of individual actions and reactions. Individual freedom is contained within our reaction to the events we are exposed to. ICBCI suggests that there are three main kinds of reactions available to individuals in the context of the group: integrating, failing to differentiate, or failing to link. The first is detailed in the previous section and within the balance sphere of Fig. 1, the latter two below, and within the red arrowed semicircles of Fig. 1.

4.2 Rigid Imbalance

ICBCI suggests that we view imbalance within the context of the balance tool engaged to support sustainable growth. Thus, instead of talking generally of rigidity as one may in a clinical context, the goal for the learning facilitator is to suggest potential corrective action for the sake of the group purpose. Thus, individuals are not 'rigid' (or 'chaotic'), but they may express 'rigid reactions' to transition tools between the four state spheres outlined above, or at any point in time (e.g., withdrawing from group activity because a reflection activity didn't go their way, see Fig. 1).

On the most basic level, rigid reactions - being the result of a failure to differentiate — indicate that the individual doesn't have access to sufficient information that affords seeing various perspectives, attributes, or qualities of a particular object. This can result in hyper-focus on a single quality, as often happens with negative information [25], or a lack of attention to any quality, as none are defined uniquely enough for the individual to differentiate them from other experience, to truly notice and value them.

As a brief sketch of the use of Fig. 1 in this context, an individual rigidly reacting to the transition from introduction to conflict states will withdraw from group activity. This can have a multitude of overlapping causes, but central to all of them will be a lack of differentiating self from other, or a previous task/group from the current task/group. Specifically, they may not trust the other individuals in the group and see them as too threatening to the self to engage in conflict (failure to differentiate self from other), or they may be 'holding on' to norms of another group/task/time/space that they preferred over the current one (failure to differentiate previous goods of enjoyment, growth, etc. from previous norms). As learning facilitators, these rigidly reactive expressions should remind us of balancing tools (e.g., Did we differentiate our purpose and norms enough for the whole group at this point? What can I do to understand what this individual is rigidly 'holding on' to?). While a full explication of Fig. 1 is beyond the scope of this paper, schematically, one can follow the logic by observing the outer ring of suggested facilitator actions that can help each individual integrate back into the group, as well as for the group to reach out to the individual. In this case, through an open and honest non-clinical socioemotional diagnosis with the individual (e.g., Want to share what's going on?), and the negotiation of an acceptable participation plan on their part (which may require revisiting established group norms).

4.3 Chaotic Imbalance

Chaotic imbalance, on the other hand, results from an individual's failure to link differentiated perspectives, attributes, or qualities of a particular object together into an integrated whole. In other words, the individual is operating from a context-poor mental environment, where the relevant experiences they have had in the past and present moment are not sufficiently overlapped to see the broader context or concept that binds them. Individuals in this 'experientially blind' state have an unresolved need for sense, for resolving conceptual conflict, or enabling pattern recognition in the moment and proceed in a predictable direction: tangential or disruptive actions designed to instigate their environment, or a 'chaotic reaction' [23].

Concretely, imagine an individual within a group transitioning from Creation to Identity, in the process of establishing meaning (e.g., "Wow, we put on a great conference") contributes, what is charitably termed, 'unintegrated group meaning' (e.g., "We are all cantaloupes in the wash!"). This chaotic reaction to a group member's heartfelt meaning statement is not only unclear, but most likely derails group processes. As learning facilitators, we have an opportunity to create space that enables the group to actively listen to what the individual truly means (e.g., "That conference was like taking a cantaloupe to the laundromat! Noisy, bumpy, messy, a waste of my time!"). This clarification, while conflicting with an already expressed perspective, is an opportunity for differentiation and linkage, rather than unresolved cantaloupe calling.

5 How the ICBCI Model Guides Facilitation

In this final section, the global axes of the ICBCI model are introduced. Each axis is derived from educational psychology [25], social psychology [3], or social cognitive affective neuroscience [15, 23] and describes the impact leaders and facilitators have on group development and individual growth within that group's context. It is helpful to see each axis as a string looped around the centralized leader (or facilitator) at any point in time, bending each axis with the direction of leader (or facilitator) actions.

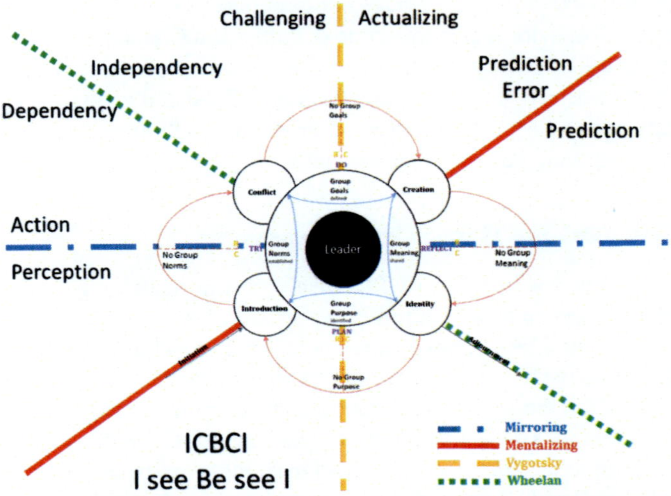

Fig. 2. ICBCI with action tendency axes moderated by leader/facilitator activity.

5.1 Facilitating Balance Through Proximal Development

Lev Vygotsky's conceptualization of the zone of proximal development (ZPD) defines three zones of capacity: (1) what an individual can do on their own, (2) what they can do with help, and (3) what they cannot (yet) do, even with help [26]. In ICBCI, the north-south axis itself defines no. (2), the ZPD, and marks the distance between opportunities for actualization, beginning on the border between nos. (1) and (2), corresponding to the right side of Fig. 2, and moments of potentially traumatic challenge, on the border between nos. (2) and (3), depicted on the left side of Fig. 2. How much a leader or learning facilitator engages in Introduction and Conflict determines the space available for actualizing vs. challenge. For example, the better foundation set for a group through productive introduction and conflict resulting in agreed norms and goals, the more likely helping behavior, and the more likely individuals are to engage in actualizing opportunities like participating in a discussion or learning a skill. Without balanced norms and/or goals, risks of trauma increase if leaders take on challenges that the group is not appropriately organized to meet.

5.2 Facilitating Balance Through Mirroring

The last three decades have produced an explosive amount of research into mirror neurons and the perception-action links they seem to track [15, 19]. At the heart of this active arena is the hypothesis that neurobiological mechanisms exist for recreating externally and internally perceived actions as internal simulations (mental perceptions) in neural 'convergence zones,' and it is this activity that scaffolds learning [27]. The skillful learning facilitator seizes opportunities to explicitly connect actions with perceptions, as when groups report who they are, how they are feeling, or what caused them to act in the way that they did. Leaders also encourage or discourage this mirroring behavior through leaning their own behavior into more action (Conflict and Creation) or more perception (Introduction and Identity) as needed throughout a group process. By maintaining a balance between action and perception in each individual and across the group, learning facilitators sustainably scaffold potential growth through intra- and inter-individual awareness of underlying perception-action links.

5.3 Facilitating Balance Through Interdependency

Wheelan, in her nuanced updating of Tuckman's general model, highlights the importance of power dynamics and leader roles within the group [3]. ICBCI integrates this work by defining an axis between dependency of the individuals in the group on the group itself (and the group itself on the leader/facilitator) and independence of each individual within the group (and the group within its context). Interdependency, thus, emerges on the mid-line of this axis, at the most crucial moments of Conflict and Identity states: turning/tipping points, and the identification of meaningful purpose, respectively. That is to say, balanced and productive conflict necessitates an interdependent perspective, after which independent action is more likely to be fruitful for the group. Similarly, maintaining an interdependent perspective during the identification of meaningful purpose likely creates productive dependency, or a reason to reform and sustain group activity. Both of these 'interdependent perspective moments' are the responsibility of the leader or learning facilitator, who can also tip these dependencies based on how much orienting vs. producing they lead the group to engage.

5.4 Facilitating Balance Through Correcting Prediction Errors

Finally, in a tour de force akin to Darwin's upending of biology or Einstein's of physics, Lisa Feldman Barrett has revolutionized our understanding of neurobiology and learning as a balance between simulated predictions, aimed at maintaining a metabolically cheap existence in the world, and errors in those predictions, resulting from our relatively limited direct experience of that world [23]. In ICBCI's final axis, learning is conceptualized in similar fashion, where Creation and Introduction spheres are centralized as the cornerstones of learning. When we orient our learning begins, matching our predictions of this group with this purpose in this context against the errors that inevitably surface in Conflict. How we handle those errors shapes the goals for our next Creation, our next set of predictions, evidenced by the meaning, and new concepts made explicit in Identity modes before encountering the next wave of error to update these, as well, ad infinitum. Leaders and learning facilitators do well when bringing group awareness to these transitions as they occur. They also bend this axis through their own actions, creating more prediction utterances when leaning more into conflict than noticing errors needed to correct them, or creating too much error correcting when leaning more into identity states than productive predictions (e.g., hypotheses, ideas, claims) needed to update the group's overall working model.

6 Conclusion

ICBCI argues for a complex dynamic systems approach to facilitation of group and individual learning, and in doing so, it provides a deep and broad foundation for facilitators themselves to test their own predictions and update them through processing errors, all while maintaining a space of learning for their group and the individuals within. While this paper does not fully do justice either to the complexity approach or the facilitation technique, it aims to provide a brief introduction to a useful model that synthesizes large bodies of foundational work in social, cognitive, and educational psychology into an empirically testable set of hypotheses concerning group development, individual learning, and the context of each within the other. Researchers are encouraged to test if the presence of balance tools facilitates learning and development, and if leader/facilitator actions adjust the axes and state space likelihoods as proposed. Facilitators are encouraged to attune their actions to balancing purpose, norms, goals, and meaning across individuals together in spacetime to maintain a group, and its ability to learn.

References

1. Lave, J.: The practice of learning. In: Contemporary Theories of Learning: Learning Theorists… in their Own Words, pp. 200–208 (2009)
2. Tuckman, B.: Developmental sequence in small groups. Psychol. Bull. **63**(6), 384–399 (1965)
3. Wheelan, S.: Group Processes: A Developmental Perspective. Allyn & Bacon, Boston (1994)
4. Kolb, D.: Experience as the Source of Learning and Development. Prentice Hall, Upper Sadle River (1984)
5. Bowlby, J.: Attachment and Loss: Attachment, vol. 1. Basic Books, New York (1969)
6. Siegel, D.: Toward an interpersonal neurobiology of the developing mind: attachment relationships, "mindsight", and neural integration. Infant Ment. Health J. **22**(1–2), 67–94 (2001)
7. Gorard, S., See, B.H., Siddiqui, N.: What is the evidence on the best way to get evidence into education? Rev. Educ. **8**(2), 570–610 (2020)
8. Larson, N., McLarnon, M., O'Neill, T.: Challenging the "static" quo: trajectories of engagement in team processes towards a deadline. J. Appl. Psychol. **105**(10), 1145–1163 (2020)
9. Glassner, A., Back, S.: Three "gogies": pedagogy, andragogy, heutagogy. In: Glassner, A., Back, S. (eds.) Exploring Heutagogy in Higher Education: Academia Meets the Zeitgeist, pp. 59–74. Springer, Singapore (2020)
10. Siegel, D.: Aware: the science and practice of presence. Neuropsychotherapist **6**(9), 24–35 (2020)
11. Thelen, E.: Dynamic systems theory and the complexity of change. Psychoanal. Dialog. **15**(2), 255–283 (2005)
12. Kaplan, A., Garner, J.: Steps for applying the complex dynamical systems approach in educational research: a guide for the perplexed scholar. J. Exp. Educ. **88**(3), 486–502 (2020)
13. Maslow, A.H.: A theory of human motivation. Psychol. Rev. **50**(4), 370–396 (1943)

14. Rotter, J.B., Mulry, R.C.: Internal versus external control of reinforcement and decision time. J. Personal. Soc. Psychol. **2**(4), 598–604 (1965)
15. Rizzolatti, G.: The mirror neuron system and its function in humans. Anat. Embryol. **210**(5–6), 419–421 (2005)
16. Immordino-Yang, M.H., Darling-Hammond, L., Krone, C.: The brain basis for integrated social, emotional, and academic development. In: How emotions and Social Relationships Drive Learning. The Aspen Institute (2018)
17. Immordino-Yang, M.H.: The smoke around mirror neurons: goals as sociocultural and emotional organizers of perception and action in learning. Mind Brain Educ. **2**(2), 67–73 (2008)
18. Csikszentmihalyi, M.: Flow: The Psychology of Optimal Experience. Harper & Row, New York (1990)
19. Immordino-Yang, M.H., Gotlieb, R.: Embodied brains, social minds, cultural meaning: integrating neuroscientific and educational research on social-affective development. Am. Educ. Res. J. **54**(1s), 344–367 (2017)
20. Williams, A.: Critical reflective practice: exploring a reflective group forum through the use of Bion's theory of group processes. Reflective Pract. **14**(1), 75–87 (2013)
21. Chi, M., Wylie, R.: The ICAP framework: linking cognitive engagement to active learning outcomes. Educ. Psychologist **49**(4), 219–243 (2014)
22. Kaplan, A., Garner, J.: A complex dynamic systems perspective on identity and its development: the dynamic systems model of role identity. Dev. Psychol. **53**(11), 2036–2051 (2017)
23. Barrett, L.F.: How Emotions are Made: The Secret Life of the Brain. Harcourt, Houghton Mifflin (2017)
24. Haradkiewicz, J., Elliot, A.: The joint effects of target and purpose goals on intrinsic motivation: a mediational analysis. Pers. Soc. Psychol. Bull. **24**(7), 675–689 (1998)
25. Clore, G.L., Huntsinger, J.R.: How emotions inform judgment and regulate thought. Trends Cogn. Sci. **11**(9), 393–399 (2007)
26. Vygotsky, L.: Mind in Society: The development of higher psychological processes. Harvard University Press (1980)
27. Damasio, A.: The brain binds entities and events by multiregional activation from convergence zones. Neural Comput. **1**(1), 123–132 (1989)

GRAF: A System for the Assessment, Accreditation and Representation of Competency Learning in Online Higher Education

Cristina Girona(✉), Lluís Pastor, Xavier Mas, and Josep-Antoni Martínez-Aceituno

Universitat Oberta de Catalunya, 156 08018 Barcelona, Rambla del Poblenou, Spain
cgirona@uoc.edu

Abstract. The evaluation and accreditation of competencies is one of the challenges that higher education must address in the 21st century. GRAF is a system of evaluation and accreditation of curricular competencies based on the integration of pedagogical engineering methodologies with data architecture design and technological applications, especially tailored to online higher education. From the teaching point of view, it provides a model, a methodology, and a tool for designing competency-based programmes and assessing the evidence of competency learning generated by the students in the teaching and learning process throughout the semester. From the student's perspective, it enhances conventional evaluation by accompanying marks with educational feedback on the level of competency achieved, and provides a graphic representation of progress. Specific tools and applications have been developed for each phase of work: for the design stage, the Competency Design Tool; for the evaluation and monitoring stage, the Competency Assessment Tool; and for the graphic representation stage, the Competency Graph Report. The Universitat Oberta de Catalunya (UOC) has spent two years developing GRAF. It has been applied to five programmes in different areas: an MBA, two master's degrees, a postgraduate course, and a university extension course. A total of 56 compulsory subjects and 293 students have been involved in these programmes. Due to the success of the initiative, the UOC has decided on a widespread application of GRAF to other programmes. We are interested in sharing the experience and lessons learned in order to improve GRAF.

Keywords: Competency assessment report · Competency-based design · Competency accreditation

1 Introduction

For some years now, design of higher education programmes has been oriented to developing personal and professional competencies. Despite efforts to date, graduates still have problems clearly identifying their stage of competency development on finishing a university programme. At the Universitat Oberta de Catalunya (UOC), as at many other universities, the training design of degree programmes is reflected in a competency map

D. Guralnick et al. (Eds.): TLIC 2021, LNNS 349, pp. 111–121, 2022.
https://doi.org/10.1007/978-3-030-90677-1_11

or matrix and specified in the proposed learning activities and methodologies. However, during the student teaching and learning process, formative and continual competency assessment becomes elusive, eclipsed by qualifications, marks, and academic transcripts, which hardly reflect competency learning and acquisition. This acquisition is not always evident to students or assessors.

The project presented here is an approach to competency design and assessment and a graphic, tailored representation of each student's competency development. A new competency assessment experience at the UOC needed to provide added value and an advance in habitual assessment practices up to now. The GRAF[1] project aims to specify the assessment process for interdisciplinary and specific competencies in degree programmes. It provides a visible place where professors can assess evidence of learning from a perspective that complements traditional assessment: assessing how well ways of doing things, acting, and behaving in professional and work contexts related to the degree, expressed in competencies, have been acquired. The project was also motivated by the desire to cover certain demands from the labour market and some of the needs expressed by universities. An academic transcript shows academic results. A graphic representation of competency development showing leadership capacity, teamwork in multidisciplinary environments, and effective communication, to give just a few examples, allows students to show and demonstrate their skills in a useful, valuable, and visual format.

Apart from providing a tool for, and graphic representation of, competency assessment, GRAF also offers a comprehensive competency design system for a full university programme. The competency work is interlaced evenly over all courses in a degree programme and specified in each of the related learning activities.

2 The State of the Art in Higher Education Competency Learning Assessment and Accreditation

Universities and higher education institutions are working to improve processes of competency assessment and their graphic visualization. GRAF is an example of this, together with other experiences, such as the University of Canberra (New Zealand) and Deakin University (Australia), or the result of the joint development among university members of the Europortfolio (European Network of ePortfolio Experts & Practitioners) European project [5, 6, 13]. In the next section, we will consider what this involves and what difficulties it entails for teachers and institutions.

Competency assessment and design of the learning process are two sides of the same coin. We can only assess the elements we decide to include in the course curriculum. This might seem obvious, but it is the cornerstone of competency assessment and also the main stumbling block for many initiatives.

Assessing competency learning means moving from a concept of content-based learning, i.e., syllabuses, toward a competency-based concept, or the skills needed to carry out an activity in the professional or academic context [18]. This fact, which in many cases might require a change in the institution's teaching culture, greatly increases complexity throughout the chain of processes involved in learning design and assessment.

[1] GRAF (from "Gráfico", in Spanish) stands for "Graph": Competency Graph Report.

Positioning ourselves in this competency design framework requires adopting an interdisciplinary approach. Skills and knowledge are no longer organized along the lines of old knowledge area taxonomies, bringing about a complex interactive relationship, designed to develop functions, actions, and behaviours in real contexts. Thus, design for competencies involves the effort of deconstructing reality to transfer its fragments to a training context, without losing sight of what connects the fragments to each other and their original context.

This approach has a number of methodological implications, which affect, to varying degrees, the value chain in the training process, from programme design to the accreditation system, without forgetting the design of training activities or assessment methods and instruments. Consequently, certain elements recur when discussing competency assessment: providing a competency map with levels of specificity; rethinking learning activities; using objective and scalable assessment instruments, such as rubrics; and designing types of accreditation suitable for the nature of competencies, such as visual systems or micro-credentials. These issues are discussed below.

2.1 Competency Map

A number of experiences have used competency maps from which learning and assessment processes are designed [2, 16]. Such maps help specify competencies to a degree at which they become workable. This means identifying dimensions or levels of acquisition [15] or learning results [1, 17] on which training and assessment activities and dynamics are based.

2.2 Learning Activities and Assessment Instruments for Competencies

Conventional training activities are not always suitable for assessing competencies. Activities are required whose design is based on complex learning situations, where knowledge and skills have to be mobilized and integrated [14], and are assessable using instruments such as rubrics, which provide rigour, objectivity, and clarity to the process and facilitate the work of the assessing professor [7].

2.3 Accreditation Systems that Recognize the Student's Competency Acquisition Progress

A competency is a personal capacity in permanent development. Helping the students to be aware of their acquisition process, whether encouraging self-reflection, providing self-diagnosis tools or showing the roadmap of the process itself, is a goal that is present in a number of experiences [8] (González-Martínez et al. [10]). Some of them focus on using visual metaphors as a strategy to convey this information to the student in an explicit and interpretable way, both at the end and throughout the training process [3, 11].

Finally, as it is possible to deduce, the application of a system of competency assessment requires the performance of an exhaustive pedagogical and methodological design that affects the entire training process in all the essential elements. We are not talking about an educational innovation that modifies certain steps of the value chain, but about a cross-disciplinary and interdisciplinary training engineering challenge.

3 Description of the System

At the UOC, we have designed a system, model, methodology, and tools for the design of competency-based training programmes and for assessing evidence of competency learning. The purpose of the system is for students to improve their competitiveness and employability in the labour market with a report accrediting their acquisition level of curricular competencies [4]. Students can then make their reports public and shareable in professional environments or to employers who might place a value on them.

In recent years, higher education institutions have begun to build their own data architecture to monitor their processes and give them visibility [19]. Our experience consists of a system based on integrating pedagogical engineering methodologies with data architecture design and technological applications, especially tailored to online higher education.

The **objectives** GRAF hopes to achieve are:

- The generation of a "Personal Competency Report", which provides visibility to student assessments by showing acquisition levels for each competency dimension and related activities and courses. The report includes a customizable graphic representation of the student's competency acquisition progress within the framework of a training programme.
- The creation of an agile, rubric-based grading system for learning activities, allowing professors to assess competencies and provide formative feedback on training, while minimizing their workload in the assessment process. The system is compatible with the traditional credit rating system and the UOC educational model.
- An accurate competency-based design for study programmes in which learning activities are linked to competency dimensions and facilitate grading and feedback [9].

The system is based on the following pedagogic principles:

- Assessing competencies involves more than correcting activities; the results students obtain for an activity need to be observed beyond the habitual grading system. This means assessing the degree of learning acquisition and orienting it to the progressive development of competencies.
- Constructive alignment of competencies, dimensions, learning results, and activities is the necessary condition for system consistency.
- Learning assessment is evidence-based. Evidence is produced by students over the course of the programme through the learning activities.

- Students receive tailored, continuous feedback on their learning and competency acquisition [12], based on the descriptions in the rubrics. The rubric is the instrument used to design and assess students' output. It describes the competency quality standards or degrees of development as defined in the training design phase to make them assessable. The rubrics are shared with students from the start, thereby informing them of what they are expected to learn.

Based on these foundations, with a painstaking pedagogic design, work then proceeds in three phases: (1) The training design phase for the competency-based programmes; (2) the student learning assessment phase; and (3) the graphic competency report generation phase. For each phase, GRAF has developed three different tools whose operation is interconnected in the virtual classroom:

- The Competency Design Tool (CDT): this enables professors to draw up competencies, dimensions and assessment rubrics, design the competency map and match up courses, competencies, dimensions and activities.
- The Competency Assessment Tool (CAT): this facilitates assessment and shows the assessment results for activities and competencies while generating tailored feedback for the student.
- The Graphic Competency Report (GCR): this reports the degree of competency acquisition and presents it graphically.

The GRAF technological design is thus consistent with its underlying pedagogic concept.

4 Application in the Context of the UOC

4.1 Application Phases

GRAF has been applied at the UOC since 2018, involving a number of different implementation phases. During the first four pilot phases (see Table 1), work processes, adaptation of competency designs, and pedagogical engineering to the characteristics of each programme and the tool development and implementation were carried out in test mode. Thus, in the first pilot tests, we focused solely on design and assessment of programme interdisciplinary competencies, while in subsequent phases, the competency map and tools also incorporated the specific and general competencies in the participating degree programmes.

Currently, GRAF includes five degree programmes. It is applied to 56 courses (242 ECTS credits), 139 professors, and 435 students[2]. The following table lists the participating programmes and the agents involved in each case.

[2] At the time of publication, the number of students was 435. Note that the previous abstract mentioned 293 students. Since then, a new master's degree has joined GRAF, hence the number has increased.

Table 1. Phases and programmes in which GRAF is applied.

Phase*	Study program	Courses	Credits	Professors**	Students**
Pilot	MBA Global Executive Education	27	80	60	105
Pilot	University Master's Degree in Strategy and Creativity in Advertising	8	60	28	75
Pilot	Assessing for learning (Postgraduate Course)	6	30	17	20
Pilot	Project Leadership and Management (Professional specialization course)	3	12	10	49
Develo-pment	University Master's Degree in Design, Visual Identity and Branding	12	60	24	186
	Total	56	242	139	435

* The Global Executive Education (MBA) programme was a pilot test in 2018–19, which has not yet reached the full development phase. The professional specialization course was a pilot test which incorporated assessment of both specific and interdisciplinary competencies.
** Professor and student numbers include all participants in each phase, regardless of whether they were also involved in other phases.

4.2 GRAF Tools

Competency Design Tool (CDT). This is a web tool that provides a design methodology for constructing a degree programme competency map. First of all, programme directors define the competencies, dimensions, and assessment rubrics. Next, directors and professors distribute the weighting for the competencies and dimensions over the degree programme, its courses, and learning activities in accordance with the work and assessment load involved. Thus, professors share in the methodological decision-making with regard to the assessment weighting, which is rated and checked from an interdisciplinary perspective covering the whole degree. The assessment weighting is expressed in percentages. The decision-making process is meticulous and based on consensus.

Professors are accompanied by an expert from the eLearn Center (eLC) in the work and are provided with guidelines and support resources produced specifically for the purpose (Fig. 1).

Competency Assessment Tool (CAT). This is an agile and efficient tool that allows professors to assess the degree of competency development, visualizing a student's complete output (deliverables) in a single space. For each student and activity, the tool shows the output or work, the dimensions assigned for observation, and the assessment rubrics. Professors make their assessment by deciding on the degree of competency development that best applies to the output or deliverable (Excellently Acquired, Acquired, Partially Acquired, Not Acquired), while automating feedback. This feedback can also be enriched by providing the student with tailored comments, broadening or clarifying descriptions from the rubric, which are then sent automatically to the student (Fig. 2).

The eLC trains professors in the use of this web tool and in assessing competencies based on evidence of learning in the UOC online environment.

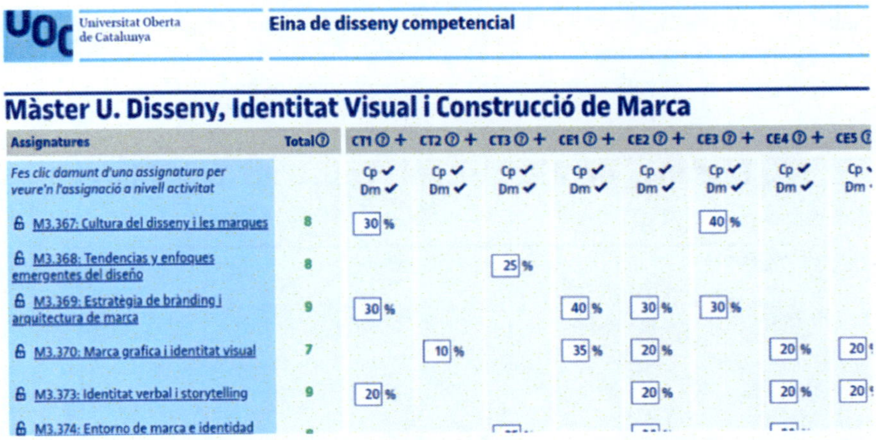

Fig. 1. Competency map for the University Master's Degree in Design, Visual Identity, and Brand Building.

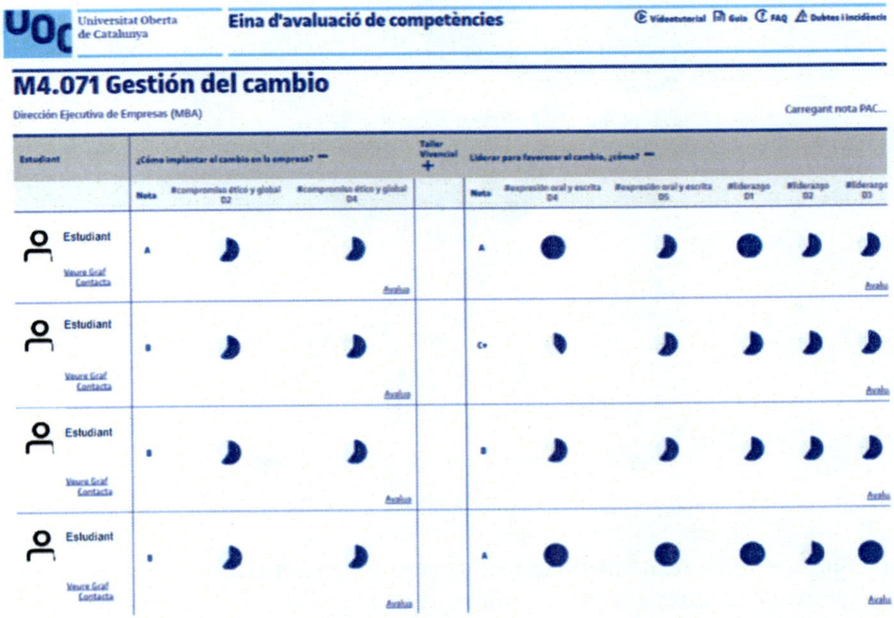

Fig. 2. Competency assessments in the assessment tool.

Graphic Competency Report (GCR). This is the result of the competency learning assessment. As students complete activities and progress in their studies, a graphic report is generated which shows their degree of acquisition for each competency. Each student has their own tailored graphic report, which they can personalize, export, and share on the web and in professional and work situations. The report accredits their competency acquisition. Currently, the report is not valid as an academic certificate issued by the UOC; however, it represents a major step towards formal accreditation, which could accompany the qualification or provide micro-credentials or badges (Fig. 3).

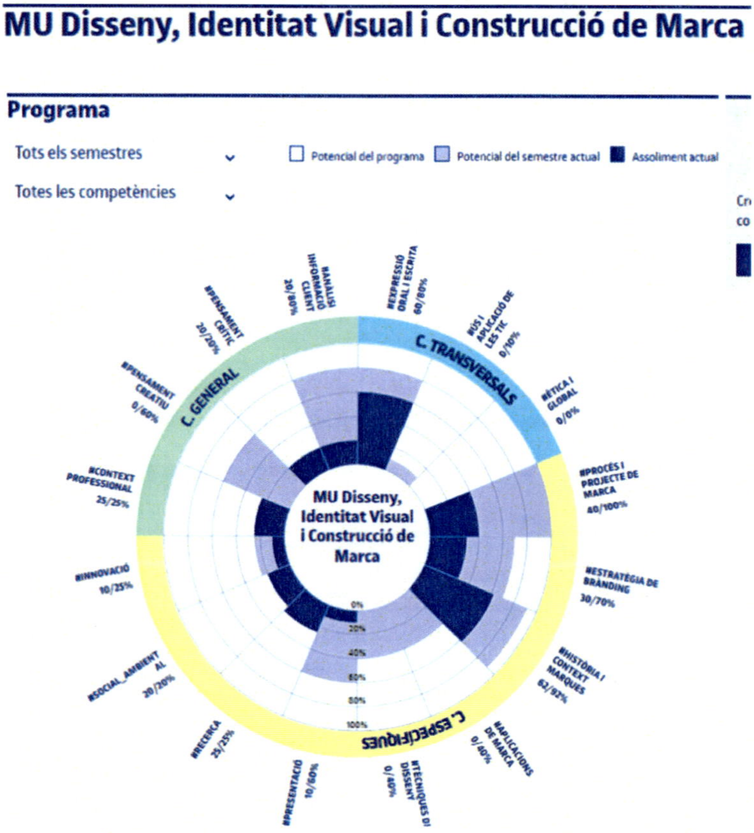

Fig. 3. A student's graphic competency report

Since GRAF started, both the methodological design and technological implementation of all its components have been regularly tested and assessed. These assessments are geared toward decisions on improving processes, tools, functionalities, and effectiveness, incorporating the results into each successive phase and programme.

Students consider this graphic report as value added to their academic transcript and state they would like to have more and better options for personalizing and exporting it. Professors feel that thanks to GRAF they now have a consensus competency map. They also feel competency assessment adds to their workload and that they need training in competencies and their assessment.

5 Conclusions

The GRAF project is an improvement in competency design, as it provides a model, methodology and tool set for competency assessment. The model, based on competencies, dimensions, and degrees of competency development, and the methodology, based on rubrics, are useful, coherent, and flexible for application to a wide range of university degree programmes. Thus, the defined model and methodology can be applied to both short programmes (specializations) and longer programmes (bachelor's, master's, and post-graduate degrees). The methodological architecture means the model cannot be applied by just one professor or part of the teaching staff from the programme. As it is based on global degree programme design, applied to all courses, it requires commitment, involvement, and participation from all course faculty. Ensuring such collective alignment requires specific commitment in terms of institutional strategy to boost its application. Thus GRAF clearly shows the need for institutions to revise their competency design and how it affects the production of degree reports, course design, and, finally, implementation in the classroom and in assessment. For the professors, it means more time spent on the design phase and providing a more detailed classroom assessment. Minimizing this extra workload requires, firstly, specific tools to automate processes and, secondly, expert advice on teaching methodology for competency design and assessment. The CDT solves the problems detected in the initial project phases and provides a robust working environment to record and trace the data on which the model is based. The CAT permits integration of pre-existing assessment tools from before the GRAF project and unifies all the required assessment actions in a single environment. Development of these two tools has produced sufficient automation to make the project sustainable in terms of professors' workload. With regard to the GCR, based on the opinions of the agents involved, it is considered value added to the academic transcript and programme. In conclusion, the GRAF tools represent an advance in competency design and assessment for professors and facilitates qualitative feedback and visualization of the competency development process for students.

The GRAF approach represents an advance towards the application of big data in training processes in higher education. It is also a starting point to promote and consolidate a culture of higher quality evaluation and personalization with great power of transformation.

Future lines of work in the project involve approaching how the model can incorporate evidence from outside the academic world, from students' personal and professional lives. Currently, the assessment and graphic competency representation provided by the project cannot be included in the official academic transcript due to legal limitations. Further work is required to ensure public legislators make the regulatory changes so this type of assessment can be included in official transcripts. At the same time, beyond this goal, the definition of a micro-credential system linked to competency development needs to be explored to provide students with a range of accreditation instruments for each of their competencies. Along these lines, from the technological perspective, blockchain could prove highly relevant in constructing a robust and reliable competency accreditation system, recognized by a range of social agents and employers. Finally, with regard to project scalability and its transfer to more degree programmes, training and advisory strategies, and actions need to be designed so that professors do not require teams of experts in methodology providing the intense, tailored accompaniment that the project has so far needed in its initial implementation phases.

References

1. ANECA: Guía de apoyo para la redacción, puesta en práctica y evaluación de los resultados del aprendizaje (2013). http://www.aneca.es/Sala-de-prensa/Noticias/2013/ANECA-presenta-la-Guia-para-la-redaccion-y-evaluacion-de-los-resultados-del-aprendizaje
2. Aref, S.: Curriculum mapping in higher education: A case study and proposed content scope and sequence mapping tool. J. Furth. High. Educ. **40**(5), 1469–9486 (2016). https://doi.org/10.1080/0309877X.2014.1000278e476
3. Blasco, L., Minguillón, J.: Using graphs for exposing the underlying competence design of academic degrees. In: Proceedings of the EuroVis 2017 Eurographics/IEEE VGTC Conference on Visualization (posters), pp. 73–76 (2017). http://openaccess.uoc.edu/webapps/o2/bitstream/10609/64566/1/lblasco-eurovis2017.pdf
4. Brown, M., et al.: 2020 EDUCAUSE Horizon Report: Teaching and learning edition. EDUCAUSE (2020). https://library.educause.edu/media/files/library/2020/3/2020_horizon_report_pdf.pdf?l=en&hash=08A92C17998E8113BCB15DCA7BA1F467F303BA80
5. Deakin University: Homepage/Professional Practice credentials. https://www.deakin.edu.au/credentials. Accessed 25 Jan 2021
6. Europortfolio: Homepage/Projects. http://www.eportfolio.eu/projects. Accessed 01 Feb 2021
7. Expósito-Langa, M., Nicolau-Juliá, D., Tomás-Miquel, J.V.: La evaluación por competencias en los grados. Desarrollo de una experiencia en el Grado en Administración y Dirección de Empresas mediante el uso y validación de rúbrica. Revista Complutense de Educación **28**(4), 1155–1171 (2017). https://doi.org/10.5209/RCED.51618
8. Gallardo-Fuentes, F., López-Pastor, V.M., Carter-Tuhillier, B.: Effects of the application of a formative and shared assessment system on the self-perception of competences acquired in pre-service teacher education. Estudios Pedagogicos **44**(2), 55–77 (2018). https://doi.org/10.4067/S0718-07052018000200055
9. Girona, C., Martínez, J., Pastor, L., Merino, M.: Assessing competencies and graphing students progress in online higher education. In: ICDE Lifelong Learning Summit, Lillehammer, Norway (2019)
10. González-Martínez, J., Esteve-Mon, F.M., Larraz, V., Espuny, C., y Gisbert-Cervera, M.: INCOTIC 2.0. Una nueva herramienta para la autoevaluación de la competencia digital del alumnado universitario. Profesorado. Revista de Currículum y Formación de Profesorado **22**(4), 133–152 (2018). https://doi.org/10.30827/profesorado.v22i4.8401
11. Grann, J., Bushway, D.: Competency map: visualizing student learning to promote student success. In: Proceedings of the Fourth International Conference on Learning Analytics And Knowledge (LAK'14), pp. 168–172. ACM, New York (2014). http://dx.doi.org.biblioteca-uoc.idm.oclc.org/10.1145/2567574.2567622
12. Guàrdia, L., Clougher, D., Anderson, T., Maina, M.: IDEAS for transforming higher education: an overview of ongoing trends and challenges. Int. Rev. Res. Open Distrib. Learn. **22**(2), 166–184 (2021). https://doi.org/10.19173/irrodl.v22i2.5206
13. Hoeppner, K.: Visualizing competency assessment. DEANZ 2016 conference, Hamilton, New Zealand (2016). [vídeo on You Tube]. https://youtu.be/JXJ5qu7Tjew
14. Hrich, N., Lazaar, M., Khaldi, M.: Improving cognitive decision-making into adaptive educational systems through a diagnosis tool based on the competency approach. Int. J. Emerg. Technol. Learn. **14**(7), 226–235 (2019). https://doi.org/10.3991/ijet.v14i07.9870

15. Marko, I., Pikabea, I., Altuna, J., Eizagirre, A., Perez-Sostoa, V.: Proposal for the development of transversal competences in the degree of pedagogy. A case study. Revista Complutense de Educación **30**(2), 381–398 (2019). https://doi.org/10.5209/RCED.57490
16. Neville-Norton, M., Cantwell, S.: Curriculum mapping in nursing education: a case study for collaborative curriculum design and program quality assurance. Teach. Learn. Nurs. **14**(2), 88–93 (2019). https://doi.org/10.1016/j.teln.2018.12.001
17. Proitz, T.S., Havnes, A., Briggs, M., Scott, I.: Learning outcomes in professional contexts in higher education. Eur. J. Educ. **52**(1), 31–43 (2017). https://doi.org/10.1111/ejed.12207
18. Riesco, M.: El enfoque por competencias en el EEES y sus implicaciones en la enseñanza y el aprendizaje. Tendencias Pedagógicas **13**, 79–105 (2008)
19. Williamson, B.: La arquitectura oculta de la educación superior: construir una infraestructura de *big data* para la 'universidad inteligente'. In: Castañeda, L., Selwyn, N. (eds.) Reiniciando la universidad. Buscando un modelo de universidad en tiempos digitales. Editorial UOC, Barcelona (2019)

A Lean Six Sigma Certification MOOC: Balancing Standard Material and Individual Guidance to Help Participants Implement Real Projects

Reiner Hutwelker[1(✉)] and Holly Ott[2]

[1] Technical University of Munich (TUM), Munich, Germany
reiner.hutwelker@tum.de
[2] Rosenheim Technical University of Applied Sciences and Technical University of Munich, Munich, Germany
holly.ott@tum.de

Abstract. Lean and Six Sigma are management systems that aim to improve the development and operation of business processes. Lean and Six Sigma tasks and tools structure the phases: Define, Measure, Analyse, Improve, Control (DMAIC). Their theoretical mastery and practical application leads to *Yellow*, *Green*, and *Black Belt* certification—monetarily rewarded by organizations worldwide. This makes Lean Six Sigma relevant for employees and attractive even for students. Our Lean Six Sigma MOOCs offer certification programs for the *Yellow* and the *Green Belt*—with more than 300k participants since 2015. In our *Yellow Belt*, the fundamentals of Lean and Six Sigma are taught and tested automatically online. In our *Green Belt*, the learner must implement a real project. The project can be learner-specific or predefined by us, is guided by a sequence of tasks and tools, and is supported by coaching and lectures. To balance our concept—wide range of project topics + digital project guidance + individual support—we develop and operate our MOOCs as business processes (Sect. 2), and outline Six Sigma in our context (Sect. 1). This is a report on the development and operation of a MOOC, not a research paper.

Keywords: MOOC · Six sigma · Development · Operation · Project guidance

1 Context and Framework of Our Lean Six Sigma MOOCs

1.1 Examples of Lean Six Sigma Certification Project Topics

False dimensions in construction drawings, *Rapid wear of bristles in street sweepers*, *Pinholes in packaging foils*, *Long cycle times for customizing cloud software packages*, *Abandonment rate in online payment process*, *Misaligned punch holes in adjustable furniture*, *Failure of serviced modules on oil rigs*, *Crumbly consistency of industrial pizza dough*, *Littered garbage and blocked sewage pipes in cities*—these examples of certification projects from our participants indicate the variety of topics that can be addressed with Lean Six Sigma and their specific relevance to real-world issues. And they indicate our challenge to support these projects and to evaluate their results.

D. Guralnick et al. (Eds.): TLIC 2021, LNNS 349, pp. 122–132, 2022.
https://doi.org/10.1007/978-3-030-90677-1_12

1.2 Subject of Lean and Six Sigma

Six Sigma—our focus here—is a scientifically based approach for problem solving in business processes. The core of Six Sigma is the Define, Measure, Analyse, Improve, Control (DMAIC) method. Each of its five phases is structured by qualitative and statistical tools. In improvement projects, these tools are applied along the DMAIC phases, to *Define* the problem, to *Measure* the problem and its potential causes, to test the related hypotheses and *Analyse* the root causes, to *Improve* the outputs by eliminating the root causes in the process, and to finally *Control* the sustainability of what has been achieved. Lean includes principles, and procedures for increasing the efficiency of processes. Six Sigma is comparable to diagnosing and treating a specific disease, while Lean strengthens the body's overall fitness.

1.3 History of Six Sigma

Six Sigma was developed at Motorola in 1987 [1] and initially applied to manufacturing. It became popular in 1996 when it was implemented throughout the company at General Electric. Since then, companies worldwide installed Six Sigma as a management system for business processes, to reduce costs and to increase customer satisfaction. To this end, companies train 5%–10% of their employees to become Six Sigma experts: The *Yellow*, *Green*, and *Black Belts*. A reputable certificate increases the chances of application and is often monetarily rewarded. This makes Six Sigma relevant for every employee and attractive also for students.

1.4 Heterogeneity of Six Sigma

Along with the training needs of companies, the supply of corresponding Lean Six Sigma training courses, coaching, and certificates has increased worldwide. Their quality varies greatly because each provider designs the content, methods, objectives, and certification individually. Wikipedia summarizes: "Criteria for Green Belt and Black Belt certification vary … There is no standard certification body") [2].

After > 30 years of Six Sigma there are different views in literature and practice on the purpose of the DMAIC phases and thus on the tasks and relevant tools. This lack of clarity is unsatisfactory for the understanding of Six Sigma and leads to uncertainty in its application. Therefore, we orient our concept to the specifications of established institutions and go beyond them with our own developments (Sect. 2.1).

1.5 Attractivity of Six Sigma

The keyword *Six Sigma Training* is searched for significantly more frequently on the web than the currently popular *Scrum Training* (Fig. 1a). The comparison is appropriate because the approaches have symmetrical properties: *Six Sigma* integrates analytical methods, principles, and tools to sustainably improve the results of business processes in service and production by simply structured and organized projects. *Scrum* integrates agile methods, principles, and tools to incrementally improve the results of development projects in IT by simply structured and organized processes.

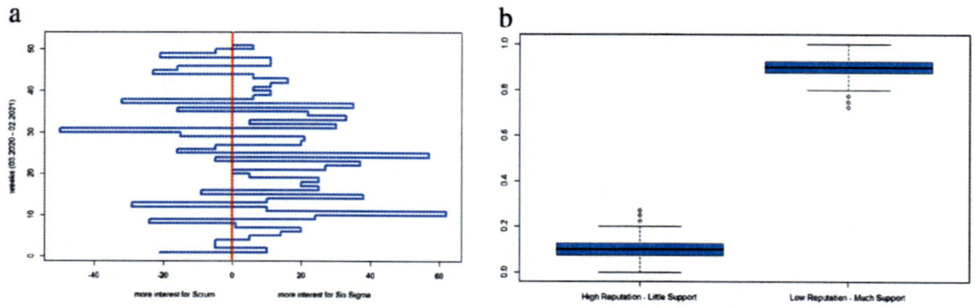

Fig. 1. a) Interest in *Scrum* vs. *Six Sigma Training* indicated by the frequency of respective keyword searches in the last year; left swing: more frequent search for Scrum; right swing: more frequent search for Six Sigma (google trends, N unknown) – b) Attitude of our students: 90% prefer: *Course at university with low reputation that cares about my needs*; 10% prefer: *Course university with low reputation that does not care about my needs* (N = 40).

1.6 Learner Needs

A survey in our Facebook group showed a clear preference of 90% for the alternative: *Course at a little-known university that takes care of my individual needs* versus the symmetric case with 10% preference: *Course at a well-known university that does not care about my individual needs.*

Our anecdotal experience with other MOOC's shows a wide range in how learners are treated and supported. Most MOOCs ask to "introduce yourself," often without notice from course instructors, and orphaned threads in discussion forums. We want the students in our MOOCs to be perceived as individuals and support their needs. This has a strong impact on our support effort, which we aim to optimize by applying Lean and Six Sigma in the development and operation of our courses.

2 MOOC Development and Operation as a Six Sigma Project

The development and the operation of a MOOC are both processes, determined by input-activity-output webs, like all processes. Thus, we apply the DMAIC method and its tools for our courses as well.

Every Six Sigma project starts with a *problem*, as a deviation from a given target or with the *challenge* of reaching a new target. Both challenge and problem are always related to a certain (intermediate) output, e.g., a video in the course or an answer to a question in a discussion. Every output is characterized by a bundle of attributes. The importance of the attributes (e.g., audience retention with videos; effort for the answers) is determined by the requirements of customer or the management. It can be specified by a target value and a tolerance range. Attributes together with the degree of their actual deviation (e.g., < 50%.; > 3min) determine the severity of a problem and, thus, the satisfaction of the customer and the costs for the company.

This formalization, structuring, and parameterization of information enables us to capture and define problems correctly at their core (Table 1), as well as further project related data. However, the extent of this formalization, the application range of suitable projects, and our support effort are interrelated.

Table 1. Definition of problems, based on outputs, their attributes, and their deviations.

Output	Attribute	Deviation
Video	Audience-Retention	<50%
Answer	Effort	>3min

- High standardization reduces the application range and our support effort
- Low standardization increases the application range and our support effort

Therefore, we continuously optimize the balance between standardized and individual guidance to keep the range of suitable projects and operate the course effectively and efficiently.

2.1 Development of Our MOOCs

Basic Teaching Objectives. Our basic teaching objectives are aligned with the American Society for Quality (ASQ) and our certification is aligned with the principles of the International Society of Six Sigma Professionals (ISSSP). With this conformance, we ensure global acceptance of our certificate and certification.

Table 2. Certification requirements of the ASQ and TUM for the six sigma green belt.

<table>
<tr><th colspan="2" rowspan="2">TUM Lean Six Sigma Green Belt Certification</th><th colspan="3">Requirements</th><th colspan="2">Online Support</th><th rowspan="2">Target Group</th></tr>
<tr><th>Theory Exam</th><th>Work Experience</th><th>Certification Project</th><th>Lectures</th><th>Coaching</th></tr>
<tr><td>edX</td><td>TUM</td><td rowspan="3">TUM
Lean Six Sigma
Yellow Belt
Teaching Objectives

(based on the
ASQ Green Belt Body of knowledge)</td><td rowspan="3">no</td><td>predefined
Standard-Project</td><td>yes</td><td>no</td><td>Students & Employees</td></tr>
<tr><td>TUM</td><td>edX</td><td>company specific
Business-Project</td><td>optional</td><td>individual</td><td>Employees</td></tr>
<tr><td>TUM</td><td>edX</td><td>selected
Award-Project</td><td>optional</td><td>group</td><td>Students & Employees</td></tr>
<tr><td colspan="2">American Society for Quality ASQ</td><td>ASQ Green Belt Body of Knowledge</td><td>3 years, full time, under supervision of a BB</td><td>no</td><td>no</td><td>no</td><td>Employees</td></tr>
</table>

The teaching objectives of our Yellow Belt certificate include the ASQ's Body of Knowledge for their Six Sigma Green Belt (Table 2). In addition, the ASQ requires a minimum of 3 years of professional experience under the supervision of a Black Belt. We do not require this because it makes the education dependent on a Six Sigma initiative at the company, putting other companies and students at a disadvantage. Instead, we require the implementation of an individual certification project for the Green Belt—as recommended by the ISSSP.

We mapped these conditions as requirements of the learner. The related Kano-Chart (Fig. 2) formalizes all resulting challenges/problems, the desired/actual state, and the consequences on learner satisfaction if a challenge is not met/the problem occurs:

1. **YB-Teaching-Objectives - ASQ-Conformity < 95%**
   Meaning: We consider it a problem if our: YB Teaching Goals (output) are less than 95% (deviation) conform to the ASQ (attribute). Please note that the first paragraph of a section or subsection is not indented.
   This is a basic requirement (Must-Be) for us because our learners expect the worldwide acceptance of the certificate as a matter of course, just as one expects a telephone to enable phone calls. If the requirement is met, the learner is indifferent but will be angry if our certificate is not accepted by a company.
2. **GB-Certification-Project - Implementation - Failed**
   Meaning: We consider it a problem if one of our learners fails (deviation) with the implementation (attribute) of the Certification-Project (object) or terminates it.
   This is a performance requirement (More-Is-Better) for us because the learner has demonstrated the fundamental competencies with the pre-acquired Yellow Belt certificate. The better the performance of a learner, the more satisfaction will result, and the failure of a project would be a problem for the learner and for us.

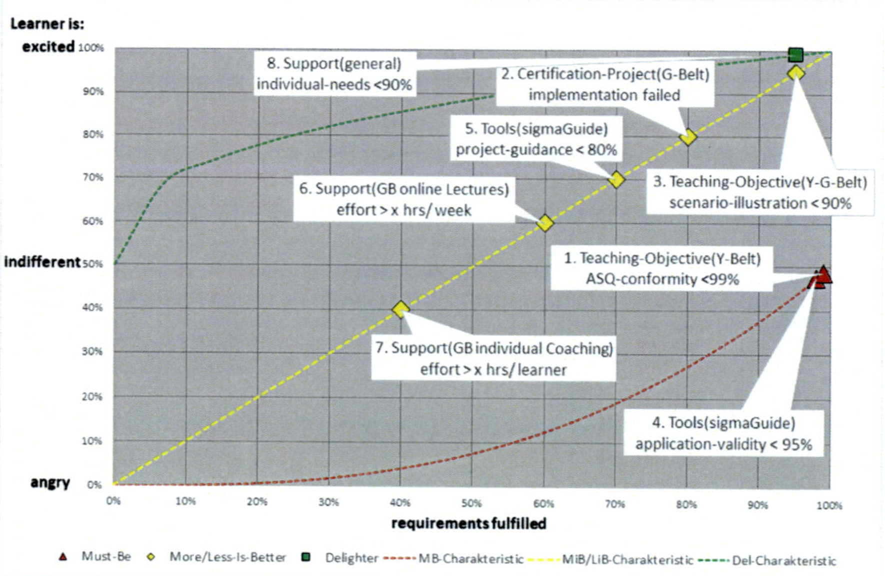

Fig. 2. Kano-Chart representing selected challenges/problems in the development/operation of our MOOC according to their requirement characteristic (Must-Be; More/Less-Is-Better; Delighter) and the resulting learner satisfaction, if requirements are (not) met.

To go beyond acceptance, we enriched these basic teaching objectives with scientific facts and practical principles, developed motivating formats and software tools, and published them in four courses on the edX E-learning platform [3].

Concept of our Yellow Belt Course Material. Our Yellow Belt teaching material addresses the competence levels: *remember*, *understand*, and *apply*. The challenge here was to illustrate more than 90% of the material in easy-to-understand, consistent, and practical scenarios (Fig. 2). That applies to 30 h video lectures, over 500 practice and homework problems, interactive exercises, case studies, and mini projects. The consistent attractive illustration is a Must-Be requirement for us because we assume that learner expect excellent material at TU Munich and would already be angry to go through an average course.

Table 3. Adaptations and developments to transform the project into a standardized process.

Functional-Model	- sharpening the constraints and purpose of the DMAIC phases and - deriving the appropriate tasks and their sequence for the project
	- development of an integrated process-problem-solution model with - formalized components (e.g. Problem:= Object-Attribute-Deviation)
sigmaGuide	For each tool: - inputs defined - application formalized and controlled - outputs defined qualitatively and quantitatively
	Seamless chronological sequencing of 30 tools along the DMAIC phases, allowing the output of one tool to become the input of the next tool
example	**guidance by controlled dialog:** - ask a question - receive the answer - make the answer to the subject of the next question - ask the question ...
Docu-mentation	Project-Story-Book template for: - documentation of project results and - grading the learning success

Automated Guidance. sigmaGuide is a model-based app that guides the learner through the DMAIC phases of the project (Table 3), by a sequence of 30, chronologically ordered and seamlessly connected tools. The input of each tool is defined, just as the input for a hammer or a screwdriver is defined. The output of each tool becomes the input of the next tool. To ensure the validity of the inputs and outputs, the requested information is hierarchically structured and requested in a controlled dialog. This concept allows the project to be treated as a standardized process.

Example: The first tool—Project-Topic in the Define phase—is intended for everyone. It serves to translate an observed weakness of a process or a deviation of its output into a statement about the topic and its suitability for Six Sigma.

Question 1/7 asks for the description of the observed weakness in own words (Fig. 3): "Some of our videos show in audience retention data: a) a short learner watch-time and b) an unusual watch pattern."

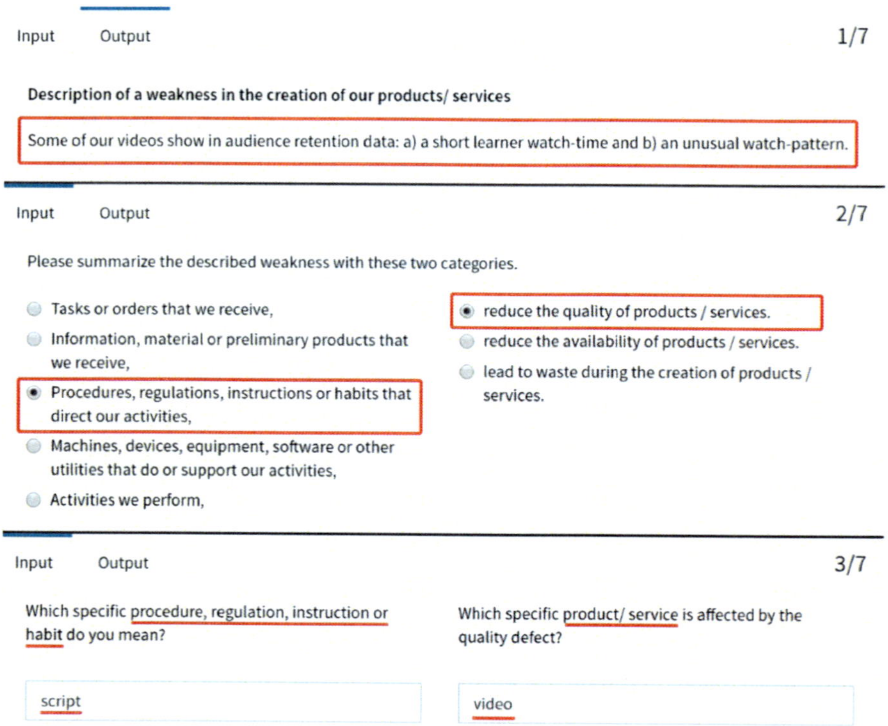

Fig. 3. Selected questions and answers in the first tool: Project-Topic of the project guideline *sigmaGuide*, to identify a Six Sigma project by a controlled dialogue.

Question 2/7 offers options a) for the *cause* of the weakness and b) for the *type of the resulting output problem* (quality, availability, consumption). The combined half-sentences extract a cause-problem relationship of the described weaknesses—e.g., *Procedures … habits that direct our activities, reduce the quality of products / services.*

Question 3/7 integrates the answers from 2/7 and asks for concretization: Which specific *procedure* and *product/ service* …? Example: *script* and *video.*

A report summarizes and interprets the answers (Fig. 4), serving to discuss the topic with a potential project Sponsor. If the Project-Topic is to be further investigated, then the Project-Definition is created in the next tool, preparing it as a Six Sigma Project.

Standardized Project Documentation. The results of each tool must be documented in the Project-Story-Book (Table 3), a template with > 60 slots for project results. Here, the learner inserts the outputs of tools, identifies the important results, interprets them, and draws conclusions. This way, each learner documents the project and its mastery.

The tension between automated guidance and individual support is reflected in the proportion of independent and dependent learners. Approximately 10% of learners successfully complete the project through automated guidance alone. In total, 5% rely on extensive individual support, meaning extra effort for us, represented in the Kano-Chart (Fig. 2).

Relevance and Suitability of the Project-Topic

The SCRIPT we use or apply (cause), decreases the quality of VIDEO (problem).

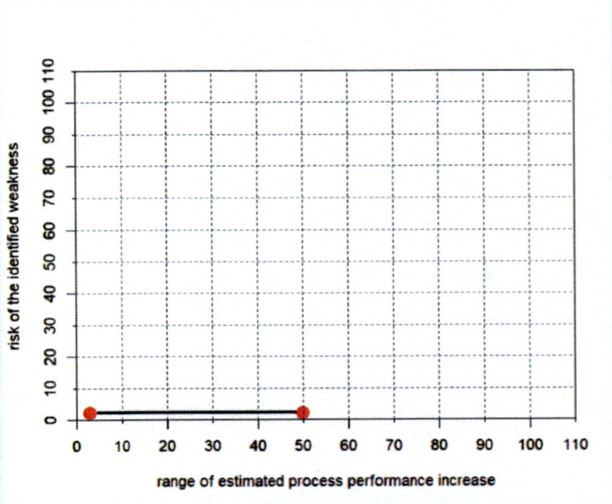

Risk of the identified weakness: 2.2 % (resulting from: cause occurrence, problem detectability, effect severity)

Expected increase of process performance:

- lower limit: 3 % (with 30 % of the solution implemented and min. expected performance increase)

- upper limit: 50 % (with 100% of the solution implemented and max. expected performance increase)

Individual data can be collected on the cause (SCRIPT) and on the problem (VIDEO). This data situation is typical for SIX SIGMA projects.

Fig. 4. Selected part of the report on the relevance and suitability of the project topic.

2.2 Peration of Our MOOCs

Support for General Inquiries. Own errors cannot be excluded in a new course. Many learners are very attentive and inform us. We benefit from this communication as we not only correct mistakes but also receive suggestions for adding new material.

In addition, we receive many questions about the course, certification, organization, technology, and career. Figure 1b shows that students have a clear preference for providers who care about their personal needs. Therefore, we have included the challenge/problem: *4. Support Individual-Needs > 95%* in our support concept—as a *Delighter* requirement (Fig. 2). Basically, of course, answering queries is a *Must-Be* and the learner would be upset if we did not answer them. However, we place great emphasis on the content and character of our response because the learner does not expect this and is therefore likely to be excited.

This brings us into the domain of call centers, which we repeatedly support with our certification projects. Analogously, we differentiate the general inquiries according to these criteria: *Source* (potential learner, enrolled learner, HR representative), *Channel* (edX, TUM, Facebook, LinkedIn), *Medium* (discussion board, course eMail, personal eMail, media post, media message), and *Topic* (course content, certification, organization, technology, career).

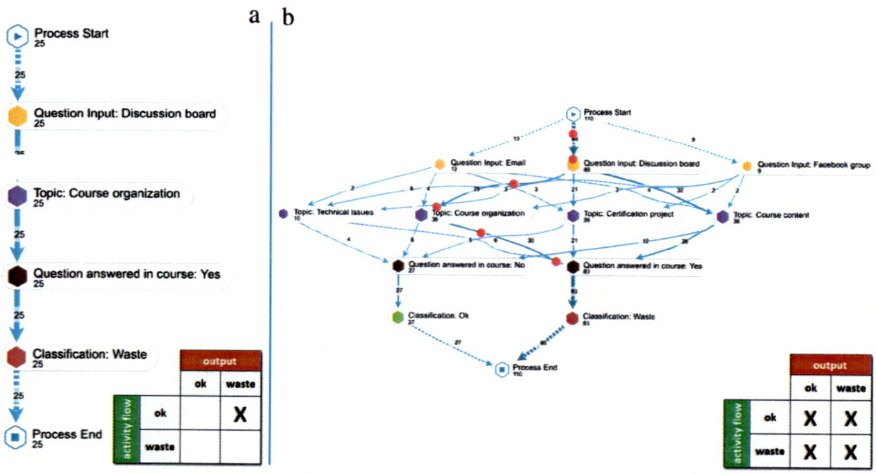

Fig. 5. Automatically generated view and evaluation of our support processes for general inquiries **a)** for the intended flow of activities with wasted outputs and **b)** for all recorded process variants, with quantified waste and value added in each case (data modified).

Although most of our answers are already included in the course material or our homepage, and the inquirer could easily detect them, we follow a central perspective of Lean Management: *Any additional request from the customer indicates a problem with our product/service.* To meet this *Delighter* requires great effort from us. (Fig. 2). In addition, the workload of our team increases if, for example, a request first must be forwarded internally before it reaches the right recipient. If you think lean, it's waste.

Therefore, we take samples to measure our process performance and to detect our problems by measuring—e.g., *Which text was sent (standardized, customized)? How long did the answer take? How often was the request forwarded internally?*

To analyse this data, we use process mining and employ a tool [4] that was developed by former students of TUM and is worldwide used today in production and service. The process mining charts (Fig. 5) are generated from our (modified) sample data.

The activity flow in Fig. 5a shows an intended, waste-free process, from the requests in the discussion board to our responses. But because our answer is already included in information about our course, the repetition of our redundant output is waste.

Figure 5b shows—superimposed—all 24 process variants, representing all cases of our sample. The prioritization of the variants according to their costs and savings potential enables us to make targeted improvements to our workflow and results.

Support for Certification Projects. The need for individual support varies greatly between our three different paths for certification (Table 2).

Typical is the certification of employees via individual business projects, selected and steered by a manager in the company. We coach these projects individually on the methodology and review the documentation of each project phase. Our study of Six Sigma success factors in N = 78 German companies [5] shows that this support is a determinant of project success.

Unique is our option for students and job seekers to acquire our certification by a predefined standard project: *Reduce the littering of public places in your hometown and control the sustainability of your measures*. It follows the United Nations Sustainable Development Goal 11: Make cities and human settlements inclusive, safe, resilient, and sustainable [5]. The topic sounds unusual but is suitable. And the projects results demonstrate relevant effects on the environment. To support these predefined projects, we offer weekly online lectures for our students.

Furthermore, we offer the annual *Global Outreach Green Belt Award* (GoGB) for our Yellow Belts to apply with a project idea that follows one of the United Nations Sustainable Development Goals [6]—like our predefined *littering project*. Award 2019: We supported a Green Belt project in a rural hospital (India) to reduce the waiting time of patients, the effort in the laboratory and to increase the efficiency of the treatment. Award 2020: We are supporting a Green Belt project in Lagos (Nigeria), to reduce litter on the streets and sewers to curb flooding in the city. Starting this year, we will provide group coaching to 3 selected projects in the final round for the award.

Our support activities for these three blended formats differ significantly:

- MOOC + wide range business project + individual coaching, fee-based
- MOOC + predefined standard project + open online lectures, fee-based and the
- MOOC + wide range UNGoal project + group coaching, free of charge (GoGB)

Targeted Improvement of the Course Material. Further reducing our support effort to guide and grade learners is a *Less-Is-Better* requirement for us (Fig. 2). To keep our learner requirements and maintain the level of project results, we reduce our effort indirectly by further improving the course material and further automating the project guidance. We distinguish between effort to:

- compensate general *laziness*, represented by our repetition of redundant information
- close specific knowledge gaps, represented by repeated errors of learners

The course content is entirely covered by videos and additionally mirrored in the related textbook [5]. Analyzing the watch-behavior therefore is key for identifying causes of learners' *laziness* and knowledge gaps (Fig. 2). Control charts (Fig. 6) for audience retention—e.g., offered by YouTube, can reveal these signals and patterns:

1. Exit at start
2. Repeated entry into the video at a specific point in time (watch-time > 100%)
3. Intervals above or below average watch-time
4. Outliers above or below watch-time control limits
5. Shifts in watch-time
6. Trends in watch-time
7. Entry at end

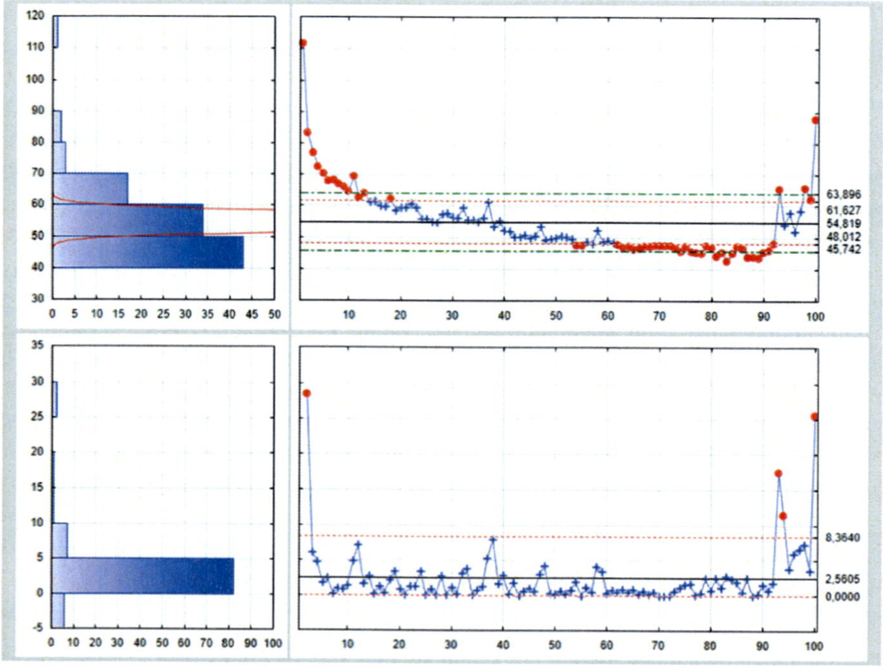

Fig. 6. Control-Chart for a video with patterns and signals in watch-behavior.

These signals in the chart serve to detect the underlying root causes of specific knowledge gaps in our scripts and audio-visual presentations. We also detect what learners like in our videos and transfer best practice on video updates.

Sender-receiver analyses of MOOCs offer an advantage over classroom lectures. And they help us balancing our standard material and individual guidance to help participants implement wide range of improvement projects, suitable for Lean Six Sigma certification.

Acknowledgments. We would like to thank Martin Grunow and Bernhard Kraus for the frame, Tina Koumpouli for her professional management, and Franka Kunft, and Karla Gabriela Urquia Rivas for their continuous support – Panorama!

References

1. Harry, M.J.: The Nature of Six Sigma Quality. Motorola University Press (1988)
2. Wikipedia - https://en.wikipedia.org/wiki/Six_Sigma#Certification
3. edX - https://www.edx.org/search?q=six%20sigma
4. Celonis - https://www.celonis.com/academic-signup
5. Hutwelker, R.: Six Sigma Green Certification Project. Springer, Heidelberg (2019)
6. United Nations Sustainability Development Goals - https://sdgs.un.org/goals

Effectiveness of an Intelligent Question Answering System for Teaching Financial Literacy: A Pilot Study

J. D. Jayaraman[1](✉) and John Black[2]

[1] New Jersey City University, Jersey City, NJ 07305, USA
jjayaraman@njcu.edu
[2] Teachers College, Columbia University, New York, NY 10027, USA

Abstract. Question answering systems have shown promise as a learning tool in several fields but are yet to be used in the financial literacy domain. The objective of this research is to build an intelligent question answering (IQA) system using state-of-the-art natural language processing technology and test its effectiveness in helping students learn personal finance. The purpose of the system is twofold—one as an intelligent system for practicing and two as a scaffolding tool for finding answers to questions. The IQA system consists of three modules—the question answering module, the evaluation module, and the feedback module—and is built using Google's state-of-the-art language model for Natural Language Processing: Bidirectional Encoder Representations from Transformers (BERT). The primary research question we attempt to answer is whether the use of the IQA system results in increased learning in the domain of personal finance and if so how. We find significant learning gains (effect size = 1.89) with the affordances of the IQA system contributing to the gains. Students felt that the IQA system helped them learn better mainly by allowing them to practice at their own pace and by providing them with indications of where they had gone wrong and allowing them to read, think, and re-attempt the answer. These promising results suggest that an IQA system could be effective as a support tool for improving learning of personal finance.

Keywords: Intelligent question answering · Natural language processing · Financial literacy

1 Introduction

Intelligent chatbots and question answering (QA) systems are an innovative technology that has been incorporated into the teaching/learning experience recently. Various types of chatbots/QA systems have been used as pedagogical agents in digital learning environments referred to as Intelligent Tutoring Systems [1]. Chatbots have shown promise as a learning tool in several fields and have been found to be particularly effective in language learning [2]. Though chatbots/QA systems have been used in various fields such as teaching math, science, and language, they have not yet been used in the domain

D. Guralnick et al. (Eds.): TLIC 2021, LNNS 349, pp. 133–140, 2022.
https://doi.org/10.1007/978-3-030-90677-1_13

of teaching financial literacy/personal finance. Financial literacy is an oft ignored but important domain. Lack of financial literacy has been linked to bad financial decisions that imperil one's future and cause significant harm to the economy [3]. The U.S. has one of the lowest levels of financial literacy among developed countries [4]. Thus, new and innovative methods of teaching financial literacy need to be explored. The objective of this research is to build an intelligent question answering (IQA) system using state-of-the-art natural language processing technology and test its effectiveness in helping students learn personal finance. The system is envisioned as an aid to the teacher and not as a standalone tutoring system for learning the content. The purpose of the system is two-fold—one, as an intelligent and adaptive system for practicing what the students learn in class, and two, as a scaffolding tool that can be used by students to find answers to questions that they may have of the content. To the best of our knowledge, this is the first such attempt in the domain of financial literacy/economics. We proceed to outline the theoretical framework, research questions, methodology, and present the results.

2 Theoretical Framework

Our study is rooted in the theory of skill acquisition [5, 6]. According to this theory, in the first phase of skill acquisition (cognitive phase), a declarative representation of the procedure is established and translated into behavior by interpretative processes. The next phase (associative phase) is characterized by the proceduralization of the skill, which leads to automatization. In this phase, much practice is necessary with no need for close supervision other than corrective feedback. Since each student differs in their skill development, the practice needs to be individualized. An IQA system is ideally suited for providing the student with individualized practice. Based on the skills acquisition theory that emphasizes individualized practice, we have built and trained our IQA system to provide practice that tries to cater individually to each student's needs.

Our study is also based on the theory behind instructional scaffolding. The idea of scaffolding was first introduced by Bruner and refers to the "steps taken to reduce the degrees of freedom in carrying out some task so that the child can concentrate on the difficult skill she is in the process of acquiring" [7]. Scaffolding is also related to Vygotsky's Zone of Proximal Development [8] and is often used interchangeably. The IQA system scaffolds the student's learning by allowing them to use the system to find answers to questions that may arise while learning the content.

3 Research Questions

The questions we attempt to answer with this research are:

1. How can we build a cutting-edge IQA system for practice and scaffolding in the domain of financial literacy?
2. Does the use of the IQA system for practice and scaffolding result in increased learning (content knowledge) in the domain of financial literacy?
3. How does the use of the IQA system for practice and scaffolding result in increased learning (content knowledge)?

Based on the theories we have described above, we hypothesize that the IQA system when used by students for practice and scaffolding will increase learning (content knowledge) in the domain of financial literacy. Since this is the first attempt that we are aware of at answering the above questions, we contribute to the literature by detailing the architecture, build process, and effectiveness of an IQA system in the domain of financial literacy.

4 Methodology

4.1 System Description

Our IQA system consists of three modules—the QA module, the evaluation module, and the feedback module. The IQA system is built using Google's state-of-the-art language model for Natural Language Processing: Bidirectional Encoder Representations from Transformers (BERT) [9]. We leveraged a BERT-based implementation of a closed domain QA system called cdQA-Suite. This library comes with a tool for further training the pre-trained BERT QA model to our domain and a simple web-based user interface. The BERT QA model is pre-trained on the Stanford Question Answering Dataset (SQuAD) [10]. For this pilot study, we focused on the topic of consumer credit (credit cards, mortgages, etc.) within the broader domain of financial literacy and trained the BERT QA model with a dataset of 150 items curated by creating questions and answers on the topic of consumer credit from various standard textbooks and websites on personal finance and leveraging the first author's expertise as a personal finance teacher (example of training data is shown in Table 1). The students used a simple web based front end to answer and ask questions.

Table 1. Simple training data.

Paragraph	Question	Answer
Credit is an arrangement to receive cash, goods, or services now and pay for them in the future. Consumer credit refers to the use of credit for personal needs (except a home mortgage) by individuals and families, in contrast to credit used for business purposes. When you use credit, you satisfy needs today and pay for this satisfaction in the future. Using credit increases the amount of money a person can spend to purchase goods and services now. But the trade-off is that it decreases the amount of money that will be available to spend in the future	What is consumer credit?	Consumer credit refers to the use of credit for personal needs
	What is credit?	Credit is an arrangement to receive cash, goods, or services now and pay for them in the future
	What is the tradeoff to using credit?	The trade-off to using credit is that it decreases the amount of money that will be available to spend in the future

The evaluation module compares the student's answer to the answer that the QA module generates and calculates a BERTScore [11]. BERTScore is a text matching evaluation metric based on pre-trained BERT contextual embeddings, which closely correlates with human judgments. The BERTScore is between 0 and 1, with a higher score indicating a better answer. The BERTScore is displayed on the screen as the percentage score for the answer the student provided.

The feedback module is currently a rule-based system. For this pilot study, we have used simple rules and feedback, but we are in the process of improving the feedback and making it "intelligent" by using the BERT framework. The rules we have used are simple—if the BERTScore is above 0.8, we provide congratulatory feedback and move on to the next question; if the score is between 0.6 and 0.8, we ask the student to improve the current answer and highlight the part of their answer that differs considerably from the correct answer (this information can be provided since the cdQA output highlights the parts of the answer that matches closely with the correct answer); and if the score is below 0.6, we ask the student to redo the answer and provide a set of links to read. As part of every feedback, the students are reminded that they could also ask questions of the system and get answers to improve their understanding. In the next phase of development, we will improve the feedback module to provide more specific feedback. We are creating a training dataset of student answers and appropriate specific feedback. We will train the BERT QA model on this training dataset and use the trained model in the feedback module. Moreover, as the system gets used, we will capture the student answers, and we will manually alter the system generated feedback to a more ideal feedback and continue to expand the feedback training dataset, thus, improving the performance of the feedback module.

The architecture/workflow for the system is shown in Fig. 1. The student starts the practice by clicking on a "Start Practice" button on the web browser interface. The system posts a question randomly chosen from a list of 150 questions. The questions are grouped by five levels of difficulty. The initial questions are always chosen randomly from the questions at the lowest level of difficulty, after which the system is adaptive. The adaptiveness of the system is currently rule based. If the student gets a 0.8 or above BERTScore three times in a row, the next practice question is randomly chosen from the next level of difficulty and so on until the last level is reached. The student then types in the answer and clicks submit, at which point the system evaluates the answer and provides feedback, along with the BERTScore displayed as a percentage. The student can then either move to the next question, modify the answer, or redo the answer depending on the feedback. The adaptiveness of the system can be greatly improved by adopting an Item Response Theory (IRT) model. We are currently in the preliminary stages of building an IRT model to enhance the adaptiveness of our system.

At any point during the practice or otherwise, the student can click the "Ask Question" button on the interface to ask any question of the system other than the one that is being worked on. The answers provided in this QA mode are meant to scaffold the student's learning.

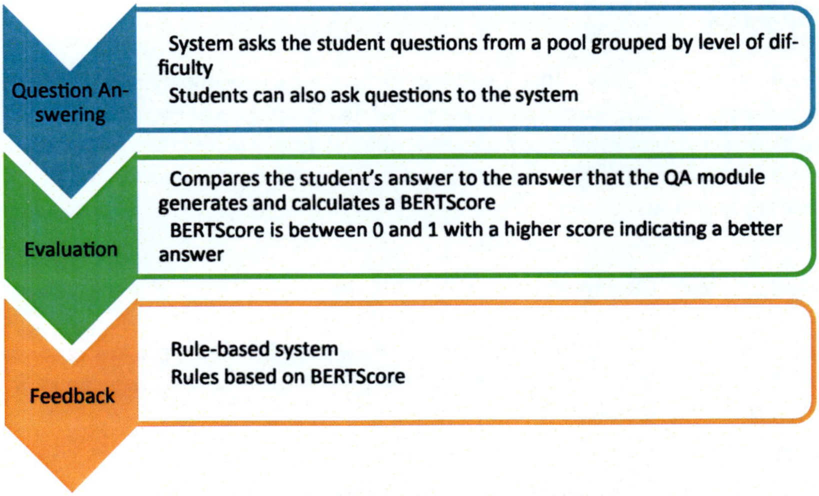

Fig. 1. System architecture workflow.

4.2 Research Design

We pilot tested our IQA system with 15 students in an undergraduate financial literacy class typically taken by freshmen at a four-year university in the northeastern United States. We used a one group pretest posttest design for the pilot study. We designed a 15-question instrument to measure learning pre and post using the IQA system. The questions were taken from assessments in standard textbooks. A short survey (Table 4) assessed the effectiveness of the IQA system as a support tool for learning. We also interviewed the students to understand how the IQA system helped them learn and also how we can improve the system for the randomized control trial we will be conducting as the next phase of this research.

4.3 Procedure

The students attended a two-hour lecture on consumer credit that included a short tutorial on how to use the IQA system. After the lecture, they were assigned homework (due in a week before the next class) to practice using the IQA system. The students had to answer at least 30 questions with a BERTScore of 0.8 or higher, and of the 30, they needed to have at least 5 questions at level 4 or higher (a counter on the UI showed them how many questions they had answered at each difficulty level). The pretest was administered after the lecture but before students started using the IQA system. The posttest and the survey were administered when they met back for class the next week, before that day's lecture started. The students also participated in a short interview to understand how the IQA system helped them learn and how it could be improved.

5 Analysis

Since this was a pilot study with a small sample size, we used nonparametric statistics to analyze the results. The learning was measured using percentage correct scores (PCS). A Wilcoxon signed-rank test (nonparametric equivalent of a paired t test) was conducted to assess the difference in pre and post means. Mean scores are reported for the Likert type survey questions.

6 Results and Discussion

Table 2 shows the mean and standard deviation for the PCS on the pre and post assessment. The mean PCS increased by about 19%. The effect size for the mean PCS was large at 1.89.

Table 2. Mean increase in percentage correct scores (PCS) scores. $n = 15$.

	Pre mean PCS	Post mean PCS	Difference	Effect size
Experimental group	54.55% (0.111)	73.64% (0.090)	19.09%	1.89

Note: Standard deviations are reported in brackets. Effect size is Cohen's d

These findings are further substantiated by the results of the Wilcoxon signed-rank test (Table 3), which showed that there was a statistically significant ($p < 0.01$) difference in the pre- and post-test paired sample mean PCS.

Table 3. Wilcoxon Signed Rank Test for difference in pre and post percentage correct scores (PCS). $n = 15$.

	V Stat	P Value
Experimental group	1.67	0.0092**

* $p < 0.1$, ** $p < 0.05$, *** $p < 0.01$

The mean scores of the survey questions (Table 4) showed that the IQA system was overall effective in helping students learn better.

Table 4. Survey results. n = 15.

Survey question	Mean score
Did the IQA system based homework help you learn better as compared to usual homework (1 – Not at all, 5 – Definitely yes)?	4.0
How accurate did you think were the answers provided by the IQA system (1 – Not accurate, 5 – Highly accurate)?	3.8
How accurate did you think the evaluation of your answers was (1 – not accurate, 5 – Highly accurate)?	4.5
How good was the feedback provided by the QA system (1 - Not good, 5 - Very good)?	3.0
Open ended questions	
How did the QA system help you learn?	
How can the QA system be improved to better help you learn?	

Though the IQA system provided good answers to questions that students needed help with, the survey showed that it could be improved. The survey also showed that the feedback module could be significantly improved. This was expected because the feedback we provided just included some links to relevant reading material and did not point out how exactly the answer could be improved. The ongoing improvements to the feedback module described in the system description will address this issue and make the feedback very specific to the answer given.

The interviews showed that the students felt that the IQA system helped them learn better mainly by allowing them to practice at their own pace and by providing them with indications of where they had gone wrong and allowing them to read, think, and re-attempt the answer. They said that the answers provided when they asked questions of the system also helped them get a better understanding of the subject matter. They also mentioned that they liked the fact that they could ask questions without fear of what others would think, unlike in a classroom. They mentioned that the answers sometimes did not make complete sense, though, and could be improved. They also felt that the feedback could be more detailed and specific.

There are several limitations to our pilot study. A randomized control trial on a much larger sample with an experimental group using the IQA system and a control group using standard paper and pencil practice questions needs to be conducted to conclude that the IQA system produces greater learning gains. The pilot study showed that we need to improve the feedback module, and we need to further train the QA module to provide better answers.

7 Conclusion

To the best of our knowledge, this research is the first attempt at building and testing an IQA system to support the teaching/learning of personal finance. The results of the pilot study showed large significant learning gains and the affordances of the IQA system contributing to the large learning gains. These promising results suggest that an IQA system could be effective as a support tool for improving learning of personal finance. But further research needs to be conducted to improve the system and conduct a randomized control trial to conclusively prove that the IQA system produces greater learning gains than standard non-technology-enabled practice and scaffolding.

References

1. Laurillard, D.: Rethinking university teaching: A conversational framework for the effective use of learning technologies. Routledge (2013)
2. Coniam, D.: The linguistic accuracy of chatbots: usability from an ESL perspective. Text Talk **34**(5), 545–567 (2014)
3. Cole, S., Sampson, T., Zia, B.: Prices or knowledge? What drives demand for financial services in emerging markets? J. Financ. **66**(6), 1933–1967 (2011)
4. OECD.: PISA 2015 Results (Volume IV): Students' Financial Literacy. OECD Publishing, Paris (2017)
5. Anderson, J.R.: Acquisition of cognitive skill. Psychol. Rev. **89**(4), 369 (1982)
6. Anderson, J.R., Fincham, J.M., Douglass, S.: The role of examples and rules in the acquisition of a cognitive skill. J. Exp. Psychol. Learn. Mem. Cogn. **23**(4), 932 (1997)
7. Bruner, J.S.: The role of dialogue in language acquisition. In: Sinclair, A., Jarvelle, R.J., Levelt, W.J.M. (eds.) The Child's Concept of Language. Springer-Verlag, New York (1978)
8. Vygotsky, L.S.: Mind in Society: The Development of Higher Psychological Processes. Harvard University Press, Cambridge (1978)
9. Devlin, J., Chang, M.W., Lee, K., Toutanova, K.: Bert: pre-training of deep bidirectional transformers for language understanding. arXiv preprint arXiv:1810.04805 (2018)
10. Rajpurkar, P., Zhang, J., Lopyrev, K., Liang, P.: Squad: 100,000+ questions for machine comprehension of text. arXiv preprint arXiv:1606.05250 (2016)
11. Zhang, T., Kishore, V., Wu, F., Weinberger, K.Q., Artzi, Y.: Bertscore: Evaluating text generation with bert. arXiv preprint arXiv:1904.09675 (2019)

Escapeling: A Gamified, AI-Supported Chatbot for Collaborative Language Practice

Christian Johnson(✉), Maksim Urazov, and Emma Zanoli

Osnabrück University, 49090 Osnabrück, Germany
{cjohnson,murazov,ezanoli}@uni-osnabrueck.de

Abstract. Escapeling is a Telegram bot aimed at improving English skills outside of a classroom setting. Intermediate language learners can collaboratively practice in a gamified escape-room scenario wherein they must complete several tasks. The three currently available tasks cover some of the most important language learning aspects: improving grammar knowledge via sentence correction, enhancing vocabulary via word guessing and encouraging writing practice in a discussion task. To complete the tasks, the users are required to cooperate by reaching common decisions or prompting their peers. Engagement with the bot is stimulated by in-game achievements and narrative components, supporting the long-term goal of alien ship escape. The bot's adaptive AI-based module tracks individual user performance and changes the task difficulty accordingly. Initial user acceptance tests suggest the efficiency of AI-assisted adaptive language learning as an aid to traditional learning approaches. Work on Escapeling is still in progress.

Keywords: Mobile learning · Adaptive learning · Gamification

1 Introduction

This article describes a conversational language learning application developed to appraise the influence of group learning and gamification on enhancing motivation and engagement in remote learning scenarios. The Escapeling application is targeted at secondary school English as a foreign language (EFL) students, integrating with the Telegram messaging application to provide several collaborative remote language practice tasks.

Recent events demonstrate the need for inventive solutions to the challenge of retaining student engagement in remote learning scenarios. Specifically in the case of second-language learning, we hypothesize that collaborative group practice is important to address this need, which can be facilitated through the intervention of mobile applications. It has been shown that portable devices such as tablets and smartphones are increasingly preferred modes of interaction, and they decrease the barrier to application entry [6]. While many applications offer mobile learning scenarios for individual users, we motivate our approach through the combined emphasis on collaborative mobile group learning and narrative-contextualized tasks supported with conversational agents.

D. Guralnick et al. (Eds.): TLIC 2021, LNNS 349, pp. 141–148, 2022.
https://doi.org/10.1007/978-3-030-90677-1_14

Conversational language learning scenarios can be considered a form of metaphorical user interaction, which are widely considered good practice for interaction design [17], potentially mitigating the alienating effects of remote learning. By encouraging users to practice language skills in a fictional task akin to a real-world communicative scenario, we hypothesize that the application will leverage the critical phenomena of "engagement" and "collaboration" to improve learning outcomes [8].

The application emphasizes a group rather than individual learning scenario following the assumptions that collaboration has been shown to foster cooperative interdependence towards providing learning support and enhancing engagement, and that these phenomena are relevant for language learning as a complex social skill.

Cooperative or collaborative learning refers to a set of classroom techniques that encourage learner interdependence as a route to cognitive and social development [11]. As Escapeling's learning activities emphasize group member interdependence towards achieving common goals, the application encourages inter-user support [13], namely that users of a higher skill level have the opportunity to act as a learning resource for other users.

As a result of this implicit support mechanism, users have access to supportive information as a component of the 4C/ID-model of complex learning [20], while simultaneously addressing challenges to engagement through social interaction [1]. The aforementioned phenomena play prominent roles in the vocabulary guessing task.

Additionally, it is known that narrative elements correlate with improved learning, since narrative-based contextualization [2] and salient "escape room" scenarios "provide an enjoyable experience that immerses students as active participants in the learning environment" [9]. The application's narrative is a science-fictional escape room setting where users must cooperatively solve language-practice tasks to escape from alien captors.

This article contributes a potentially novel application addressing the challenges of remote language learning, accompanied with investigation and discussion of the relative impact of the application's features on learner engagement and language outcome. In the following sections, we describe the features and evaluation of the application's user experience.

2 Application Features

The Escapeling application is delivered in the form of a bot through the Telegram messaging application. The bot provides guidance throughout a science-fictional narrative learning environment. Upon first contact with the application, the bot sends both textual and audiovisual messages to introduce the narrative elements of the learning scenario. After this introduction, every user is given the chance to play a collaborative language learning game. To this end, they are redirected to a group session, where they collaborate with other users on solving a number of language tasks. At the time of writing, our application offers three such tasks: sentence correction, word guessing, and discussion. During each session, the group collaborates in an effort to escape from their captors, and their performance on the proposed language tasks impacts their chance of success. User performance metadata are collected throughout the whole process in order to adapt task difficulty to user proficiency, leveraging "human-in-the-loop" AI methodologies.

In the following sections, we will expand on each of the aforementioned elements in turn.

2.1 Pedagogical and Gamified Elements

The application offers three language practice tasks to be completed collaboratively in a group of intermediate English learners.

The sentence correction task is primarily focused on grammar. In this task, one user is randomly selected to review the correctness of one sentence, displayed in the chat for all users. The user first decides whether the sentence is correct; if it is incorrect, they need to identify the word that causes the mistake and provide a correct alternative. Even though a single user is responsible for one sentence at a time, cooperation among users is encouraged by giving the group the opportunity to discuss the given sentence. The task is successfully completed if a group collects all pieces of a "codeword," distributed when a single task iteration is successfully completed.

The word guessing task is primarily focused on vocabulary learning. In turn, each user is given a word, and their objective is to enable the other users to guess the given word by explaining its meaning. In case someone doesn't know the meaning of the word they received, they can ask for an alternative word. As with the sentence correction task, users must collect pieces of a codeword by correctly guessing all the words assigned to them.

The discussion task is primarily focused on written language production through a group discussion. The discussion is first prompted by a short text on a given topic, and it is then guided by subsequent questions about the text. Group performance is evaluated via a number of metrics meant to assess user participation. In that sense, comparatively less stress is put on grammar and vocabulary, while more emphasis is given to encouraging contributions from all users.

Multiple studies suggest that gamified components can improve students' performance and encourage extra interaction with learning materials [21]. Escapeling combines multiple gamification strategies working towards the common goal of regular, long-term, and fruitful interaction with the bot. Its learning tasks are supplemented by gamified elements, such as badges, activity streaks, unlockable content elements, and public achievement display.

While completing the tasks, users can earn Streak and Code Cracking achievements. Streaks are awarded every 2nd, 3rd, 5th, etc. consecutive day the users interact with the bot, with the distance between streaks gradually increasing. They indicate user consistency and facilitate the habit of regular interaction with the application. Code Cracking badges are awarded every 1st, 4th, 8th, etc. unlocked code digit and indicate the overall time spent interacting with the bot, encouraging long-term interaction, even after breaks. The digits themselves serve as Escapeling's unlockable content in its simplest form, creating a sense of progression towards the common goal of alien ship escape. Each new achievement takes longer to complete, making them rare and more difficult, which was found to be a successful means of keeping the users motivated over time [10].

Achievements can be viewed by the user individually in a private chat with the bot, but new achievements also appear in the group chat when a task is finished and can be seen by the whole group. This stimulates competition and activates the users' natural social comparison mechanisms, yielding an additional emotional response, which, in turn, has been shown to motivate further interaction with the application and better learning performance [21].

2.2 Adaptive and AI-Supported Elements

We describe the application as "AI-supported" because, although we leverage AI methodologies to achieve user-specific task recommendations, we do not aim for a completely automated adaptive learning solution. We, instead, opt for a "human-in-the-loop" concept, which suggests possible "paths" to users before allowing them to choose their ultimate task content.

It is demonstrated that adaptive user models incorporating student or teacher feedback can be effective compromises to resolve the challenges of exhaustively automated adaptive processes [7]. In this spirit, the Escapeling application collects data about students' task performance in the form of task iteration metadata, as well as task-item-specific performance results, which together inform a proficiency-based, hierarchical user model intended to select task items of optimal difficulty for the elected student of a given task iteration.

Figure 1 outlines the user model, which situates several grammar and vocabulary subskills beneath two independent higher-order language skills. Multiple factor analyses of university and K12 language assessments confirm the fit of independent rather than correlated factor models, motivating our decision to avoid a correlational assumption in the user model [18, 19].

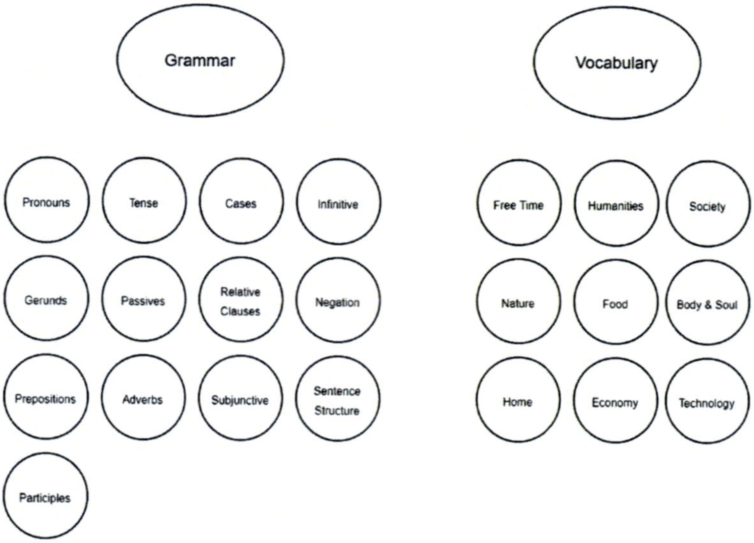

Fig. 1. Application hierarchical user model with language sub-skills.

Subskill categories for the grammar and vocabulary skills are selected so as to correspond to grammatical categories represented in the SCoRE corpus [4] and semantic word groups represented in the English WordNet corpus [15], respectively, as these are the source corpora from which the task data are drawn. The unscored discussion task leverages data from public news sources. To coordinate task items with user performance, we assign scalar difficulty values to each task item. Difficulty values for sentences are provided by the SCoRE corpus, while values for vocabulary items are programmatically assigned using a small neural network difficulty classifier trained on word-difficulty-relevant features [3, 12, 14].

Performance metadata such as duration, number of messages, and number of successful sub-iteration tasks are collected during play and used to modulate each user's sub-skill values, which probabilistically determine the likelihood of encountering corresponding task items. Intuitively, if a user frequently succeeds in explaining vocabulary words related to the domain "Society," similar words are less likely to appear in an effort to prioritize practice of other domains. On a group-level, updates to an elected user's values are applied as a soft-update to other users. The flowchart in Fig. 2 outlines the protocol by which performance data are collected and used to adapt the application's content.

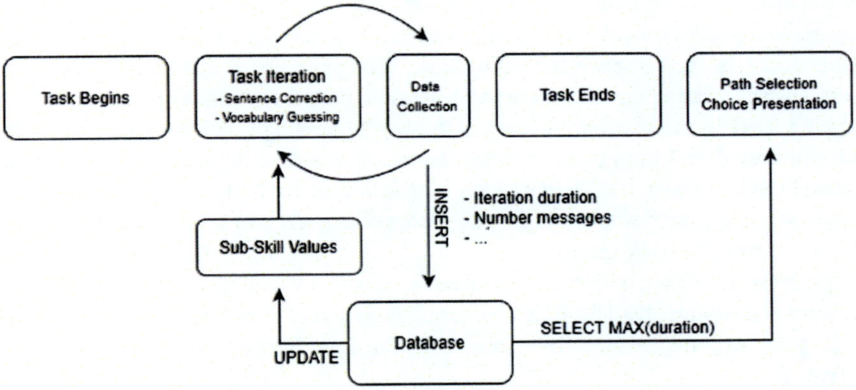

Fig. 2. Application data flow for a single group task.

The aforementioned metadata are also used to suggest the additional "paths" that allow users to semi-manually influence the application's adaptability. Users can opt for a dynamic third option, which is suggested based on their task performance from metadata excluded from the automatic probabilistic adaptation.

3 Concept Validation

The application was tested remotely by teenagers with English knowledge between pre-intermediate and upper-intermediate. Considering no intermediary adjustments planned and the limited testing capabilities due to the COVID-19 situation, the group size in the first testing phase was restricted to seven people, covering the two tasks available at that moment; the second group was restricted to five people, covering the three currently available tasks. The first single test session lasted 60 min. To ensure the results' reliability, the second test involved daily 30-min sessions over a period of two weeks.

Testers received two instruction sets—one with general information and technical requirements, distributed two days prior to the test, and the other with instructions to perform certain actions in the application, sent during the test. The test started with a background questionnaire, followed by a screen-recorded game session and another form, containing questions about the users' satisfaction, their willingness to recommend the application further, and basic task understanding.

The first test phase additionally focused on the application's performance. Assessing the performance of a learning task presented a challenge because not all common testing approaches would have worked reasonably. In contrast to most regular task scenarios, a learning task's goal is to transfer knowledge, even at the cost of increased processing effort or delays. It is, therefore, less relevant how quickly the users finish with the task, provided they learn from it.

Diah et al. (2010) previously encountered the same issue with their learning application and chose success rate, defined as the proportion of successful trials to the total amount of trials, as their base metric [5]. We adopted a similar approach. Our performance assessment was informed by two metrics: success and error rate. We defined success and partial success for the application's two tasks to be worth 1 or 0.5 points, respectively. In the sentence correction task, success required three users in the game room communicating throughout the task session; partial success required two communicating users in the session. Success in the word guessing task was defined as at least one hint posted, followed by an incorrect guess and a further description, or by a correct guess; partial success was defined as a hint followed by a guess posted. The second performance measure was the error rate, defined as irrelevant or rule-violating actions count per complete task session.

Both tasks finished with success and error rates of 1.0 and 2.0 accordingly. The players were actively involved in all task sessions, and most of their errors could be attributed to no prior experience with Telegram API, not directly concerning the application's design.

In the first test, six out of seven users reported their overall satisfaction to be 7/10 or higher. In the second test, all five users reported a satisfaction score of 7/10 or higher. In both tests, all users reported their willingness to recommend the application further to be 5/10 or higher. Lower satisfaction scores were primarily reported after the session where the testers had experienced technical problems.

The test results demonstrate that the application was well accepted by the testing group with only minor usability issues. A vast majority of participants reported their high satisfaction with the application, and all testers reported at least a certain readiness to recommend the application to others. This suggests that Escapeling has potential as a collaborative language learning environment and calls for further work with the initial concept.

4 Conclusion

Escapeling presents an innovative approach to remote language learning. Its collaborative didactic framework is built with a combination of motivating narrative, gamified elements such as achievements, unlockable content and public achievement display, and adaptive module powered by machine learning algorithms. Initial tests yielded relatively high satisfaction and performance scores that at least hint at the viability of the approach we adopted. However, the application's scope is currently limited to just three tasks that do not constitute a full-fledged language learning experience. Escapeling's ability to maintain long-term engagement and its effect on the users' progress need to be further assessed. Longitudinal testing and application development are in progress.

References

1. Bahrani, T., Sim, T.S.: Informal language learning setting: technology or social interaction? Turk. Online J. Educ. Technol.-TOJET **11**(2), 142–149 (2012)
2. Chen, Z.-H., Chen, H.H.-J., Dai, W.-J.: Using narrative-based contextual games to enhance language learning: a case study. Educ. Technol. Soc. **21**, 186–198 (2018)
3. Chen, X., Meurers, D.: Characterizing Text Difficulty with Word Frequencies, pp. 84–94 (2016). https://doi.org/10.18653/v1/W16-0509
4. Chujo, K., Oghigian, K., Akasegawa, S.: A corpus and grammatical browsing system for remedial EFL learners. In: Leńko-Szymańska, A., Boulton, A. (eds.) Multiple Affordances of Language Corpora for Data-driven Learning, pp. 109–128. John Benjamins, Amsterdam (2015)
5. Diah, N.M., Ismail, M., Ahmad, S., Dahari, M.K.: Usability testing for educational computer game using observation method. In: 2010 International Conference on Information Retrieval & Knowledge Management (CAMP) (2010)
6. Dykes, G., Knight, H.R.: Mobile Learning for Teachers in Europe: Exploring the Potential of Mobile Technologies To Support Teachers and Improve Practice. UNESCO, France (2012)
7. El-Hadad, G., Shawky, D., Badawi, A.: Balancing the role of machine learning and teacher in adaptive learning guidance system (ALGS). In: The International Conference on E-Learning in the Workplace 2020, New York (2020)
8. Franceschi, K., Lee, R., Zanakis, S., Hinds, D.: Engaging group e-Learning in virtual worlds. J. Manage. Inf. Syst. **26**(1), 73–100 (2009)
9. Fotaris, P., Mastoras, T.: Escape Rooms for Learning: A Systematic Review (2019)
10. Groening, C., Binnewies, C.: "Achievement unlocked!" - the impact of digital achievements as a gamification element on motivation and performance. Comput. Hum. Behav. **97**, 151–166 (2019). https://doi.org/10.1016/j.chb.2019.02.026
11. Johnson, D.W.: Cooperative learning: increasing college faculty instructional productivity. ASHE-ERIC Higher Education Report No. 4, 1991. ASHE-ERIC Higher Education Reports, George Washington University, One Dupont Circle, Suite 630, Washington, DC 20036–1183 (1991)
12. Leroy, G., Kauchak, D.: The effect of word familiarity on actual and perceived text difficulty. J. Am. Med. Inform. Assoc.: JAMIA **21**(e1), e169–e172 (2014). https://doi.org/10.1136/amiajnl-2013-002172
13. McGroarty, M.: The benefits of cooperative learning arrangements in second language instruction. NABE J. **13**(2), 127–143 (1989)
14. Medero, J., Ostendorf, M.: Analysis of vocabulary difficulty using Wiktionary. SLaTE: 61–64 (2009)
15. Miller, G.A.: WordNet: a lexical database for English. Commun. ACM **38**(11), 39–41 (1995)
16. Sasaki, M., Hirose, K.: Explanatory variables for efl students' expository writing. Lang. Learn. **46**, 137–168 (1996). https://doi.org/10.1111/j.1467-1770.1996.tb00643.x

17. Sharp: Interaction Design: Beyond Human Computer Interaction. Cornell. Encyclopedia of the Sciences of Learning. Springer, Boston (2006)
18. Shin, S.-K.: Did they take the same test? Examinee language proficiency and the structure of language tests. Lang. Test. **22**(1), 31–57 (2005)
19. Stricker, L., Rock, D., Lee, Y.-W.: Factor structure of the LanguEdge test across language groups. ETS Res. Rep, Ser. **2005**, i–43 (2005)
20. Van Merrienboer, J.J.G., Clark, R., Croock, M.: Blueprints for complex learning: the 4C/ID-model. Educ. Tech. Res. Dev. **50**, 39–61 (2002)
21. Zainuddin, Z., Chu, S.K., Shujahat, M., Perera, C.J.: The impact of gamification on learning and instruction: a systematic review of empirical evidence. Educ. Res. Rev. **30**, 100326 (2020). https://doi.org/10.1016/j.edurev.2020.100326

Emotional Intelligence Development in Tourism Education and Training Through Digital Technologies

Sofia Kallou[1(✉)], Aikaterini Kikilia[1], and Michail Kalogiannakis[2]

[1] Faculty of Administrative Economic and Social Sciences, Department of Tourism Management, University of West Attica, Athens, Greece
akikilia@uniwa.gr

[2] Faculty of Education, Department of Preschool Education, University of Crete, Crete, Greece
mkalogian@uoc.gr

Abstract. Emotional Intelligence has become a significant topic of interest in management and organizational psychology and behavior, providing a framework of skills and abilities that allow individuals to succeed in life and the working environment. Additionally, digital technologies have grown exponentially during the past decades, disrupting traditional approaches and changing perceptions about teaching and learning, also providing new methodologies for developing a new type of high-level skills. Furthermore, the increase of tourism economic importance for many countries and recognizing the human capital value through education and training contribute to competitiveness and productivity. This paper focuses on the critical points for developing emotional intelligence in some crucial areas of tourism education and training through digital technologies. A rapid literature review was conducted in combination with the SWOC analysis tool focusing on Emotional Intelligence development and the use of digital technologies in Tourism Education and Training. It proposes a new perception of curricula, teaching methods, infrastructure, and the educator's role for redesigning tourism education and training according to the Knowledge Society and the challenges of the 21st century.

Keywords: Emotional intelligence · Tourism education and training · Digital technologies

1 Introduction

The scientific community considers the role of emotions and their utilization in the workplace as a critical key that affects employees' professional excellence, success, and performance [1]. On the other hand, the ongoing rapid development of emerging technologies in learning leads to transforming the pedagogical strategies, design, methods, and learning content to achieve significant learning outcomes. Additionally, digital learning technologies provide new communication and collaboration forms while creating opportunities and challenges for education and lifelong development [2, 3]. This

D. Guralnick et al. (Eds.): TLIC 2021, LNNS 349, pp. 149–159, 2022.
https://doi.org/10.1007/978-3-030-90677-1_15

article analyzes the combined use of Emotional Intelligence (EI) and digital technologies in learning for tourism education and training. A rapid review analysis in combination with the SWOC analysis tool aims to outline a) the strengths, weaknesses, opportunities, and challenges of developing EI; b) the strengths, weaknesses, opportunities, and challenges of using digital technologies; and c) the key points of the development of EI in combination with new technologies, in tourism education and training. Simultaneously, the analysis results focus on the critical areas for successful application in Tourism Education and Training (TET).

2 Theoretical Background

2.1 Emotional Intelligence in the Tourism Sector

EI has become a significant topic of interest in management and organizational psychology and behavior and has been considered essential as it plays a role as the contributor to personal excellence, success, and performance [4]. EI is an ability that provides a framework of skills and abilities that allows individuals to successfully manage various emotional states and take advantage of such skills for personal development and better management of daily situations [5, 6]. Salovey and Mayer [7] are at the forefront of shaping EI theory by defining it as "the ability of a person to control and regulate his or her emotions and emotional reactions to himself and others, to be able to distinguish them, and to use the information that emerges from them to guide his thinking and actions." Goleman [1] has developed a model where the EI consists of four competencies: self-awareness, self-management, social awareness, and relationship management.

The role of EI in the working environment has increased significantly in recent years. It has been linked to leadership effectiveness, job satisfaction, increased work performance, interpersonal relationships, mutual understanding, and communication with others [8–10]. Besides, individuals with high EI levels can have more extraordinary adaptability and flexibility and deal with difficult situations that may arise in business. Moreover, these individuals seem to be more creative and energetic and show altruistic work behaviors [11, 12].

Tourism is an industry that focuses on people with visitors seeking experiences from the destinations communicating at the same time with various stakeholders. Additionally, it is the communication and contact between tourists and service employees that creates memorable experiences [13]. Prentice [14] acknowledges that, while various factors contribute to these experiences (offers, situations, personal factors), tourism employees make the difference to these visitors' experiences. Service employees with a high level of EI can understand customers' thoughts and emotions and develop empathy to respond to customers' requirements appropriately, resulting in higher levels of customer satisfaction. Understanding customers' emotions in the tourism industry quickly and correctly are crucial as service businesses' reputation depends on employees' ability to satisfy the customer's needs on time [15].

Wen et al. [16], and Jung and Yoon [17], concluded that the tourism industry could increase employees' satisfaction by developing their EI and suggest practical solutions in human resources management, as employees with a higher EI also have a high performance. According to Wolfe and Kim [18], the EI of hotel employees was a predictor of job satisfaction and performance, and also, it was a factor of reduced stress and burnout.

Studies have shown that EI is not a gift but can be learned and improved in individuals through training instead of intelligence [19]. In his research, Clarke [20] concluded that, although training can provide self-awareness as an EI ability, the process of the development of this ability continues taking place after months from the training through on-the-job learning mechanisms.

2.2 Tourism Education and Training

Globalization and rapid socio-political changes lead the tourism industry to hire and training high-quality employees. Tourism education and training are essential for developing human resources, supporting policies, and providing incentives and supplies to learners, contributing to their efficiency while giving the company a competitive advantage [21, 22]. However, in a world characterized by radical changes and innovations, tourism education did not evolve satisfactorily and remained traditional. A transformation that focuses on innovative teaching methods, curricula, and new learning environments is crucial for the evolution of tourism education and training. Additionally, focusing on developing soft skills, critical thinking, communication skills, and teamwork will also provide advantages to tourism sector students and employees [23, 24].

2.3 Digital Technologies

The ongoing rapid development of emerging technologies has brought about radical changes in education, training goals and programs, methods, societal structures, and corporate values [3, 25]. Such emerging technologies include Learning Management Systems (LMS), Moocs, Web Conferences Technologies (Zoom, Skype, Webex, Microsoft Teams, etc.), Gamification, Mobile Learning, and Virtual Technologies [2]. To overcome the challenges in digital learning environments, researchers recommend creating online courses that encourage new teaching styles and practices [26, 27], mainly based on social presence, interaction, and relationship development [28]. It is essential to point out that many advantages can be found through digital technologies, such as ubiquity, personalization, reduced costs, flexible access to lectures, time management, and convenience [29].

In the tourism sector, researchers have shown that hospitality students are enthusiastic about using online learning and enjoyed the experience compared to the traditional classes [30]. Additionally, Kim and Jeong [31] have recognized online learning as the future of hospitality and tourism education.

3 Methodology

In this paper, a rapid literature review was conducted in combination with the SWOC (Strengths, Weaknesses, Opportunities, Challenges) analysis tool. The SWOC (Fig. 1) is a strategic planning tool that is also used in the academic analysis as a robust framework to discuss and identify the strengths, weaknesses, opportunities, and challenges of the internal and external environment [32].

Internal Attributes	**STRENGTHS**	**WEAKNESSES**
External Attributes	**OPPORTUNITIES**	**CHALLENGES**

Fig. 1. The structure of SWOC Tool Analysis.

There are many methods for conducting a SWOC analysis. In the present research, the utilization of the SWOC tool will be done in three (3) axes (Fig. 2):

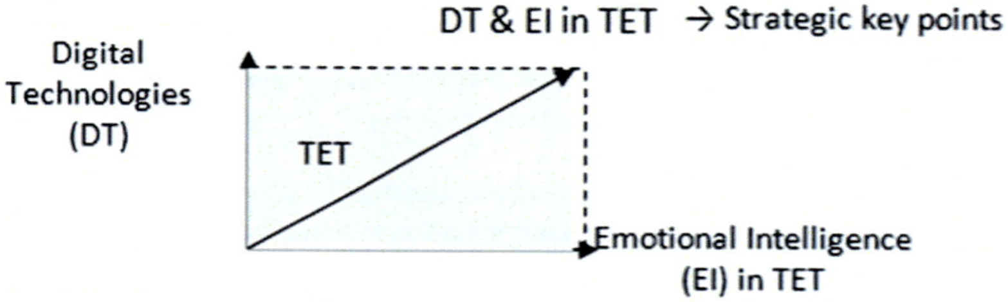

Fig. 2. Axes of analysis using SWOC.

The objective of conducting a SWOC analysis is to develop critical areas of focus for improving TET in some critical areas: curriculum, teaching techniques, infrastructure, and educators. The following figure (Fig. 3) shows the methodological framework of analysis (based on SWOC) utilized in the present research [33].

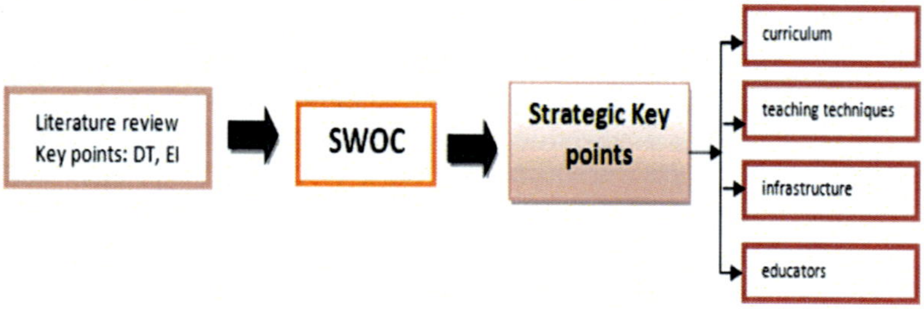

Fig. 3. Methodological Framework.

4 Analysis

TET is a field of adult education where EI skills and hard skills are highly effective. The adaption of EI informal and non-formal tourism education and the use of emerging technologies are analyzed. According to the methodological framework, the following presentation is referred to as rapid literature research findings using the SWOC analysis (Table 1 and 2).

Table 1. Findings of emotional intelligence development in TET.

STRENGTHS	WEAKNESSES
• EI can be developed through training [17] • EI is linked to leadership effectiveness, job satisfaction, performance, relationships development, reduced stress, and communication [6–8, 14–16] • Individuals with high EI show flexibility and altruistic behaviors [9, 10] • Empathy development leads to customer's satisfaction [1, 13], and enhance the tourist experience [12] • EI of hotel employees is negatively related to counterproductive work behavior [15] • EI of hotel employees has, as a consequence, stronger customer loyalty and profitability [34]	• The training of educators is costly. Additionally, EI takes time to develop [1]
OPPORTUNITIES	**CHALLENGES**
• EI as a tool for improving performance and job satisfaction [15] • EI can help employees understand the tourists' needs better, leading to tourist satisfaction [35] • Employees with high EI cope with stressful situations. Therefore, tourism organizations should offer emotional health services as a part of an employee benefits package [7, 15] • EI as a helpful marketing tool to enhance the tourist experience [12]	• EI could be used to manipulate others [34] • EI equates with more self-control, yet extreme self-control levels lead to perfectionism, low innovation, and creativity levels [34]

Table 2. Findings of using digital technologies in TET.

STRENGTHS	WEAKNESSES
• Effectiveness [36] • Ubiquity, personalization, reduced costs, flexibility, time management, convenience [24, 29] • Portability, availability [2] • Open access, free e-learning platforms, participation of a vast number of learners [37, 38] • Real-time interactions in synchronous learning [39] • Immediate feedback, exchange of emotions, social presence in synchronous learning [40] • Autonomy, commitment, performance, motivation, engagement, social skills [2, 40] • Epidemic prevention and control [41]	• Technical difficulties, availability of the platforms [42] • Less physical interaction with teachers [43] • Not immediate feedback in asynchronous communication [43]
OPPORTUNITIES	**CHALLENGES**
• Changes in knowledge and skills [17] • Creating online courses that encourage social presence and quality interaction [28] • New teaching styles emphasizing communication, interaction, and relationship development [28, 41] • Educators' training for developing effective e-learning practices [42] • Lifelong development changes educational goals, means, and corporate values [3, 25]	• Emotional problems from isolation [42] • Not well-trained educators for instruction remotely [41] • Cost of technology [37]

Finally, the strategic key points that emerged from the previous findings concerning the utilization of EI through the digital technologies in TET are shown in Table 3.

Table 3. Strategic key points.

Curricula	Infrastructure
The introduction of EI in tourism curricula should provide opportunities for the trainees to improve communication and teamwork [24, 44] Curricula should also focus on improving the emotional and social skills and coping with stress [23, 45] Critical thinking of the trainees should improve too [23, 24] The relationship between critical thinking and EI is essential in developing a balanced curriculum [46]	Digital technologies form a new framework for the redesign of education that will meet the needs and the challenges of the 21st century Digital Technologies enhance learning and teaching practices through multiple ways of representing and distributing content, communication, development of critical thinking, dynamic problem-solving tools, and new educational practices [52] Therefore, new technologies must be concluded in tourism education to be used effectively in teaching practices

(continued)

Table 3. (*continued*)

Curricula	**Infrastructure**
Teaching methods	**Educators**
Educational techniques for training in emotional skills should be preferred. These are storytelling, group work, conflict management, and brainstorming [47, 48]. Additionally, techniques to enhance communication skills [42] such as case studies, role play (promotion of experiential learning) [49] There should be a student-centered approach that enhances interest and ensure learner participation [50] A climate of freedom and security should be cultivated so that to promote learning through critical reflection [51]	Educators' training must have a mentor's central role and be equally focused on the cognitive and emotional dimension cultivating critical thinking [45, 51] The educator's goal should be creative and critical learning, leading to autonomy, personal development, and integration [48] Educators should be trained in digital teaching strategies [37] Educators should establish frequent and quality interaction with the trainees in digital environments, leading to students' satisfaction and learning [42, 43]

5 Discussion

The landscape of tourism education and training is altering dramatically [31] through the rapid development of emerging technologies and the tourism industry's requirements. Digital technologies are present in most institutions and companies, and they contribute to creating new learning environments that have opened up the development of new educational and training experiences and encourage learners' active participation and the improvement of specific skills and abilities [53]. Perceiving these changes, the tourism industry is trying to adapt, thereby looking for talented, flexible employees capable of handling others and their own emotions [15].

In this framework, the field of education and the tourism industry has to understand the need for new training programs and methods that will lead to a broader perspective of developing EI through digital technologies. Regardless of hard skills, employees need to acquire EI development, which will add massive value to the education system and the tourism industry. This will contribute to the upgrade of the provided services' quality and to further development and productivity of the tourism sector.

A development of EI through digital technologies on TET should be based on new perceptions of the academic curricula, teaching methods, infrastructure, and the educator. Notably, a curriculum and teaching methods focusing on improving emotional and social skills, communication, collaboration, interaction, and critical thinking, based on constructivism and experimental learning, should be considered. Effective teaching development of the educators that promote emotion and social skills, except cognitive skills, in connection with digital teaching strategies is necessary. Communication techniques and frequent interaction with the learners in digital environments lead to a more attractive and exciting learning process, resembling the classroom environment. Additionally, the investment in digital technologies for the redesign of tourism education and training is essential according to the Knowledge Society's requirements and the challenges of the 21st century. They define new knowledge and skills in the new working environment that is evolving.

6 Limitations

The rapid review process of this paper is not a systematic review, since it is based on limited data in order to produce information in a short period of time [54]. However, this paper is an important contribution in the field of tourism education and training that refers to a complex and multifactorial issue on which more extensive research will be conducted in the future.

7 Conclusions and Future Research

Digital technologies in recent years have become ingrained in education and training programs, enhancing the quality of the learning process and leading learners to respond to the new knowledge society needs. While new technologies are rapidly integrated with education and training, EI is becoming a critical factor that has garnered attention both in the educational and business field.

This paper analyzes EI and digital technologies in the tourism sector and develops key areas of focus for improving TET. The strategic key points that emerged from the findings concerning that the utilization of EI through the digital technologies in TET is valuable for the tourism education, as well as the tourism industry, so to be developed innovative solutions and new learning environments, facilitating the learning process, and providing a qualified of soft skills human capital. Integrating EI in academic curricula that emphasizes emotional and social competencies and innovating teaching methods adds exceptional value and provides high-quality services in the tourism sector. Additionally, qualified educators and investment in digital technologies are essential.

The development of EI in tourism education and training through digital technologies is a progression that every educational system will need to make and is also a vital opportunity for the tourism and hospitality sector to thrive and meet the needs that arise globally.

Further research should be done in using specific digital technologies or a combination of them in the learning process for the development of EI and innovative pedagogical methods that could be used to deliver knowledge in tourism education and training.

References

1. Goleman, D.: Working with Emotional Intelligence. Bantam Books, New York (1998)
2. Sofos, A., Kostas, A., Parashou, V.: Online Distance Learning From Theory to Practice, Greek Academic Electronic Textbooks and Aids (2015)
3. Jimogiannis, A.: Digital Technologies and Learning of the 21st Century. Kritiki Publications, Athens (2019)

4. Van Rooy, D.L., Viswesvaran, C.: Assessing emotional intelligence in adults: a review of the most popular measures. In: Bar-On, R., Maree, J.G., Elias, M. (eds.) Educating People to Be Emotionally Intelligent, pp. 259–72, Westport, CT (2007)
5. Drigas, A., Papoutsi, C.: Emotional intelligence as an important asset for HR in organizations: leaders and employees. Int. J. Adv. Corp. Learn. **12**(1), 58–66 (2019)
6. Jordanoglou, D.: Human Resource Management in Organizations, New Trends and Practices. Kritiki Publications, Athens (2008)
7. Mayer, J., Salovey, P.: Emotional Intelligence. Imagination, Cognition and Personality, 9/90 (1990)
8. Samanta, I., Kallou, S.: The role of emotional intelligence in counterproductive work behavior. Eur. Bus. Manag. **6**(2), 20–27 (2020). https://doi.org/10.11648/j.ebm.20200602.11
9. Lee, J.H., Ok, C.: Reducing burnout and enhancing job satisfaction: critical role of hotel employees' emotional intelligence and emotional labor. Int. J. Hosp. Manage. **31**(4), 1101–1112 (2012). https://doi.org/10.1016/j.ijhm.2012.01.007
10. Goleman, D., Cherniss, C.: The emotionally intelligent workplace: How to select for, measure, and improve emotional intelligence in individuals, groups, and organizations. Jossey-Bass (2001)
11. Karadimas, T., Karadimas, P.: The relationship between emotional intelligence and effective leadership. E-J. Sci. Technol. **11**(4), 55–70 (2016)
12. Carmeli, A.: The relationship between emotional intelligence and work attitudes, behavior and outcomes. An examination among senior manager. J. Manag. Psychol. **18**(8), 788–813 (2003). https://doi.org/10.1108/02683940310511881
13. Chang, K.C.: Examining the effect of tour guide performance, tourist trust, tourist satisfaction, and flow experience on tourists' shopping behaviour. Asia Pac. J. Tour. Res. **19**(2), 219–247 (2014). https://doi.org/10.1080/10941665.2012.739189
14. Prentice, C.: Enhancing the tourist experience with emotional intelligence. Tour. Rev. (2020).https://doi.org/10.1108/TR-06-2019-0254
15. Koc, E., Boz, H.: Emotions and developing emotional intelligence in tourism and hospitality businesses. In: Koc, E. (ed.) Emotional intelligence in tourism and hospitality, pp. 15–35. CABI, Wallingford (2019)
16. Wen, J., Huang, S., Hou, P.: Emotional intelligence, emotional labor, perceived organizational support, and job satisfaction: A moderated mediation model. Int. J. Hosp. Manag. **81**, 120–130 (2019). https://doi.org/10.1016/j.ijhm.2019.01.009
17. Jung, H.S., Yoon, H.H.: Why is employees' emotional intelligence important? The effects of EI on stress-coping styles and job satisfaction in the hospitality industry. Int. J. Contemp. Hosp. Manag. **28**(8), 1649–1675 (2016). https://doi.org/10.1108/IJCHM-10-2014-0509
18. Wolfe, K., Kim, H.J.: Emotional intelligence, job satisfaction, and job tenure among hotel managers. J. Hum. Resour. Hosp. Tour. **12**(2), 175–191 (2013). https://doi.org/10.1080/15332845.2013.752710
19. Salovey, P., Mayer, J.D., Caruso, D.: The positive psychology of emotional intelligence. In: Snyder, C.R., Lopez, S.J. (eds.) The Handbook of Positive Psychology, pp. 159–171. Oxford University Press, NewYork (2002)
20. Clarke, N.: The impact of a training programme designed to target the emotional intelligence abilities of project managers. Int. J. Project Manage. **28**(5), 461–468 (2010). https://doi.org/10.1016/j.ijproman.2009.08.004
21. Baum, T: Managing Human Resources for the European Tourism and Hospitality Industry: A Strategic Approach. Hapman & Hall, London (1995)
22. Harris, M.: Human Resource Management, A practical approach, 2nd Edition, Harcourt, Inc. Dryden Press, Fort Worth (2000)

23. Simantaraki, E., Dimou, I.: Public tertiary tourism education graduate competencies. Do the TEI and ASTE graduates meet the hotel industry's expectations? In: Conference: 5th International Scientific Conference Tourism Trends and Advances in the 21st Century, Rhodes, Greece (2013)
24. Hsu, C.H.: Tourism education on and beyond the horizon. Tour. Manag. Perspect. **25**, 181–183 (2018). https://doi.org/10.1016/j.tmp.2017.11.022
25. Holladay, P.: Pedagogy for online tourism classes. In: Benckendorff, P., Zehrer, A. (eds.) Handbook of teaching and learning in tourism. Edward Elgar Publishing, Cheltenham (2017)
26. Kikilia, A., Barbounaki, S.: The Evolution of E-Learning in Greek Universities and Technological Education Institutes – the Government Intervention. ICELW New York, at Columbia University (2011)
27. Papachristos, D., et al.: An educational model for asynchronous e-learning. A case study in higher technology education. Int. J. Adv. Corp. Learn. **3**(1), 32–36 (2010)
28. Bigatel, P.M., Ragan, L.C., Kenan, S., May, J., Redmond, B.F.: The identification of competencies for online teaching success. J. Asynchronous Learn. Netw. **16**(1), 59–67 (2012). https://doi.org/10.24059/olj.v16i1.215
29. Aithal, P.S., Aithal, S.: Impact of online education on higher education system. Int. J. Eng. Res. Modern Educ. **I**(I), 225–235 (2016)
30. Annaraud, K., Singh, D.: Perceptions of hospitality faculty and students of massive open online courses (MOOCs). J. Hosp. Tour. Educ. **29**(2), 82–90 (2017). https://doi.org/10.1080/10963758.2017.1297714
31. Kim, J., Jeong, M.: Research on hospitality and tourism education: now and future. Tour. Manag. Perspect. **25**, 119–122 (2018). https://doi.org/10.1016/j.tmp.2017.11.025
32. Henzi, D., Davis, E., Jasinevicius, R., Hendricson, W.: In the students' own words: what are the strengths and weaknesses of the dental school curriculum. J. Dent. Educ. **71**(5), 632–645 (2007)
33. Sindhu, S., Nehra, V., Luthra, S.: Solar energy deployment for sustainable future of India: hybrid SWOC - AHP analysis. Renew. Sustain. Energy Rev. **72**(2017), 1138–1151 (2017). https://doi.org/10.1016/j.rser.2016.10.033
34. Harvard Business Review: https://hbr.org/sponsored/2019/08/the-ei-advantage-driving-innovation-and-business-success-through-the-power-of-emotional-intelligence?utm_source=Team+A+Ultimate+Trade+NEW&utm_campaign=ec2258d13aFour_Seasons_Announce_Resort_Napa_Valley_COPY_01&utm_medium=email&utm_term=0_06ef706b4d-ec2258d13a-297548727. Accesed 15 Nov 2020
35. Cavelzani, A.S., Lee, I.A., Locatelli, V., Monti, G., Villamira, M.A.: Emotional Intelligence and tourist services: the tour operator as a mediator between tourists and residents. Int. J. Hosp. Tour. Adm. **4**(4), 2 (2003). https://doi.org/10.1300/J149v04n04_01
36. Mollenkopf, D., Vu, P., Crow, S., Black, C.: Does online learning deliver? A comparison of student teacher outcomes from candidates in face-to-face and online program pathways. Online J. Dist. Learn. Adm. **20**(1) (2017)
37. Dhawan, S.: Online learning: A panacea in the time of COVID-19 crisis. J. Educ. Technol. Syst. **49**(1), 5–22 (2020). https://doi.org/10.1177/2F0047239520934018
38. Chew, S.W., Cheng, I.L., Chen, N.S.: Yet another perspectives about designing and implementing a MOOC. In: Jemni, M., Kinshuk, K.M. (eds.) Open Education: From OERs to MOOCs. LNET, pp. 117–133. Springer, Heidelberg (2017). https://doi.org/10.1007/978-3-662-52925-6_6
39. Lowenthall, P., Snelson, C., Dunlop, J.: Live synchronous web meetings in asynchronous online courses: reconceptualizing virtual office hours. Online Learn. J. **21**(4), 177–194 (2017)
40. Kalogiannakis, M., Papadakis, S., Zourmpakis, A.-I.: Gamification in science education. A systematic review of the literature. Educ. Sci. **11**(1), 22 (2021). https://doi.org/10.3390/educsci11010022

41. Qiu, H., Li, Q., Li, C.: How technology facilitates tourism education in COVID-19: case study of Nankai University. J. Hosp. Leisure Sport Tour. (2020). https://doi.org/10.1016/j.jhlste.2020.100288
42. Chen, T., Peng, L., Yin, X., Rong, J., Yang, J., Cong, G.: Analysis of user satisfaction with online education platforms in China during the COVID-19 pandemic. Healthcare **8**(3), 200 (2020). https://doi.org/10.3390/healthcare8030200
43. Skordis-Worrall, J., Batura, N., Haghparast-Bidgoli, H., Hughes, J.: Learning online: a case study exploring student perceptions and experience of a course in economic evaluation. Int. J. Teach. Learn. High. Educ. **27**(3), 413–422 (2015)
44. Dirkx, J.M.: The meaning and role of emotions in adult learning. New Dir. Adult Continuing Educ. **120**, 7–18 (2008). https://doi.org/10.1002/ace.311
45. Goleman, D.: Leadership: The power of emotional intelligence. Northampton, MA (2011)
46. Kaya, H., Senyuva, E., Bodur, G.: The relationship between critical thinking and emotional intelligence in nursing students: a longitudinal study. Nurse Educ. Today **68**, 26–32 (2018). https://doi.org/10.1016/j.nedt.2018.05.024
47. Dirkx, J.M.: Knowing the self through fantasy: toward a mytho-poetic view of transformative learning. In: Kimmel, J.C. (ed.) Proceedings of the 39th Annual Adult Education Research Conference, University of Incarnate Word and Texas A & M University, San Antonio (1998α)
48. Knowles, M.S., Holton, E.F., III., Swanson, R.A.: The Adult Learner. Routledge, London (2012)
49. Karalis, T., Papageorgiou, H.: Design, implementation and evaluation of lifelong learning programs. Labor Inst. - GSEE, Athens (2011)
50. Kasworm, C.E.: Emotional challenges of adult learners in higher education. New Dir. Adult Continuing Educ. **120**, 27–34 (2008). https://doi.org/10.1002/ace.313
51. Kokkos, A.: Critical Thinking: A Critical Issue. Adult Education: International Approaches and Greek Routes, pp. 65–93, Metaichmio, Athens (2010)
52. Tsiotakis, P., Jimoyiannis, A.: Critical factors towards analysing teacher's presence in on line learning communities. Internet High. Educ. **28**, 45–58 (2016). https://doi.org/10.1016/j.iheduc.2015.09.002
53. Salcido-Cibrián, L.J., Ramos, N.S., Jiménez, Ó., Blanca, M.J.: Mindfulness to regulate emotions: the mindfulness and emotional intelligence program (PINEP) and its adaptation to a virtual learning platform. Complement. Ther. Clin. Pract. **36**, 176–180 (2019). https://doi.org/10.1016/j.ctcp.2019.07.003
54. Tricco, A.C., Antony, J., Zarin, W., Strifler, L., Ghassemi, M., Ivory, J., et al.: A scoping review of rapid review methods. BMC Med. **13**, 224 (2015)

Mindfulness Intervention Courses in STEM Education: A Qualitative Assessment

Eunmi Kim[1(✉)], Cheon Woo Moon[2], Sangseong Kim[3], Erva Ozkan[1], and Tim Lomas[4]

[1] Center for Contemplative Science, KAIST, Daejon, Republic of Korea
eunmikim@kaist.ac.kr
[2] Department of Chemical and Biomolecular Engineering, KAIST, Daejon, Republic of Korea
[3] Department of Mathematical Sciences, KAIST, Daejon, Republic of Korea
[4] School of Psychology, The University of East London, Stratford Campus, London, UK

Abstract. Mindfulness-based interventions (MBIs) have been applied in many contexts, including educational sectors of K-12 and various graduate schools, such as medical and law schools. Research shows that highly competitive students are likely to benefit from mindfulness practices. However, few STEM-focused colleges have been able to assess its value and apply mindfulness practices. This study presents a case report of an MBI course offered by the Korea Advanced Institute of Science and Technology (KAIST). We devised eight questions in three categories to investigate how each individual experiences the MBI course, how they enjoy the academic course, and the suitability of the MBI curriculum in STEM education. Qualitative assessments of feedback featuring questions and answers were analyzed using grounded theory. The results demonstrate three central phenomena: a) gradual changes and development in students' emotional intelligence, b) development in physical awareness, and c) enhanced sense of joy during the session. These results imply that even during the COVID-19 pandemic, international students—being isolated in their home countries and taking the course online—would likely still benefit from the intrinsic effect of MBIs. It also implies that MBI courses could be recommended as mandatory classes for all KAIST students, where feasible. Nevertheless, further research is needed to fully explore the impact of such programs, both online and in-person, including the use of self-report scales of common measures of mental health to contribute empirical data to the literature.

Keywords: Academic curriculum · Grounded theory · Mindfulness-Based Interventions (MBIs) · Online learning · STEM education

1 Introduction

1.1 Mindfulness in the Learning System

Mindfulness was originally defined as "paying attention in a particular way: on purpose, in the present moment, and non-judgmentally" [1]. Since then, mindfulness has emerged as an essential core element of a wide range of disciplines. The applicability

D. Guralnick et al. (Eds.): TLIC 2021, LNNS 349, pp. 160–169, 2022.
https://doi.org/10.1007/978-3-030-90677-1_16

of mindfulness—in any condition and context or in any population at any age—has led to the introduction of mindfulness-based interventions (MBIs). A growing body of literature from the past few decades recognizes the importance of MBIs in education as a consequence of challenges such as social-emotional difficulties, as well as physical and mental illnesses [2].

The Need for Mindfulness in Education. Studies show that the pressure of achieving academic standards and high marks while facing academic competition has caused depression and high levels of stress and anxiety among students for many years [3–5]. Existing research illustrates the association between these mental disorders and a low grade point average, decreased participation in campus activities, and relationship difficulties [4].

Another well-recognized issue associated with stress, anxiety, and depression is sleep deprivation and insomnia; these symptoms have similar results, such as a decrease in grade point average in academic studies and suicide attempts [6–8]. The widespread occurrence of these critical issues among students necessitates a better approach to preparing the young generation to live quality lives and contribute to the future of the world as mindful leaders.

MBIs in Higher Education Systems. Since Jon Kabat-Zinn started mindfulness-based stress reduction (MBSR) at the University of Massachusetts Medical School in 1979, numerous contemplative practices have been fostered and researched in both academic and clinical settings [9]. These are aimed at enhancing the well-being of their community members (e.g., mindfulness at work, presented at Harvard University [10], compassion cultivation training at Stanford University [11], and Koru mindfulness at Massachusetts Institute of Technology [12, 13]). Overall, it seems that mindfulness courses will be offered to students at more institutions worldwide.

Mindfulness in education has been applied in many sectors of K-12 education and in various graduate schools, such as medical and law schools [14, 15]. MBSR, mindfulness-based cognitive therapy, and mindfulness-based therapy are notable examples of MBIs. Although mindfulness is practiced in most universities' well-being or life centers, it is also an accredited course in some universities' curricula, such as Brown University in the United States and Monash University in Australia [16–18]. Moreover, Oxford University in the United Kingdom and the Lesley University in the United States have integrated mindfulness in their 2-year master's degree course [19]. Existing research highlights that highly competitive students—such as medical students—also benefit from mindfulness practices [20]. Therefore, this study aims to explore the applicability of MBIs among STEM-focused students.

Positive Outcomes of Mindfulness Practice. There is currently reasonable evidence of mindfulness playing a crucial role in mental and physical well-being, self-regulation, and a broad range of mental and emotional conditions. Further, significant progress in neuroscience and neuroimaging shows how mindfulness alters the brain [21]. These fairly strong positive impacts led scientists, clinicians, and educators to integrate mindfulness into the learning system to improve students' physical and psychological health and to contribute to students' academic success. In 2018, Chi et al. reported that MBIs are effective in reducing depression among adolescents [22], while in 2019, Bamber and Morpeth [23] showed that mindfulness reduces anxiety among college students. Law schools have also incorporated mindfulness and compassion practices, at which positive outcomes—such as improved attention, higher grades, and wiser decisions—became evident among students [24].

1.2 Difficulties Faced by Students in Korea's Higher Education System

During the past few decades, academically driven education has been representative of Korea's economic success; however, there are significant issues specific to the Korean education system that should be considered, in addition to general education problems worldwide. Korea has shown remarkable economic growth and has become a highly developed and high-income country, thanks to its meticulous national economic development plan and passion for education [25].

However, when considering Korean education from a different perspective, many negative effects have come to light. These include an extremely competitive education system, a lack of parental empathy, an academic success-oriented society, and elitism based upon academic positions [25, 26]. Parallel to the rapid progress of industrialization, higher education was universalized and recognized by the government as a strategic transition point for economic development [27].

Furthermore, there has been a strong connection between having high test scores and gaining power throughout history, and people have come to believe that academic achievement can be a step up to the upper class [28]. A previous study found that Korean families tend to recognize their children's success as their own [28]; relatedly, students perceive academic success as a large part of their self-worth. Considering all these issues together, the pressure related to academic achievement has led to frequent suicide ideation among students [28, 29]. While these difficulties are encountered at most levels of education, medical schools, law schools, and STEM-based colleges experience them more intensely. Taken together, there is a clear and urgent need to address these issues to assist current and future students and improve their mental well-being, health, and success.

1.3 Korea Advanced Institute of Science and Technology (KAIST): STEM-Focused College

KAIST, one of the top and first STEM-focused colleges in South Korea, opened in 1971 with a special legislative mandate from the Korean government and two founding missions: to educate scientists and engineers to work for the nation's industrialization and to make innovations by conducting basic and applied research that would drive the nation's economic growth [30]. To fulfill these missions, high-quality STEM education is critical for continuous national and global economic growth.

Factors that cause students to suffer from a high degree of stress, anxiety, and depression include highly competitive homogeneity, with over 80% of the residents staying in dormitories; emotional immaturity among students who transferred to KAIST from high school second grade; and the implementation of English-medium instruction policies in education [31]. Four suicides among KAIST students, committed within a span of three months, called to attention the fierce competition and stress among students in Korean society [32]. The number of suicides among students is very high, even when considering the fact that Korea is known for its high suicide ratio among OECD countries, which was 23 out of 100 in 2019 [33].

1.4 KAIST Center for Contemplative Science and KAIST Global Leadership Center Effort

The KAIST Center for Contemplative Science (KCCS) was established in 2018 as a bridge to connect the humanities and science research pools, and to contribute toward solving the problems faced by humanity. The center offers contemplative programs through six avenues: 1) The "Lecture Series" program invites experts in the field of contemplative science and practice to KAIST to share their work and build collaborative research; 2) "Step by Step" is designed to help participants learn stress management skills, wherein simple contemplative practice steps were devised in consultation with KAIST students, faculty, and administrative staff; 3) through the "All Steps" channel, KCCS offers evidence-based mindfulness and compassion training to those who have completed the "Step by Step" program—here, participants learn more enhanced resilience skills; 4) through the "Reaching Out" program, KCCS addresses the contemplative practice needs of researchers and individuals in the community; 5) the "Public Lecture" avenue is related to the center's contribution to the general public; and 6) the "Talk Concert" program shows KCCS's intention to bridge art and contemplative practices in the future [34].

The KAIST Global Leadership Center (GLC) offers various programs that enable students to take on roles of knowledge-creating leaders with an international mindset through the completion of the leadership program [35]. MBI has been offered to Korean students through the GLC channel since the spring semester of 2019. International students started their first MBI course in the 2020 fall semester, wherein the course feedback was gathered and analyzed qualitatively in this study.

2 Methods

2.1 Participants

A total of 27 undergraduate students were initially registered for the 2020 fall MBI course for international students; this was the first time an MBI course was presented at KAIST. The participant group included 16 freshmen and five sophomores; three of the participants were female. There were no additional advertisements other than the course being on the list of "Leadership and Humanity" courses. Most of the participants were international students, and the Korean students were fluent in English. Less than half stayed on the KAIST campus; the rest stayed in their own home countries because of the COVID-19 pandemic. Nine participants completed a feedback form containing eight open-ended questions devised by the researchers. There is no further information about the students who submitted the forms, as they were collected anonymously.

2.2 MBI Course

The course was offered through the KAIST GLC as one of the mandatory classes within the context of "Leadership and Humanity". The MBI course consisted of 10 ordinary classes of 2.5 h; that is a total of 25 h of practice, excluding orientation and a one-day retreat class. Compared with the usual MBSR program of 28 h with a one-day retreat class, the intervention is shorter in practice hours; however, the total duration is longer. Occasionally, there were gaps between classes longer than a week, due to the exam period or holidays. Each 2.5-h class had a unique theme, such as "coping with stress" and "mindful eating". These themes were drawn from the original MBSR program. A special activity called "mindful test taking" was also offered [36].

Starting with a brief lecture and sharing thoughts about the theme, students were instructed to follow several mindfulness practices. Some practices were closely related to the theme, such as mindful eating, walking, listening, and speaking. General practices, such as breathing meditation and yoga, were often included. After each practice that lasted for 15–30 min, the students shared their impressions of the experience in dyad or triad sessions. Each class had a 15- to 30-min break, during which students were encouraged to mindfully prepare for the next practice.

Students were required to attend at least eight of the 10 classes to pass the course. One student dropped out of the course after the "add and drop period" due to unforeseen personal circumstances. The average attendance rate was 9.17 out of 10, and 12 of the 26 students attended every class.

2.3 Procedure

At the end of the last class, the students were asked to complete the feedback form devised by the researchers. The feedback form consisted of eight questions, of which the first four were related to the changes students experienced with regard to their body, stress, and relationships with other people. The next two questions addressed the enjoyability of the course contents and interventions, such as duration and online environment. Next, we asked whether the students would recommend the course to other KAIST students, while the last question invited any additional feedback. Nine of the 26 students who completed the course submitted the qualitative survey, along with signed informed consent forms. Table 1 presents the evaluation questions.

Table 1. Qualitative assessment questions.

	Questions	Core intention (question numbers)
1	How was your relationship with your body before, during, and after the program?	Effect of participating in the course (1–4)
2	How did the program affect your symptoms of stress and anxiety?	

(*continued*)

Table 1. (*continued*)

	Questions	Core intention (question numbers)
3	How did the program affect your relationship with other people during and after the training?	
4	Were there other changes not mentioned above? (purpose of life, productivity, etc.)	
5	Was the program enjoyable? Which aspect of the program was most enjoyable to you?	Feedback on the enjoyability/durability of the course (5–6)
6	How do the logistics of the program (promotion/registration, participant management, time, duration, online/in-person) help you to keep attending and participating in class?	
7	Do you think this program will be helpful to most KAIST students? Are there particular types of students who would or would not need the program?	General suitability of the course (7)
8	Give us any additional feedback. (Optional)	Personal experience (8)

2.4 Analysis: Coding Scheme with Grounded Theory

The feedback data were analyzed using grounded theory methodology. The authors independently developed the coding schemes and performed open coding. After reaching consensus on the list of codes, the authors attempted to find a central phenomenon that penetrates as many codes as possible. After finalizing three central phenomena, the authors clarified the connection between each response and central phenomenon—whether as direct examples or as clues regarding the cause of such phenomenon—to produce further insights.

3 Results

The feedback was collected from nine responders upon their completion of the MBI course. After the keywords were reviewed, they were categorized into three central phenomena: a) gradual changes and development in their emotional intelligence, b) physical awareness, and c) feeling of joy during the session.

First, there were various responses related to emotional intelligence: self-realization, emotional balance, stress management skills, healthier communication and relationships with individuals and the community, and increased productivity. Interestingly, four of the participants reported increased productivity, but with different causes: fewer distractions, unwinding comfortably during practice and being awake when doing the task, the exchange of useful information with their peers, and enhanced focus.

Second, mainly through Questions 1 and 4, seven respondents mentioned physical awareness—two of them experienced deep focus on each part of their body, while the other five became more frequently aware and cared for their body. Answers to Question 2 showed that all of the respondents experienced physical relaxation during mindfulness practices and learned to implement them in their daily lives to control stress and anxiety levels and boost productivity.

Third, responses to Questions 5 and 6 showed that every respondent enjoyed the course; however, their favorite aspect of the course varied. Four respondents valued group activities and free sharing, during which they could meet new friends and share their experiences during the practice and other personal stories. Three respondents reported valuing everyone's active participation and the presence of other students performing the same practice simultaneously. Two respondents mentioned that they would prefer the course's in-person protocol, because it would allow more interaction and fun. One valued the relaxing atmosphere, and the other two enjoyed the various mindfulness practices the most.

Alongside the three central phenomena, one aspect stood out. In answer to Question 7, all the respondents agreed that the course was recommendable to every KAIST student. Their supporting reasons, mainly the interactive aspect of the course, tended to match their source of satisfaction from the course, which shows that they were able to realize their needs.

There were other feedback points for Questions 6 and 7. For example, considering the class duration of 2.5 h, one respondent was comfortable with it, while two preferred shorter classes. Moreover, one participant pointed out that the course will only be helpful to those who are interested and cooperative, and that there were some who were not. Another participant noted that the course successfully motivated regular attendance, which is key to its success; thus, it is helpful.

4 Conclusion

Existing literature reports a strong relationship between Korea's rigorous education system and emotional exhaustion or mental disorders, such as depression, stress, and anxiety, among students. This study poses the question of how we can address these existing problems in a better and sustainable way to improve students' success and leadership skills. The feedback, which was collected from students and analyzed using a grounded theory approach.

The three most prominent results that emerged from the grounded theory analysis are as follows: the KAIST students who took part in the MBI course made progress in emotional intelligence, including intrinsic and extrinsic awareness; the logistics of the active participation-based MBI course were completely satisfying for students and may be requested by them; and long-term MBI courses lead to physical changes—such as body awareness—in students.

The most important finding of this study is that, even through online classes, STEM-focused KAIST students can benefit from MBIs. In particular, they feel connectedness and openness among themselves, even if they never see each other, being from different countries and having different cultures. The serial mindfulness practice led to several changes in students' behaviors toward others. Some comments in the feedback included the following: "I would like to invite all students with different personalities to join our class to know each other's characteristics better, so that we can live together well" and "I'd like to recommend this course to my peers".

Based on our results, we propose the necessity of mandatory mindfulness courses in the curriculum of KAIST to contribute to students' academic achievement and performance as mindful leaders. There is a growing need to educate mindful leaders in many areas. For example, Harvard Business School encourages "these future leaders to see success as making a positive difference in the lives of their colleagues, their organizations, their families, and society as a whole" [37].

We further suggest enhancing the applicability of the contemplative programs offered by the KCCS. College students may have a better chance of learning mindfulness through mandatory classes. As we consider MBIs to also have a big impact on students' academic performance with better standards of living, MBIs should be considered mandatory courses.

The present study has certain limitations: small number of responses and the absence of a comparison group, as well as possible bias of the population. Quantitative data such as perceived stress scale or mindfulness scales would improve the quality of creditability of the study. However, considering that this is the very first MBI course offered at KAIST for international students, there is the possibility of a better study design in future.

Future research should compare online and in-person protocols, conduct pre- and post-surveys on mindfulness, self-compassion, and leadership. Randomized controlled trials should be studied with various depth and diversity of populations. Clinical studies could be done with students who suffer from depression, anxiety, and high levels of stress. Considering the research-driven academic environment, graduate students should be included as participants as well.

References

1. Kabat-Zinn, J.: Full Catastrophe Living, Revised Edition: How to Cope with Stress, Pain and Illness Using Mindfulness Meditation, xxvii, Revised Hachette, UK (2020)
2. Desmond, C.T., Kinder, W., Hanich, L.B., Chukwu, O.C.B.: Handbook of Mindfulness-Based Programs: Mindfulness Interventions from Education to Health and Therapy, 1st edn. Routledge, UK (2020)
3. Cui, Z., Zhang, Y., Ma, L.: Mindfulness-based interventions for prevention of depressive symptoms in university students: a meta-analytic review. Mindfulness **10**, 2209–2224 (2019)
4. Maynard, B.R., Solis, M.R., Miller, V.L., Brendel, K.E.: Mindfulness-based interventions for improving cognition, academic achievement, behavior, and socioemotional functioning of primary and secondary school students. Campbell Syst. Rev. **147**(2017)
5. Pascoe, M.C., Hetrick, S.E., Parker, A.G.: The impact of stress on students in secondary school and higher education. Int. J. Adolesc. Youth 25(1) (2020)
6. Hershner, S.D., Chervin, R.D.: Causes and consequences of sleepiness among college students. Nat. Sci Sleep **6**, 73–84 (2014)
7. Choueiry, N., Salamoun, T., Jabbour, H., Osta, N. E., Hajj, A., Khabbaz, L. R.: Insomnia and relationship with anxiety in university students: a cross-sectional designed study. PLoS ONE **11**(2) (2016)

8. Nepon, J., Belik, S.L., Bolton, J., Sareen, J.: The relationship between anxiety disorders and suicide attempts: findings from the national epidemiologic survey on alcohol and related conditions. Depress. Anxiety **27**(9) (2010)
9. Barnes, N., Hattan, P., Black, D.S., Schuman-Olivier, Z.: An examination of mindfulness-based programs in US medical schools. Mindfulness **8**(2), 489–494 (2017)
10. Harvard Human Resources. https://hr.harvard.edu/mindfulness. Accessed 25 Feb 2021
11. The Center for Compassion and Altruism Research and Education. http://ccare.stanford.edu/education/about-compassion-training/. Accessed 25 Feb 2021
12. MIT Medical. https://medical.mit.edu/community/stress-reduction. Accessed 25 Feb 2021
13. Barbezat, D.P., Bush, M.: Contemplative Practices in Higher Education: Powerful Methods to Transform Teaching and Learning. Wiley, Hoboken (2013)
14. Broderick, P.C.: Learning to Breathe: A Mindfulness Curriculum for Adolescents to Cultivate Emotion Regulation, Attention, and Performance. New Harbinger Publications. Oakland (2013)
15. Schonert-Reichl, K.A., Roeser, R.W.: Handbook of Mindfulness in Education: Integrating Theory and Research into Practice. Springer, Berlin (2016). https://doi.org/10.1007/978-1-4939-3506-2
16. Cieslak, K.J., et al.: An environmental scan of mindfulness-based interventions on university and college campuses: a research note. J. Sociol. Soc. Welf. **43**(4), Article 7 (2016)
17. Brown School of Public Health. https://www.brown.edu/public-health/mindfulness/programs/mindfulness-schools-and-higher-education. Accessed 25 Feb 2021
18. Monash University Health and Wellbeing. https://www.monash.edu/health/mindfulness/mindfulness-at-monash. Accessed 25 Feb 2021
19. de Bruin, E.I., Meppelink, R., Bögels, S.M.: Mindfulness in higher education: awareness and attention in university students increase during and after participation in a mindfulness curriculum course. Mindfulness **6**(5), 1137–1142 (2014). https://doi.org/10.1007/s12671-014-0364-5
20. Dobkin, P.L., Hutchinson, T.A.: Teaching mindfulness in medical school: where are we now and where are we going? Med. Educ. **47**(8), 768–779 (2013)
21. Tang, Y.Y., Hölzel, B.K., Posner, M.I.: The neuroscience of mindfulness meditation. Nat. Rev. Neurosci. **16**(4), 213–225 (2015)
22. Chi, X., Bo, A., Liu, T., Zhang, P., Chi, I.: Effects of mindfulness-based stress reduction on depression in adolescents and young adults: a systematic review and meta-analysis. Front. Psychol. **9**, 1034 (2018)
23. Bamber, M.D., Morpeth, E.: Effects of mindfulness meditation on college student anxiety: a meta-analysis. Mindfulness **10**(2), 203–214 (2019)
24. Bush, M.: Mindfulness in higher education. Contemp. Buddh. **12**(1), 183–197 (2011)
25. Lee, J.K.: Lee educational fever and South Korean higher education. Revista Electronica de Investigacion Educativa **8**(1) (2006)
26. Lee, J.K.: Korean experience and achievement in higher education. SNU J. Educ. Res (11), 1–23 (2001)
27. Kim, J., Kim, H.S.: Globalization and Access to Higher Education in Korea. Fairness in Access to Higher Education in a Global Perspective. Sense Publishers, Rotterdam (2013)
28. So, K., Kang, J.: Curriculum reform in Korea: issues and challenges for twenty-first century learning. Asia Pac. Educ. Res. **23**(4), 795–803 (2013). https://doi.org/10.1007/s40299-013-0161-2
29. Beach, J.M.: Children Dying Inside A Critical Analysis of Education in South Korea. Createspace Independent Publishers, USA (2011)
30. KAIST PR Materials, Vision 2031. https://kaist.ac.kr/en/html/kaist/012104.html. Accessed 10 Mar 21

31. Kim, E.G., Kweon, S.O., Kim, J.: Korean engineering students' perceptions of English-medium instruction (EMI) and L1 use in EMI classes. J. Multiling. Multicult. Dev. **38**(2), 130–145 (2017)
32. Miller, K.: South Korea: student stress fuels suicides as standards rise. In: Understanding Global Higher Education, pp. 231–233. Sense Publishers, Rotterdam (2017)
33. OECD, Suicide rates. https://data.oecd.org/healthstat/suicide-rates.htm. Accessed 03 Mar 2021
34. KAIST Center for Contemplative Science. http://kccs.kaist.ac.kr/. Accessed 03 Mar 2021
35. KAIST Global Leadership Center. https://leadership.kaist.ac.kr/sub0101. Accessed 18 Mar 2021
36. Biegel, G.M.: The Stress Reduction Workbook for Teens: Mindfulness Skills to Help You Deal with Stress, 67–68. New Harbinger Publications, California (2017)
37. George, B.: Mindfulness helps you become a better leader. Harv. Bus. Rev. **26**(10), 21–32 (2012)

Digital Project-Based Learning in the Higher Education Sector

Christoph Knoblauch(✉)

Ludwigsburg University of Education, Ludwigsburg, Germany
christoph.knoblauch@ph-ludwigsburg.de

Abstract. This paper discusses evaluation findings from a digital project-based course in the higher education sector. The empirical findings focus on students' attitudes, practices, and preferences towards digital project-based learning. The evaluated course was developed and taught in the context of the Master's program for Education at Ludwigsburg University of Education (LUE). It aims at the autonomous project-based planning, execution, and critical reflection of an individual empirical study carried out by the participating students. This paper presents a discussion of the structure, the digital methodology, and the outcomes of the course, using digital qualitative interviews for the evaluation research. By doing so, the study investigates how students deal with a digital project-based learning environment and how these results can be constructively implemented to improve future digital or blended-learning scenarios in the higher education sector.

Keywords: Digital project-based learning · Higher education sector · Empirical qualitative evaluation

1 Background and Context of the Digital Project-Based Course

The process of learning and teaching in digital settings has become increasingly important and has changed considerably in the face of new challenges. Autonomous learning and working with the assistance of new information technologies have become integral features of societies and their educational systems all over the world. The world of learners and teachers has become more autonomous and at the same time more interconnected as a result of new challenges such as globalization and digitalization. As reflected in the research by many academics whose interests lie within the scope of project-based, autonomous, digital, and blended learning. In this regard, digital and project-based learning are discussed as some of the most important ways to cope with new challenges of a globalized world and to develop and foster competencies in autonomous learning in the higher education sector [12, 17]. Digital and project-based learning approaches seem to have proven particularly relevant in this context, for they can offer tools for communication, autonomous learning, collaboration, and shared learning between students and

D. Guralnick et al. (Eds.): TLIC 2021, LNNS 349, pp. 170–179, 2022.
https://doi.org/10.1007/978-3-030-90677-1_17

teachers [14, 16]. Consequently, there is a strong demand for the development of digital project-based courses in an increasingly autonomous and digital learning environment. The higher education sector plays a major role in these developments and must deal with unknown and known challenges in order to prepare students for learning and teaching in more autonomous, project-based and digital contexts [10]. Against this background, however, well-researched resources offering guidance on how to combine digital and project-based learning and how to design adequate courses for the higher education sector remain rare. This study reports on the development, implementation and evaluation of a course in the Master's program "Teacher Education"[1] at Ludwigsburg University of Education (LUE) in Germany. Participants were students enrolled in teacher education programs at LUE in the summer term of 2020. The key question of the project is how students cope with project-based learning courses in a digital setting. Therefore, students' attitudes, practices, and preferences concerning digital project-based learning and factors as competence development and autonomous learning play major roles in the design of the course and its evaluation. Digital project-based learning shows the potential to help students foster independent learning skills and develop learning competence in individual online learning contexts [11]. The course discussed, therefore, shows a mix of synchronous and asynchronous course sessions [5]. In addition to this, the participating students reserve several weeks within the semester for the project-based, autonomous development, and execution of a small-scale empirical study.

The focus of this paper is to discuss the structure and the evaluation of the course in order to present results on the possible impacts of digital project-based learning in the higher education sector. Therefore, the paper discusses digital project-based learning scenarios that support the participating students in a two-fold way: On the one hand, participants should develop their individual learning skills in project-based digital contexts. On the other hand, future teachers should become aware and capable of digital and project-based skills that could transfer to their own future teaching in a classroom setting. In order to facilitate an insight into the learning strategies of students in the described settings, the paper discusses a qualitative study focusing on the attitudes and preferences of students to project-based learning in digital contexts. By doing so, the study assesses current practice and analyzes the course mentioned in detail.

2 Project-Based Learning in Digital Contexts: Description of the Course Design

2.1 A Subsection Sample

This course served as the basis of the study and was conducted at LUE in the summer semester of 2020—in the middle of the COVID-19 pandemic. The main foci of the course are (a) the discussion of empirical studies, (b) empirical research methods, and (c) the autonomous implementation of a small-scale empirical study in the field of education. Thus, the digital project-based character of the course aims for the interconnection of students with their future work area, their future colleagues and their future

[1] https://www.ph-ludwigsburg.de/7684+M5054de7a952.html (access on 2021/4/8).

students through an empirical project in the field of education. The combination of digital and project-based learning offers many options for digital learning scenarios in the higher education sector and has a special focus on the development and fostering of autonomous learning [1, 8]. Approximately 50% of the course were carried out in a digital asynchronous way, using learning management systems such as Moodle, whereas the other 50% of the digital course sessions were conducted in a synchronous way using mostly videoconferences.

The asynchronous sessions were based on a mix of learning arrangements consisting of readings, audio presentations, videos, podcasts, interactive forums, and chats. To these ends, (1) pertinent presentations with audio commentary were created; (2) a podcast with experts was made available; (3) common issues were presented through a selection of pertinent literature; (4) videos of professors and students working in empirical research were provided; and (5) digital forums were established in order to offer an ongoing interactive exchange of ideas.

The synchronous sessions mostly consisted of guided discourse offering the chance to reflect on the contents of the course and discussions of the various empirical studies designed and carried out by the students. The video conferences were based on student questions and reports offering group discussions and individual sessions in break-out-rooms.

Against the backdrop of the course structure, the idea of digital learning is understood as a form of 'blended learning', which combines individual eLearning, interactive digital collaboration, and autonomous project-based learning in various ways [9]. The empirical focus of the course aims at helping students to learn about empirical methods and to practice empirical research in order to prepare for their Master's Thesis and their future in teaching. Students can gain up to three credit points (as defined by the European Credit Transfer and Accumulation System [ECTS]) for the course and can use their empirical projects as a basis for their Master Thesis.

3 Evaluation of the Course

The evaluation was designed as a qualitative study, focusing on the feedback of the participating students of the course. It uses a creative qualitative design [6], which is conducted digitally [7]. The study is intended to create a flexible research process, to produce various findings and to meet the COVID-19 regulations of the time. In the context of the complex research focus—students' attitudes and preferences to a digital project-based course and possible developments of competencies in learning—semi-structured qualitative interviews were carried out in an asynchronous digital way. This way of interviewing students is a rather new technique in qualitative research, which encourages respondents to reflect on their answers by allowing them to structure their ideas and responses beforehand [15]. In this particular study, students had the opportunity to record their answers as audio files independently and, therefore, can take as much time for reflection as they individually considered appropriate. The completed audio files were then sent to the research team via a digital transfer system. Drawbacks of this technique, such as a lack of face-to-face interaction and constructive dialogue, have been taken into account and play a role in the analysis of the data. Even though there is rare guidance on the implementation of the described technique of asynchronous digital qualitative interviews, there has to be emphasis that—apart from the innovative way of data generation—the study follows established structures of qualitative research and analysis in manifold ways [4]. Using this method of qualitative interviews, the study looks for data, which offers a comprehensive view through individual, reflected, and subjective feedback, which is in this study recorded asynchronously and individually by the respondents.

The qualitative semi-structured questionnaire, which was sent to the students via email, starts with a short salutatory address and some operation guidelines on how to produce the audio file. The operation guidelines emphasize that the respondents can take as much time for reflection as needed.

The semi-structured questionnaire shows seven categories that were developed in a deductive procedure [3]: (1) The individual project of the respondent, (2) expectations of the respondent about the course, (3) content of the seminar, (4) structure of the course, (5) digital learning, (6) comments about the course, and (7) an open section for various feedback. These deductive categories were built based on the discussion of former research, and evaluations in the field of education and beyond [10, 13]. Each category includes several key questions, which are supported by specifying impulses or further questions. The questionnaire mainly focuses on the digital and the project-based character of the course, asking students about their expectations, attitudes, preferences, and possible development of competencies. The questions furthermore focus on the use of the provided online content, various options for interaction, attitudes towards autonomous learning, reflections on learning strategies, and perceptions of change within learning individual strategies and attitudes.

4 "I Can Do This": Findings of the Qualitative Study

The analysis and discussion of the data show eight combined categories, which are based on the deductive categories and new inductive impulses found within the data: (1) motivations for AE: enrollment in a digital project-based course, (3) descriptions of the students' projects, (4) project-based character of the course, (4) project implementation in digital contexts, (5) digital character of the course, (6) individual learning progress, (7) impact on further studies and future work of the students, and (8) critical perspectives on the course structure, implementation, and content. The findings within the data are structured and discussed according to these categories. However, some answers and reflections within the data show links to more than one category and are therefore discussed under the perspectives of different categories.

Motivations for AE: Enrollment in a Digital Project-Based Course. For many students, empirical research in the field of education and the prospect of the implementation of a small-scale empirical project seems to be a strong motivation for AE: enrollment in the course. In general, education is often discussed as an empirical discipline in the conducted interviews and many students seem to see a benefit in the implementation of an individual research project connected to their studies and their future work. The idea of autonomous and project-based work also seems to be a motivation, as some students report that they missed working independently in their studies so far.

Another factor seems to be the interdisciplinary approach of the course. Many students report that they wanted to develop interdisciplinary competencies in order to implement research in different subjects. Furthermore, the project-based character seems to be a motivation as it is discussed as a hands-on approach, which can produce insights with a practical orientation. Other students want to learn about various projects in different fields of education in order to gain a more profound insight into specific themes. Of course, there are also very straightforward reasons: Students report that they want to prepare their Master Theses and gain Credit Points (ECTS) in order to finish their studies.

Project Description. The participating student projects all focus on education and implement an empirical study: Qualitative interviews with students and teachers from different school sectors (primary and secondary school) and the higher education sector were conducted, in order to gain insights into various fields of education. The projects deal with student and teacher attitudes and preferences towards various topics in primary and secondary schools and the higher education sector. Some projects focus rather on subject-specific research issues, while others focus on interdisciplinary questions.

There are participants, for example, who explore the question, "Do students in the higher education sector feel well prepared for their future work as teachers in schools?" Other projects focus on the question of why pupils in schools like or dislike certain subjects, how pupils work with pictures in school lessons, and how they use rituals in school. In general, the projects show a large variety of topics in the field of education, mirroring the various interests of students in this field.

Project-Based Character of the Course. The reactions to the project-based character of the course are mainly positive. The interviewed students value the possibility to pursue their individual interests and thereby develop competencies within their chosen fields of education. The possibility of individual time management and the chance to design and carry out an individual project self responsibly also seems to be appreciated by the interviewed students. However, many students emphasize that individual time management has to be accompanied by a solid course structure, which sets a framework to the various projects.

Especially, in the first stage of the course, strong support through the lecturer seems to be very important for the interviewed students. In this context, the importance of (1) a reliable structure and time frame, (2) the development of a knowledge foundation through pertinent literature and the discussion of 'good practice' projects, and (3) the close mentoring of the design of the individual projects are emphasized as very important moments. The interviews show that students value a fine balance of autonomous project-based learning, which is accompanied by a reliable course structure and individual support, especially when the project is conducted: *"...so I could learn and work at my own pace, within a sound course structure and with the option to receive individual support for my project"*. In order to accomplish this balance, it is crucial to plan the course with its different stages at the very beginning of the semester in close collaboration with all students. In this way, it is possible to create a reliable structure for all participants.

In this context, many answers also emphasize the importance of realizing individual ideas and projects and the connected processes of learning: *"...and I also realized what could have been better (within a particular project phase) and I improved step by step and got more and more professional in my role as an interviewer"*. Against this backdrop, many students also report that they developed confidence throughout their project-based work: *"...the fact that I worked on my own project and I could actually see: I can do this!"* Additionally, some students appreciate that the project-based character of the course leads to a real involvement with various topics and thereby helps to understand rather abstract scientific themes. *"...we were able to reach certain goals by implementing our individual project. Therefore, the learning processes are more effective and bigger – on top of that, the learning is sustainable and interdisciplinary"*.

Furthermore, the digital project-based design of the course is valued as a fitting response to the challenges of a fully digital semester. However, many students report that they miss the presentation and discussion of the various projects in a face-to-face setting. Some students also find it hard to develop an individual idea for a project, as the possibilities in the field of education seem to be endless. A closer individual mentoring at the beginning of the course could have been helpful in these cases.

Project Implementation in Digital Contexts. The interviewed students used different strategies and processes for the implementation of their projects. While some students conducted their projects digitally, others used blended forms of online and face-to-face meetings, and some students conducted their projects face-to-face only.

Students who chose a digital way mostly report about positive effects such as (1) the possibility to record sessions via video conferencing software, (2) the distribution of digital materials before and during sessions, (3) the opportunity to gather and share information online and simultaneously, (4) the chance to meet without using a car or public transport, and (5) the time-saving aspects of digital communication.

In contrast, students who chose face-to-face project sessions emphasize the necessity of face-to-face interaction with children in primary school or early childhood education, mostly because children in this age group cannot handle the required digital devices on their own. Additionally, some students report that they experienced an unobstructed flow of information and communication in their face-to-face sessions and that the use of some materials—e.g., pictures or items—may have a better effect.

Digital Character of the Course. In general, the interviewed students show a high acceptance of digital learning. The fact that the course offers a variety of different digital learning methods is highly appreciated in many answers. Against this backdrop, the students especially mentioned the learning platform as pivotal, as well as the mix of synchronous and asynchronous sessions, and the possibility of revising digital input such as podcasts. In addition to this, online forums for asynchronous discussions and Power-Point-Presentations with audio comments were discussed as highly effective tools: *"The presentation with audio comments are very comfortable because one doesn't have to make notes under time constraints and one can really think things through"*.

Furthermore, many students liked to work with digital content, like interactive websites for empirical research and appreciated Q&A Forums, which were available any time. The flexibility of a digital-project-based course and the benefits that come with this flexibility were also discussed intensely. Some students reported that they do not have to commute anymore; others find it easier to combine children and studies or, in general, manage their life. Overall, the answers show a high level of satisfaction with the digital character of the course: *"I hope we can use many of these methods in future courses again"*. In this context, however, many students again report that they miss the interaction with teachers and other students in face-to-face situations: *"(In face-to-face interaction) ...you could have in-depth talks with other students in the course or maybe afterwards with some coffee"*.

Individual Learning Progress. Project-based learning encourages independent learning. All interviewed students report that they developed new skills independently while working on their projects: *"I think I really learned a lot in this course because I taught myself. I developed competencies on different levels independently"*. Many students report that the planning and execution of their project helped them to develop new competencies and deepen their knowledge in various areas like Grounded Theory, Qualitative Interviews, or time management. Some students value the factor of autonomous work as a special property of the course: *"Autonomous working, the development of one's own ideas and the responsibility that comes with the execution of the project. This was missing in my studies"*.

Against this backdrop, many students report that they want to use their new competencies as interdisciplinary tools for other subjects and their future work: *"I can use my knowledge about empirical research for my master thesis and maybe for other subjects"*. One student reports that she gained a deeper insight into the subject by realizing how empirical studies in this specific subject are deployed and discussed. In general, the individual learning progress is mainly associated with the project-based and autonomous character of the course, which seems to help students by the development of new competencies in various disciplines: *"...and in my opinion, you only learn things, when you actually do things. This is why I appreciated the projects so much"*.

Further Studies and Future Work of the Students. The students discuss various effects of the course on their further studies and their future work: the development of new interests, more profound insights in chosen topics, connections to different topics and fields of education, first steps in the preparation of their master thesis, and motivation to develop their projects in the future.

Many students especially value the discussion of empirical studies and tools of empirical research as helpful preparations for their future work. The reflection of models for competence measurement and students learning progress serve as examples. Some students argue that they could improve their teaching by interviewing their students in new empirical projects. In addition to this, students emphasize that (1) the communication with children and teachers as future partners in the field of education and (2) the development of interdisciplinary competencies through the implementation of various processes within their projects were of great importance for them.

Critical Perspectives on the Course Structure, Implementation, and Content. Many students report that they missed a profound presentation and discussion of the various projects in class. The asynchronous presentations of the individual projects with videos and podcasts at the end of the course proved to be very effective, but seem to be unsatisfying in terms of personal involvement and the appreciation of individual efforts. For future courses, several synchronous sessions should be arranged at the end of the semester in order to attach more importance to the presentation and discussion of individual projects.

Some students struggle with the development of an individual project as the possibilities in the field of education seem to be endless. In future courses, closer mentoring at the beginning could help to solve this problem. Another option would be a stronger and closer collaboration between the students. This could lead to project teams, which work together on a joint question.

While most students approve the balance of (1) content-based sessions and (2) autonomous project-based learning, some students would like to develop a greater knowledge about (empirical) project-based work first. Optional sessions for a more profound discussion of project-based work and empirical methods could be helpful for future courses.

Some students report that the interaction between students in the course was rather difficult. Even though it was always possible to interact via online forums, chats and video calls, these students argue that the digital character of the course inhibited interaction: *"Interaction with others was rather difficult, but it is always like this in online courses"*. As many students appreciate the autonomous interaction via open forums and chats, an option for future courses could implement video forums in theme-centered groups without a teacher. In this context, the idea of closer collaboration between students was mentioned several times: Teams, which discuss common methods and projects or even work on the same project, are certainly a constructive and promising idea for future digital project-based courses.

5 Options for the Future: *"I Hope We Can Use Many of These Methods in Future Courses Again"*

Digital project-based courses in the field of education can generate a high level of motivation in students and encourage them to do autonomous and self-responsible work. In this context, the implementation of empirical research within the projects can lead to the development of interdisciplinary competencies, a deeper insight into chosen topics, and many-sided connections to future work fields. Against this backdrop, digital project-based learning has to offer a fine balance of autonomous project-based learning, which is accompanied by a reliable course structure, scientific content, and individual support. Customized Presentations with audio comments for this course and its projects and the use of online forums for asynchronous discussions proved to be very effective ways in the digital context of the course. Videos and podcasts from experts in the field also proved to be very effective.

However, the digital character of the course shows a lack of collaboration between the participating students. The implemented projects could not all be presented and discussed in a satisfying way, as no face-to-face meetings were possible. For future courses, more collaboration among students could be promoted via digital break-out-rooms and synchronous sessions without the teacher. Furthermore, collaboration could be developed in terms of research methods, common topics, and the implementation of projects.

Digital project-based courses require a structure, which enables students (1) to find themes and develop projects individually through online content and scientific perspectives; (2) to share and discuss their ideas online; (3) to develop shared knowledge through collaboration; and (4) to develop new content and tasks for themselves and others. The course designers should present a solid course structure, which describes the balance between autonomous work and collaborative learning. This course structure should also support students with a manageable and reliable schedule.

In the study, students comment on the process of autonomous work as being a crucial factor in their learning experiences. By following individual interests and ideas, especially in a project-based setting, students can develop and foster competencies through active and project-based learning processes—especially students in the higher education sector can gain from these processes, as they can (1) set appropriate goals autonomously, (2) work and learn within a reliable structure, (3) have frequent opportunities for individual and collaborative revision, and (4) connect with future work fields. These processes support one another as they aim at the development and fostering of competencies and help students become more aware of their learning activities and future work fields.[2]

These encounters, which are made possible by the digital and project-based design of the course, might be regarded as transformative as they meet the challenges of the time constructively and support students in their ways of becoming autonomous experts in their chosen fields of education.

References

1. Akinwamide, T., Adedara, O.: Facilitating autonomy and creativity in second language learning through cyber-tasks, hyperlinks and net-surfing. ELT **5**(6), 36–42 (2012)
2. Barron, B., Schwartz, D., Vye, N., Moore, A., Petrosino, A., Zech, L., et al.: Doing with understanding. lessons from research on problem- and project-based learning. J. Learn. Sci. **7**(3/4), 271–311 (1998)
3. Berg, B., Lune, H.: Qualitative Research Methods for the Social Sciences, 9th edn. Pearson, Boston (2017)
4. Ehlers, U.: Qualitative online Befragungen. In: Mikos, L., Wegener, C. (ed.) Qualitative Medienforschung Ein Handbuch von Mikos, Lothar, Wegener, Claudia, pp. 327–339 (2017)
5. Farros, J.M., Shawler, L.A., Gatzunis, K.S., Weiss, M.J.: The effect of synchronous discussion sessions in an asynchronous course. J. Behav. Educ., 1–13
6. Flick, U.: Interviews in der qualitativen Evaluationsforschung. In: Flick, U. (ed.) Qualitative Evaluationsforschung. Konzepte - Methoden -Umsetzung. Orig.-Ausg. Rowohlt-Taschenbuch-Verl., Reinbek bei Hamburg, pp. 214–233 (2006)
7. Gnambs, T., Batinic, B.: Qualitative online-Forschung. In: Mey, G., Mruck, K. (ed.) Handbuch Qualitative Forschung in der Psychologie. VS Verlag für Sozialwissenschaften, Wiesbaden, pp. 320–332 (2010)
8. Gudjons, H.: Handlungsorientiert lehren und lernen: Schüleraktivierung - Selbsttätigkeit - Projektarbeit (8. Aufl Ausg). Klinkhardt, Bad Heilbrunn (2015)
9. Hochschulforum Digitalisierung: The Digital Turn – Pathways for Higher Education in the Digital Age, Berlin (30) (2017)

[2] cf. Baron, B., 1998:273.

10. Knoblauch, C., Keßler, J.U., Jakobi, M.: Schools of education as agents of change: coping with diversity in India and Germany through a collaborative, interactive and blended-learning environment – a pre-test study. In: Auer, M., May, D. (eds.) REV 2020. AISC, vol. 1231, pp. 866–876. Springer, Cham (2021). https://doi.org/10.1007/978-3-030-52575-0_71
11. Knoblauch, C., Keßler, J., Jäger, N.: Embracing diversity with blended learning. brainstorming across borders yields new insights at home: EAIE Forum, Spring 2020, pp. 15–17 (2020)
12. McGuinness, C., Fulton, C.: Digital literacy in higher education. A case study of student engagement with e-tutorials using blended learning. JITE: IIP **18**, 1–28 (2019)
13. Mertens, D.: Research and Evaluation in Education and Psychology. Integrating Diversity with Quantitative, Qualitative, and Mixed Methods, 3. edn. SAGE, Thousand Oaks (2010)
14. Moran, K.: Taking a leap of faith. redefining teaching and learning in higher education through project-based learning. Interdiscip. J. Probl. Based Learn. **8**(2) (2014)
15. Salmons, J.: Qualitative Online Interviews. Strategies, Design, and Skills, 2nd edn. SAGE, Los Angeles (2015)
16. So, H., Bonk, C.: Examining the roles of blended learning approaches in computer-supported collaborative learning (CSCL) environments: a delphi study. Educ. Technol. Soc. **13**(3), 189–200 (2010)
17. Stefanou, C., Stolk, J., Prince, M., Chen, J., Lord, S.: Self-regulation and autonomy in problem-and project-based learning environments. Act. Learn. High. Educ. **14**(2), 109–122 (2013)

Gamification of Project Business Studies

Matti Koivisto(✉)

South-Eastern Finland, University of Applied Sciences, 50100 Mikkeli, FI, Finland
matti.koivisto@xamk.fi

Abstract. During recent years, gamification of education has received increased attention and interest. Although critics argue that gamification derails the focus of learning and increases stress and competition, most scholars see that benefits outweigh the risks. They believe that gamification increases student engagement and reinforces their problem-solving skills, collaboration, and communication. In this paper, gamification is applied to skills required in the management of the company's project business. In project business, the focus is not on individual projects, but on the organization's project portfolio. The project portfolio refers to the company's simultaneous projects and project opportunities that have common strategic goals and that are competing for the same resources. In the empirical part of the study, post-graduate students' feedback on the project portfolio management (PPM) exercise was investigated using both manual and machine learning-based content and sentiment analysis. The results of the experiment indicate that student sentiment toward the PPM workshop was positive and that gamification seems to be a good method of learning project business skills, especially in the creation of the management and decision-making system and in managing a diverse project portfolio.

Keywords: Gamification of education · Project business · Collaborative learning

1 Introduction

Demand for project management skills has never been higher. Organizations are constantly creating innovations, and projects represent the main method of bringing us products or services that have not existed before. The increased popularity has led to more complicated projects and, above all, has increased the number of simultaneous projects. Managing a large number of projects—also known as project portfolio management (PPM)—is not an easy task because it involves many challenging decision-making tasks, including resource allocation, scheduling, risk and financial management, and project prioritization. A new project environment with the large project portfolios and the increased importance of project business requires wider skills and new learning methods.

The aim of this paper is to analyze the suitability of gamification for learning project business skills. Our paper is organized in the following way. In Sect. 2, we briefly review gamification of education and project business, with an aim to develop a model for learning skills required in project business. The model serves as a framework to

D. Guralnick et al. (Eds.): TLIC 2021, LNNS 349, pp. 180–190, 2022.
https://doi.org/10.1007/978-3-030-90677-1_18

analyze student sentiment toward a PPM exercise carried out in a Finnish university of applied sciences. The study on the gamified PPM workshop is described in Sect. 3. In the study, the student feedback is analyzed using both content and sentiment analysis. To limit unintended errors, machine learning was used to verify the feasibility of the model and the correctness of the manual sentiment analysis. Finally, in Sect. 4 we discuss the results and draw the final conclusions.

2 Literature Review and Model Development

2.1 Gamification in Education

In recent years, games and game-like elements have been introduced to several domains, including entertainment, business, and education. In education, practitioners have applied two different approaches: Game-Based Learning (GBL) and gamification. In GBL, the game is the starting point, and at a simple level, GBL has been defined as "learning that is facilitated by the use of a game" [1]. Gamification, instead, has been defined as "the use of game design elements in non-game contexts" [2] and "the process of making activities more game-like" [3]. Both of these definitions emphasize that, in gamification, the focus in not on the game, but on the learning process.

Gamification of education typically aims to improve students' motivation, engagement [4], participation, and learning outcomes [5], and it combines play-like simulation, functional proficiency, and social interaction with learning [6]. Many scholars have reported the positive effects of gamification on learning outcomes at different education levels and subjects [7]. Naturally also some critical views have been reported [8, 9], but most of the empirical studies on gamification in higher education have provided positive outcomes [10].

Researchers have modeled gamification of learning in different context (e.g., in MOOCs [11] and in eLearning [12]). For our purpose, a model for virtual team collaborative learning (LIC gamification model) based on gamification of education, collaborative learning, virtual teams, and technology [13] is especially interesting. The model consists of the following three main parts: a learner as a player, an instructor as a coach, and a classroom as an arena.

2.2 Project Management and Project Business

Project management is the application of knowledge, skills, tools, and techniques to a project's activities to meet its requirements and to reach its completion [14]. Project business, on the other hand, goes beyond a single project and it is defined as "the part of business that relates directly or indirectly to projects, with a purpose to achieve the objectives of a firm or several firms" [15]. In project business, many projects are managed simultaneously, with the aim to fulfill the organization's strategic business objectives. Managing a diverse project portfolio is not an easy task because PPM involves many challenging decision-making tasks, including resource allocation, scheduling of the individual tasks, risk management and project prioritization. To solve this problem, organizations have created different kinds of PPM tools for increasing efficiency and productivity.

Researchers have paid a lot of attention to the success factors of a single project, but project business success has received far less interest. Scholars have analyzed project business from many related perspectives, such as trust [16], outsourcing [17], and project selection in the project portfolio [18]. However, Artto et al. introduced a framework that can be used to analyze the success factors of project business [19]. This model is presented in Fig. 1. The model has the following four elements: a management system, financial management, PPM, and management of customer and supplier networks.

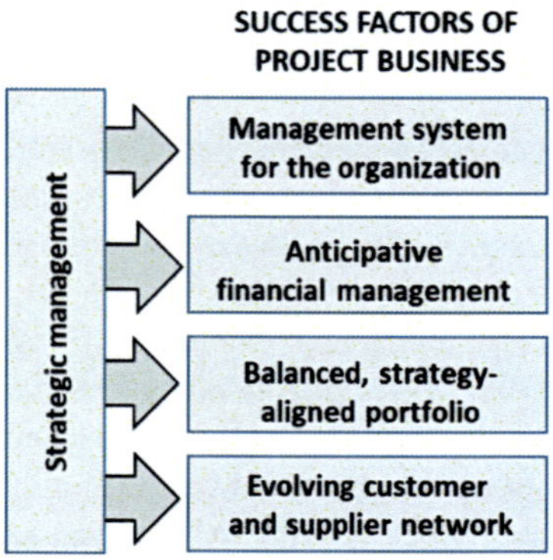

Fig. 1. Project business success factors [19].

2.3 Framework for Gamified Learning of Project Business Skills

In the previous sections, we introduced some models for gamified learning and the key success factors of the project business. Based on these discussions, a theoretical framework for gamified project business education is now created. The model (see Fig. 2) has two parts: (1) project business and (2) gamification, representing the subject of the learning and the learning method, respectively. The project business side of the model is based on the four key success factors introduced earlier, namely an organization's management system, financial management, portfolio management, and customer and supplier management.

The gamification side, instead, has three elements: a learner and learning, a learning session and environment, and finally an instructor and instructing. Here, the learner is seen both as an individual and as a member of a team, because learning takes place in a team and project business, and portfolio management contains decision making both alone and with others. The learning session refers to all activities of the exercise, and environment contains both physical and virtual learning environments. Finally, the instructor's role contains all the teaching and supporting functions, as well as materials provided before, during, and after the learning session.

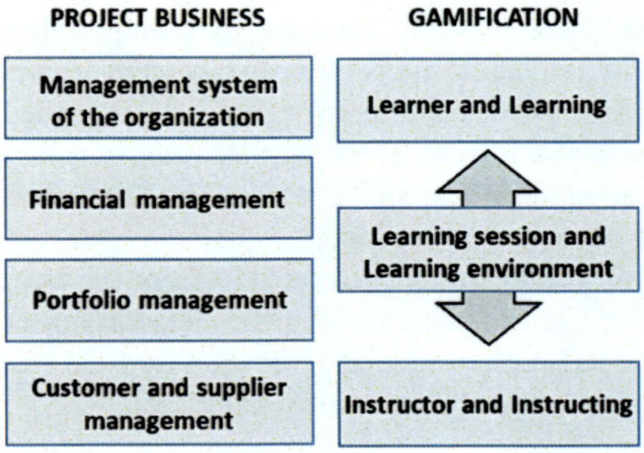

Fig. 2. Framework for gamified learning of project business.

3 Study Design

3.1 Research Methods and Data Collection

As mentioned earlier, the aim of the study is to analyze the suitability of gamification for project business education. To collect some empirical data, we organized a small-scale learning experiment in a Finnish university of applied sciences. In the experiment, a group of post-graduate engineering students (N = 42) took part in a PPM exercise. In the exercise, they worked in teams of four or five students, and they managed a project portfolio of a fictional company. During the exercise, students carried out typical portfolio management tasks like allocating resources, scheduling tasks, creating reports, and prioritizing projects with PPM software. All participants were adult learners with full-time jobs and at least some work experience in the project organizations.

After a six-hour exercise, students gave written feedback about the learning session. The total number of the feedback sentences was 358. The sentences were analyzed in two different ways. First, content analysis was used to assess message characteristics systematically. Second, sentiment analysis was applied to interpret and classify students' emotions towards the different parts of the model. Because of the researchers' human nature, both content and sentiment analysis are prone to the researcher bias. To avoid any unintended errors in the research process, we used machine learning to verify both the feasibility of the theoretical framework and the correctness of the sentence classification.

3.2 Design of Gamified Learning Exercise

Design of the gamified learning exercise was based on the following five questions that designers of gamified learning have to consider [20]:

- Why is gamification used?
- What is the focus of the exercise?
- Who is the target or learner in the gamified exercise?
- How is the gamified exercise shown and presented to the learners?
- How does the gamified exercise work (steps of the game)?

Answers to the first three questions have already been presented in Sect. 2. Gamification is used to improve students' motivation, engagement, participation, and learning outcomes. The exercise focuses on learning various skills needed in project business management. The learner is, of course, the participating student, but his or her learning takes place both at an individual and at a team level because the exercise contains decision making both alone and with others.

In meaningful gamification, designers have to focus on the aspects of the underlying activity to understand where an integration of game elements makes sense [21]. In order to do that, presentation and implementation of the exercise were guided by Marczewski's general game element principles [22] as shown in Fig. 3.

GENERAL GAME ELEMENTS	IMPLEMENTATION HERE
Narrative/Story	You are a member of the management team
Theme	You must run the project business of the company
On-boaring/tutorials	Short demo by the instructor and learning videos
Signposting	Instructions for the next task
Progress/feedback	From instructor after completing of each task From software during the tasks
Curiosity	Unknown events in each task
Strategy	Decisions about the team rules and working methods
Time pressure	Limited time and feel of hurry
Scarcity	Limited resources to complete the projects
Consequences	Decisions in one project affects others
Loss aversion	Wrong resource allocation leads to lost bonuses
Investment	Students spent the whole Saturday in the workshop
Flow	Balance of skills and challenges

Fig. 3. Framework for gamified learning of project business.

The exercise consisted of the following four steps. In the first step, the instructor revealed the theme and the narrative, followed by a short introduction to the PPM software. Because the students needed guidance to the program also during the exercise, they had access to short task-specific video tutorials. In the second step, the participants added their own projects to the project portfolio and allocated resources to their project. Each team had the same projects, but all members of a team had a unique project. This way, the outcomes of the teams could be compared against each other.

Next, in step three, students had to shift their perspective from a single project to the project portfolio and its success. The scarcity of resources forced them to prioritize projects, reschedule tasks, and reallocate resources, as well as react to the new orders from the company headquarters. This step was the most essential part of the exercise, and it involved a lot of cooperation, negotiations, and decisions on team rules, utilizing methods under time pressure. After each step, all participants met in a short feedback meeting. These meetings also provided guidance to the next tasks. Finally, at the end of the workshop, the instructions for homework or step four were given.

3.3 Content Analysis

Content analysis is a popular research method used to determine the presence of certain words, word groups, or concepts in text, speech, or some other form of qualitative data.

The content analysis has different approaches and, here, a directed content analysis (DCA) is used. DCA is a structured process guided by an existing theory [23], and researchers use the key concepts of the theory as coding categories. In this case, seven elements of the model described earlier created the framework for categorization.

In this paper, content analysis was used for two different purposes. First, it was used to validate the new model introduced in Sect. 2.3. Like always, the model represents the conceptual world, and therefore, it is a more or less simplified version of the real world. The aim of model validation is to find out if the suggested model is useful, addresses the right problems, and provides accurate information about the system being modeled [24]. The main question, in this case, is how well the new model encapsulates the key elements of the gamified learning of the project business. In order to find this out, we used a simple n-gram-based content analysis for the student feedback. The analysis was done with Microsoft Azure Machine Learning Studio, and the workflow contained the typical data processing steps, including data selection, cleaning, and preprocessing before extracting the n-grams as shown in Fig. 4.

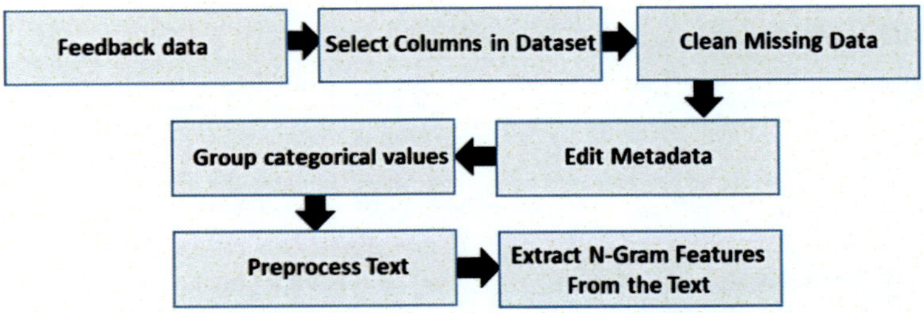

Fig. 4. Workflow to extract n-grams from the feedback data.

The analysis identified 50 key words or word groups from the text. Then, the n-grams and the items of the model were manually mapped together to find out how well the model and the content of the feedback matched. The results are shown in Table 1. From 50 n-grams, 28 were related to project business and 17 to gamification. Five n-grams were not associated to any item of the model. Bearing in mind that all models are wrong but some are useful, the results suggest that the framework contains the key elements of the phenomenon, and it can be used for analyzing gamified learning of the project business.

Second, we carried out a manual content analysis and assigned all feedback sentences to corresponding categories. As mentioned earlier, we used DCA and the new model for classification. If a comment contained opinions related both to the subject of study (project business) and the learning method (gamification), it was classified to both categories. Portfolio management and the learning session or environment were the most often commented areas, with 80 and 71 sentences, respectively. Instructor and instructions, together with customer and support networks, received the least feedback from students. More details can be found in Table 2.

Table 1. N-grams and their associations to the elements of the model.

Element of the model	n-grams	N
Management system of the organization	manager, project_manager, decision, problem, leader, operate, realization, advantage, company, operation	10
Financial management	budget, anticipate_control, anticipate_control_economy	3
Portfolio management	accordance_strategy, portfolio, resource, work, creation_portfolio, strategy_company, different_project, progress, view, point_view, point	11
Customer and supplier management	network_subcontract, customer_network_subcontract, network_subcontract_network, customer_network_subcontract_network	4
Learner and learning	group, deal, opinion, like, easily, succeed, understand, able	8
Learning session and environment	time, place, open, think, program, software	6
Instructor and instructing	support, facilitate, follow	3
Not mapped	surely, level, really, accord, possible	5

Table 2. Feedback sentences in each category

Element of the model	Number of statements
Management system of the organization	51
Financial management	40
Portfolio management	80
Customer and supplier management	26
Learner and learning	59
Learning session and environment	71
Instructor and instructing	15

3.4 Sentiment Analysis

Sentiment analysis, in general, aims to identify opinions and determine attitudes towards a particular topic. In sentiment analysis, classification can be done different ways using binary, ternary, or ordinal classification. It this study, binary classification (positive or negative) was used. Initial manual classification gave the following results: 232 positive and 79 negative sentences. Forty-seven comments did not have a clear positive or negative attitude, and they were therefore removed from the further study.

As pointed earlier, manual classification is an error-prone method. Therefore, a dictionary-based machine learning approach was used to analyze the quality of the manual classification process. The simplified machine learning workflow is shown in

Fig. 4. First, we had to preprocess the data in order to get it in correct format. Second, we divided the data into two parts—first for model creation and second for using it. In model creation, we used supervised learning with a neural network algorithm to create a binary classifier, which divided the sentences either into positive or negative categories. Our neural network algorithm used one hidden layer with 100 nodes and maximum 100 learning iterations. After running our model, we compared the outcomes of manual and machine learning classifications, and the accuracy of the machine learning model was 0.78. In other words, the algorithm classified 68 sentences differently than we had done manually. After reviewing all differently classified cases, we found out three clear errors in the original manual classification and we corrected them (Fig. 5).

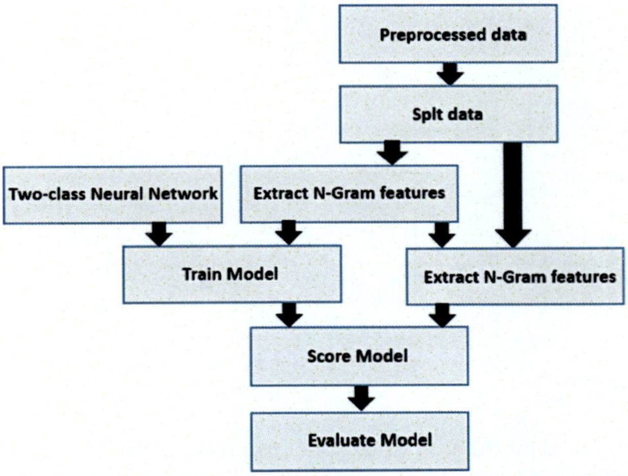

Fig. 5. Simplified machine learning based sentiment analysis workflow.

Finally, we combined the classification and sentiment results to find out the students' sentiment toward different elements of the model. In Table 3, numbers of positive and negative statements, the relative sentiment score, and qualitative sentiment evaluation are presented. Sentiment score was calculated as follows: the difference between the number of positive and negative statements is divided by the total number of statements. Qualitative sentiment assessment follows the logic presented in reference 25 for correlation coefficient interpretation. According to the results, students expressed strong positive sentiment toward the management system, portfolio management, as well as learner and learning. The only dimension receiving more negative than positive comments; thus, a moderate negative sentiment was customer and supplier management.

Table 3. Amount of positive and negative statements and sentiment of each element.

Element of the model	Positive statements	Negative statements	Sentiment score	Qualitative sentiment
Management system	48	3	0.88	Strong positive
Financial management	26	14	0.30	Weak positive
Portfolio management	75	5	0.88	Strong positive
Customer & supplier management	8	18	−0.38	Moderate neg
Learner and learning	56	3	0.90	Strong positive
Learning session and environment	41	30	0,15	Weak positive
Instructor and instructing	9	6	0.20	Weak positive
All elements	263	79	0.54	Moderate pos

4 Conclusions

The study contributed meaningful evidence in two areas. First, the study demonstrated how to support manual content and sentiment analysis with machine learning. In this study, machine learning was used to validate the model and to verify the correctness of the manual sentiment classification in a simple but successful way.

Second, the study provides valuable information on gamification of the project business education. The positive sentiment of the students' feedback clearly points out that their attitude toward the workshop was positive and that they found the exercise useful. The more detailed analysis also indicated that a gamified PPM exercise is a good and motivating method to learn how to create and operate the management system, allocate resources, schedule the tasks, and prioritize the projects. Similarly, exercises of this kind seem to be a less suitable way to study financial management and especially the customer and supplier network management.

There can be many reasons behind these findings. However, an obvious explanation can be the role of the PPM software in the general information system architecture of a company. The PPM software serves as a production information system for a project-oriented company and provides support for managerial decision-making. Organizations have typically dedicated customer and supplier management systems, as well as finance and accounting applications; therefore, these areas cannot be covered fully in an exercise utilizing PPM software. Further studies are naturally needed to find more detailed information.

Although sentiment toward learning was strongly positive, student feedback also provided some suggestions how the current learning session, and instructions could be developed further. The main sources of criticism were related to the timing. A large number of students reported that their team had to hurry in some parts of the exercise, which limited their possibility to compare different alternatives. Some students also pointed out that they would like to get familiar with the PPM software before the exercise. Based on these comments, the exercise will be, in the future, divided into three separate parts: before, during, and after the exercise. In this new implementation, students will use video tutorials to get familiar with the software before the workshop. This allows students to concentrate on decision-making and teamwork during the learning session. Finally, most of the reporting will be carried out after the learning session, offering students more time to reflect their experiences.

Acknowledgments. This work was supported by Master Talent program of Thinking Portfolio Oy.

References

1. Whitton, N.: Games-based learning. In: Seel, N.M. (eds.) Encyclopedia of the Sciences of Learning, pp. 1337–1340. Springer, Boston (2012). https://doi.org/10.1007/978-1-4419-1428-6_437
2. Deterding, S., Dixon, D., Khaled, R., Nacke, L.: From game design elements to gamefulness: defining gamification. In: 15th International Academic MindTrek Conference: Envisioning Future Media Environments, New York, pp. 9–15 (2011)
3. Werbach, K.: (Re)Defining gamification: a process approach. In: Spagnolli, A., Chittaro, L., Gamberini, L. (eds,) Persuasive Technology. PERSUASIVE 2014. LNCS, vol. 8462, pp. 266–272. Springer, Cham (2014). https://doi.org/10.1007/978-3-319-07127-5_23
4. Dichev, C., Dicheva, D.: Gamifying education: what is known, what is believed and what remains uncertain: a critical review. Int. J. Educ. Technol. High. Educ. **14**(1), 1–36 (2017). https://doi.org/10.1186/s41239-017-0042-5
5. Buckley, P., Doyle, E.: Gamification and student motivation. Interact. Learn. Environ. **24**(6), 1162–1175 (2016)
6. Sousa, M.J., Rocha, A.: Leadership styles and skills developed through game-based learning. J. Bus. Res. **94**, 360–366 (2019)
7. Legaki, N.-Z., Xi, N, Halmari, J., Karpouzis, K., Assimakopoulos, V.: The effect of challenge-based gamification on learning: an experiment in the context of statistics education. Int. J. Hum. Comput. Stud. **144** (2020)
8. Bogost, I.: Why gamification is bullshit. In: Walz, S.P., Deterding, S. (eds.) The Gameful World, pp. 65–80. The MIT Press, Cambridge (2015)
9. Morozov, E.: To Save Everything, Click Here: The Folly of Technological Solutionism. PublicAffairs, New York (2013)
10. Hung, A.C.Y.: A critique and defense of gamification. J. Interact. Online Learn. **15**, 57–72 (2017)
11. Saputro, R.E., Salam, S., Zakaria, M., Anwar, T.: A gamification framework to enhance students' intrinsic motivation on MOOC. TELKOMNIKA Telecommun. Comput. Electron. Control **17**, 170–178 (2019)

12. Rutkauskiene, D., Gudoniene, D., Maskeliunas, R., Blazauskas, T.: The gamification model for e-learning participants engagement. In: Uskov, V.L., Howlett, R.J., Jain, L.C. (eds.) Smart Education and e-Learning 2016. SIST, vol. 59, pp. 291–301. Springer, Cham (2016). https://doi.org/10.1007/978-3-319-39690-3_26
13. Srisomboon, P., Jeerungsuwan, N.: Gamification model for virtual team collaborative learning via cloud technology. In: The 12th International Conference on eLearning for Knowledge-Based Society (2015)
14. Project Management Institute: A Guide to the Project Management Body of Knowledge (PMBOK Guide), 5th edn. Newtown Square, Pennsylvania (2013)
15. Artto, K., Wikström, K.: What is project business? Int. J. Project Manage. **23**(5), 343–353 (2005)
16. Smyth, H., Gustafsson, M., Ganskau, E.: The value of trust in project business. Int. J. Project Manage. **28**(2), 117–129 (2010)
17. Sandhu, M.A., Shamsuzzoha, A., Helo, P.: Does outsourcing always work? A critical evaluation for project business success. Benchmarking Int. J. **25**(7), 2198–2215 (2018)
18. Costantion, F., Di Gravio, G., Nonino, F.: Project selection in project portfolio management: an artificial neural network model based on critical success factors. Int. J. Project Manage. **33**, 1744–1754 (2015)
19. Artto, K., Martinsuo, M., Kujala, J.: Project Business. Helsinki, Finland (2011)
20. Hallifax, S., Marty, J-C., Serna, A., Lavoué, E.: A design space for meaningful structural gamification. In: Extended Abstracts of the 2018 CHI Conference on Human Factors in Computing (2018)
21. Nicholson, S.: A user-centered theoretical framework for meaningful gamification. Games+Learning+Society 8.0, Madison, WI (2012)
22. Marczewski, A.: User types. In: Even Ninja Monkeys Like to Play: Gamification, Game Thinking and Motivational Design, 1st edn., pp. 65–80. CreateSpace Independent Publishing (2015)
23. Hsieh, H.-F., Shannon, S.: Three approaches to qualitative content analysis. Qual. Health Res. **15**, 1277–1288 (2005)
24. Macal C.: Model verification and validation workshop on "threat anticipation: social science methods and models" The University of Chicago and Argonne National Laboratory April 7–9, Chicago, IL (2005)
25. Dancey, C.P., Reidy, J.: Statistics without Maths for Psychology. Pearson Education (2007)

Mistakes: The Shadow Capital for Learning

Adamantios Koumpis(✉)

Berner Fachhochschule, 3005 Berne, Switzerland

Abstract. Mistakes, whether resulting from defective judgment, deficient knowledge and carelessness, or misconceptions and misunderstandings, are never trivial to deal with. However, if approached appropriately, they can become a source of potential corporate value so that they can facilitate future growth and development. Accounting of mistakes can prove an extremely tricky area — the same as in financial reporting, accounting is not an end on its own but a means to an end. We live with our mistakes, but we don't have to die with them. Finding the courage to change things may seem difficult, but the opportunity is big and is available anytime we decide to seize it. It is never too late to see an opportunity for change: the more we invest in explaining our mistakes to ourselves, the easier we shall get "rid" of them and minimize the risks of making the same mistakes again. Admitting mistakes is not easy but should not become unbearable for the people: it is what we learn from our mistakes and our failures that may drive us into our future successes.

Keywords: Mistakes · Shadow capital · Intellectual capital · Corporate value

1 Introduction

Many of the ideas that I present here come from discussions and exchanges I had the pleasure to have back in November 2020 with a good partner in research, Dr. Jerry Andriessen of Wise and Munro, the Netherlands. Jerry is a thought leader, and I owe him much. In an informal interaction we had, and in response to some of the ideas I am also presenting here, he responded by writing to me:

" [...] Learning is not the thing people do, it is part of something they do. Apart from simple things, learning has different outcomes for everyone. The process is different every time, the outcomes are different every time.

Learning is not an isolated activity. We learn because we interact with others. Together, we set the stage, perform the actions, the interactions, the verbalisations, the whole practice. This practice is not a constant, but is ever changing, we each tell different stories about it. Learning is an arbitrary selection of the things we do differently, compared to some earlier time.

Learning has purpose, without a purpose, there is no learning. Robots cannot learn. Machines cannot learn. They can be programmed, and they can be programmed to acquire new facts and procedures according to some algorithmic mechanism. That is not learning.

D. Guralnick et al. (Eds.): TLIC 2021, LNNS 349, pp. 191–198, 2022.
https://doi.org/10.1007/978-3-030-90677-1_19

We have forgotten how to learn, yes, because have never understood how we learn. We learn because we interact. […]" [1].

The author has spent considerable time and effort in the past years reflecting on the subject of mistakes, and a great part of the ideas we present today relates to three articles that appeared in the International Journal of Advanced Learning [2–4]. However, while getting older makes people look back and try to detect patterns of success so that they repeat them, it is also an opportunity to identify patterns of failure and try early recognise and avoid them. To this aim, we combine thought that appeared in these three previous papers but integrated from the perspective of the Learning Organisation.

We are all aware of the problems related with New Year's resolutions: most of them end in failure, increasing our disappointment levels to unmanageable levels; a matter for psychotherapists, one would one say, is whether the same reasons that appear in people's New Year's resolutions would not appear in corporate business plans and agendas. As another research partner mentioned to me once when reflecting on how people deal with their weight management problems, "the problem is not what people eat and drink between Christmas and New Year – it is rather what people eat between New Year and Christmas." However, the focus is mainly in this small one-week or ten-day period that counts for 1.9 or 2.73 of an entire year's duration, respectively.

In the next sections we present certain perspectives that might help overcome the difficulty of stigmatizing mistakes and — most importantly — those who make them. An open culture that allows people to not feel fear to fail will help us seize success.

2 Shadow Capital of Mistakes

Mistakes, whether resulting from defective judgment, deficient knowledge and carelessness, or misconceptions and misunderstandings are never trivial to deal with. However, if approached appropriately, they can become a source of potential corporate value so that they can facilitate future growth and development.

Same as with *gray problems*, which is a term used for IT problems where the causing technology is unknown or unconfirmed, mistakes are usually underestimated or not valuated in their true size.

Human nature drives people to make emotional interpretations and seek simplicity. However, mistakes are not always the results of a higher-order complexity and can afford empathy-free interpretations. The stigma of failure and the elaborations usually related to the conduct of a mistake hinder us from learning and improving ourselves and our corporate performances from our mistakes.

Same as with the accounting practices that govern corporate business transactions, an investment towards developing some common understanding on our mistakes can help improve our performance and increase our individual and corporate learning potentials.

Richard Weaver, in the book *Ideas Have Consequences* [5], presented the negative aspects and what we now perceive as the dark shadow of nominalism in the Western civilization since this doctrine gained prominence in the High Middle Ages.

Leaving back the times of absolute truth, where a mistake can be any of the ones recognised in the "official" list of the ten commandments, to the times of what Weaver calls as the Great Stereopticon, namely a metaphor that aims to denote an emergent conceptual construct that serves to manipulate people's emotions and beliefs and to separate them from their humanity via "the commodification of truth," a mistake is anything that can be presented as such, employed and capitalised as such to serve the agendas of the part that made it or any other part that has a vested interest related to it.

A CEO who fires several (tens, hundreds, or thousands) of employees to protect shareholder value in times of the COVID-19 slowdown and the fear of an approaching recession may be regarded as cruel for the means they employ, but their ends may in general be appreciated or even praised. One rarely asks whether there may have been other means to protect shareholder value than "demising" people, a euphemism that might had been used also in past times to denote physical extinction.

Raoul Vaneigem, one of the introducers of the Situationist movement in Europe, supports that nothing is sacred and everything is permitted to be said ('rien n'est sacré, tout peut se dire') [7]. However, there is a difference between something that is said—i.e., expressed with words and concepts—and something that is materialised—i.e., expressed with actions. Sadly for some of us, fortunately for the rest of the world, the world is not populated by intellectuals: the CEO who would dare to write an article sharing with a wide audience their idea about firing employees so that their company regains its strength and growth potential would face criticism and might get sacked the next day. But firing people as such may not cost them their position — at least not instantly.

This brings us to some of the remarks we made before and which relate to the fact that nobody wants to openly talk about mistakes, as they seem to diminish the value and the power of the person who makes them.

2.1 Mistakes in Human Ventures

The concept of shadow capital is the same as this of shadow banking, not an official term in the business or finance literature. Looking to the concept of shadow banking may actually offer some advantages in better understanding the concept of shadow capital. The former symbolizes one of the many failings of the financial system leading up to the global crisis. The term shadow bank was coined by economist Paul McCulley [6] and had originally a distinctly U.S. focus and referred mainly to nonbank financial institutions that engaged in what economists call maturity transformation and when commercial banks use deposits, which are normally short term, to fund loans that are longer term.

Shadow banks do something similar: they raise (that is, mostly borrow) short-term funds in the money markets and use those funds to buy assets with longer-term maturities. But because they are not subject to traditional bank regulation, they cannot (as banks can) borrow in an emergency from the Federal Reserve (the U.S. central bank) and do not have traditional depositors whose funds are covered by insurance; they are in the "shadows."

Something similar happens with our mistakes: if appropriately employed, mistakes can "finance" our future development and growth—but only if appropriately managed in terms of supporting our learning curves. We have mentioned before the role of learning from mistakes and how this results in an increase of corporate value.

Accounting of mistakes can prove an extremely tricky area — same as in financial reporting, accounting is not an end on its own but a means to an end. We live with our mistakes, but we don't have to die with them. Finding the courage to change things may seem difficult, but the opportunity is big and is available anytime we decide to seize it.

It is never too late to see an opportunity for change: the more we invest in explaining our mistakes to ourselves, the easier we shall get "rid" of them and minimize the risks of making the same mistakes again. At the very end, we humans can be inventive and creative and can always come up with some new mistakes. What may only be necessary is a methodological framework that provides the conceptual basis for reporting mistakes and including them as part of a continuous learning process. To this, we promote the use of the Triple Bottom Line (TBL) model, which we describe in the next section.

3 Triple Bottom Line for the Learning Organisation

To support the sustainability of a learning paradigm, we have explored the potential of applying the TBL model to organisations that want to build their future as learning organisations. TBL has been designed originally as an accounting framework with three components: one social, one environmental, and one related to financial and business aspects [8].

We consider this as of major importance for two reasons:

- Firstly, TBL shall support the learning process not only in meeting its business goals, but also for taking into account social and environmental aspects and considerations which are extremely relevant to all types of organizations but especially for the case of small- and medium-sized enterprises that would otherwise not afford to deal with the complexity of running a profitable business with low levels of human, technology, and technology resources while also caring to improve their learning curves.
- The application of the TBL method for helping organizations learn from their mistakes demands from us to see beyond the traditional bottom line of business to all three layers of the profits that a business can make — namely socially, environmentally, and economically. Measuring learning processes using the TBL is one of the best markers of how sustainable the learning operations of a business are and how profitable learning from their mistakes really is.

The challenge is not just technical: it requires profound adaptations in business models and the underlying business mindset. Research and innovation offers unique opportunities to reach out and engage with customers and other partners in novel ways, identifying and addressing specific areas of improvements with a view to ensuring that companies do not need to treat their mistakes with regret. Lessons we have accumulated from the software industry where companies test their software firstly within their employees (alpha testers) and then with their customers and end users (beta testing) provides an example on how one may take an inclusive approach in the offering of solutions that may still not work as planned or as expected, but exhibit a positive impact to a company's value propositions to their clients' base.

3.1 Socially Sustainable Learning

The Social "bottom line" measures "business' profits" of learning from conducted mistakes in terms of human capital, comprising also aspects related to the collective intellectual capital that is created by all members of a particular business setting or ecosystem.

Learning related to the social bottom line is increased by offering fair and beneficial working environments and employment practices, and through corporate community involvement, and it can also be measured in the impact of the individual company contributions to the local or the regional economy at large. Some aspects that matter when assessing learning related to the social bottom line are, among others, whether a particular company operates as a job-growth driver in their city and/or their province, if their employees are better situated in terms of economic and employment stability, and if they support local and regional initiatives and contribute to the overall sustainability of their community/region, as well as issues related to fair employment standards, gender equality, and employee demographics.

The basic point for the social bottom line is that, after all, if a business is not nurturing healthy and positive relationships with their community, their client base and employee pool shrinks accordingly. Learning related to the social bottom line questions the belief that the less a business pays its work force, the longer it can afford to operate. Instead, the social bottom line measures the long-term sustainability of business human capital, with the understanding that a business that is also a desirable workplace will always be able to operate into the future, since there will be a work force striving to be part of the business. Essentially, corporate interests and labour interests are seen as interdependent. It is easy to see that, due to its intangible nature, the social bottom line can be difficult to measure or quantify. However the Global Reporting Initiative (GRI, https://www.globalreporting.org/) has developed guidelines to enable businesses to report and measure their social impact.

3.2 Environmentally Sustainable Learning

The TBL approach we take for the promotion of the learning organisation takes the view that the smaller impact a business has on the environment and the fewer natural resources it consumes, the longer and more successful they will be. This definitely forms a change in the mindset of all business actors and stakeholders. Changing the way we consume and the way we produce is a recurring leitmotiv in the Circular Economy Action Plan of the European Commission where it is recognised that "products placed on EU market will be designed to last longer, to be easier to repair and upgrade, recycle and reuse." To this aim, providing incentives for product-as-a-service is one only of the plausible ways to consider the development of new business models, where companies will keep the ownership and responsibility for the product throughout its life cycle.

In the light of the above, learning that pays attention to the environmental bottom line means managing, monitoring, and reporting mistakes and learning from them in a holistic way, recognising all aspects of production, waste, emissions, and consumption. What may have been considered in the past as typically the work of an EHS department (environment – health – safety), is now taking with an increasing pace a higher priority as the most sustainable business models also make waste reduction and green policies in corporate-wide and supply chain-wide values across all levels.

Same as with learning from the social bottom line, measuring and reporting mistakes related to the environmental bottom line is certainly possible. However, depending on the nature and the size of a business, it can be a time-consuming and difficult process that asks for specific skills and capacities to be established. And it is again here that the Global Reporting Initiative offers helpful insights for metrics for measuring and reporting environmental TBL.

3.3 Economically and Business Sustainable Learning

According to the TBL approach, economic and business sustainability of learning is not simply associated with traditional financial capital. The economic and business capital of learning—or what we consider as the shadow capital of mistakes—under the TBL model should be assessed in terms of how much of an impact a particular mistake has on the wider economic and business environment of the company, its sector, a region, country, or even globally.

The core idea here is that a healthy and sustainable business that strengthens the economy it is a part of is one that will continue to succeed in the future, since it contributes to the overall economic health of all its constituent support networks, as well as the corresponding supply and value chains, and last but not least, the communities it is a part of at the local or regional level. Of course, a business needs to be aware of its traditional profits, as well, and the TBL accounts for this.

Relevant aspects to consider regarding the economic sustainability of learning are: does a company help local/regional suppliers, as well as actors from other areas (e.g., local or regional entrepreneurs, links with academia and research entities of the region), stay in business and innovate? Or do its activities put the local economy at risk? Does it pay employees and contractors enough to stimulate economic growth and spending? Or does its compensation policy shrink local economy? Does it choose materials, technologies, and know-how that are economically a good investment? Or does it buy cheaper products that leave a negative impact and raise other issues (of all three types: economic, environmental, and social) in other areas?

In case the above may seem too theoretical, one may consider the case of how Clarebout, a supposedly successful business and the largest producer of frozen potato products in Europe, is now recognised by the citizens of the region as a polluter and a poor employer [9, 10].

The case reflects rather well the shift in what citizens and the society at large perceive as positive for them in the longer run. TBL, in this respect, can be used as a tool that can help reflect all different considerations and bring them to a common bottom line.

The main reason we promote the use of the TBL method is that, in the wider business context, it provides us with a holistic understanding of all different aspects related to a broad gamut of different mistakes and mistake types, bringing them under a single common framework.

In this respect, TBL offers the necessary sustainability framework for learning from one's mistakes that can help companies capitalize on their mistakes in order to allow them to stay in business, based on their impact on the environment, the relationship to local and regional economies and the community, and their contribution to the economy. In fact, all three factors play a major role in determining if it can stay in business and generate profits and value—no single bottom line can sustain a company alone; this is a lesson also to be learned from the case of Clarebout we described above.

4 Conclusion

The idea of repentance has an important place in the Christian religion. However, repentance does not need to take an emblematic character; it is a simple process—in Greek, the term used is *μετάνοια*, which means nothing else than change of one's mind. Below, we present in the form of aphorisms aspects we have discussed in the previous sections.

Learning is a *very private* matter. While for personal matters, people are capable to show an increased degree of control over their learning processes and, thus, also on how to recognize, valuate, and capitalize on their own mistakes, for professional and working contexts, this does rarely happen.

Over the years, businesses have found it difficult to contend with and account for intangible assets (same as it is the case with mistakes).

Mistakes are precious; they are invaluable and constitute the shadow capital of any individual or company or organization.

We need to change a culture where nobody wants to openly talk about mistakes as they seem to diminish the value and the power of the person or the company who makes them.

Greatness has a price called commitment: in a living and open-to-learning organization, groups of people examine themselves, their activities, and their organisation, leading to better adaptation to changing circumstances and challenges, to enhancing their service provision as well as their own intellectual and spiritual wellbeing and that of their organisation.

Mistakes are like abandoned children—like our very own abandoned children. It is tough to try and look deep into the eyes of our mistakes.

Most business organizations still tend to assign all decision-making responsibilities to one visionary founder, powerful CEO, or senior executive—and together with them, they also assign both their expectations for unending successes and, as one may expect, the accountability for any types of possible mistakes.

The core problem with mistakes is when we are applying a bad rule about mistakes: "It may have been a good rule that kept me out of trouble when I was six years old, but it is not a good rule now that I am older. It is time to change the rule. What would be a better one? In reality, what is a mistake?".

Accounting of mistakes can prove an extremely tricky area—same as in financial reporting, accounting is not an end on its own but a means to an end. We live with our mistakes, but we don't have to die with them. Finding the courage to change things may seem difficult, but the opportunity is big and is available anytime we decide to seize it.

It is never too late to see an opportunity for change: the more we invest in explaining our mistakes to ourselves, the easier we shall get "rid" of them and minimize the risks of making the same mistakes again.

Admitting mistakes is not easy but should not become unbearable; it is what we learn from our mistakes and our failures that may drive us into our future successes.

Acknowledgments. I would like to thank Jerry Andriessen and Juliano Sales—with whom the author had the opportunity of long and fruitful discussions, where we had the opportunity to revisit recurring themes and topics. With Michael Scriney, we had the opportunity to also reflect on how humans can learn from computers and more specifically how Human Intelligence may gain from the breakthroughs of Artificial Intelligence technologies. There is still a long way to walk and it is a journey full of mistakes—those that have been already made and those that wait for us to make them. What is important is what we may learn out of them.

Authors usually thank their reviewers as an expression of gallantry. Here, the author was fortunate enough to receive a comment by one of the reviewers on the relevance of the thinking of Michael Oakeshott, drawing our attention to his book "The voice of liberal learning." The author admits that has never heard or read before the name of Oakeshott (renowned English philosopher and political theorist) nor had been exposed to any of his works. It took some good time to locate a copy of the book in a library (borrowed very few times and the last back in 2006). It is quite a pity that with such a massive exposure we experience to mega-amounts of information, most of which may be of ephemeral and doubtful of any value, some really good pieces of work and books miss the opportunity to find their way to a broader audience.

References

1. Andriessen, J.: comment on learning, personal communication with the author, November 2020
2. Koumpis, A.: Mistakes as Shadow Capital for the Learning Organization. LAP Lambert Academic Publishing, Saarbrücken (2016). ISBN13: 978–365996 2660
3. Koumpis, A.: Mistakes as shadow capital in human ventures. Int. J. Adv. Corporate Learn. (iJAC) **7**(4) (2014)
4. Koumpis, A.: Learning from mistakes results in increase of corporate value. Int. J. Adv. Corporate Learn. (iJAC) **6**(3) (2013)
5. Weaver, R.: Ideas Have Consequences, University of Chicago Press, Chicago (Expanded Edition) (2013)
6. McCulley, P.: The shadow banking system and Hyman Minsky's economic Journey, Global Central Bank Focus, May 2009
7. Vaneigem, R.: Rien n'est sacré, tout peut se dire. Réflexions sur la liberté d'expression, La Découverte (2003)
8. Slaper, T.F., Hall, T.J.: The triple bottom line: what is it and how does it work? Indiana Busin. Rev. **86**(1) (2011)
9. Lemoine, R.: En Wallonie, un petit village en lutte contre le géant de la frite, Liberation (2021). https://www.liberation.fr/international/europe/en-wallonie-un-petit-village-en-lutte-contre-le-geant-de-la-frite-20210516_52CZIWA47VFLHGLFUT7TITCMA4/
10. Boffey, D.: The chips are down: Belgium counts the cost of betting all on the potato, The Guardian (2021). https://www.theguardian.com/environment/2021/may/15/the-chips-are-down-belgium-counts-the-cost-of-betting-all-on-the-potato

Lab 21 – A Space for Learning, Sharing and Innovating

Christina Merl(✉)

TalkShop/2CG®, 1190 Vienna, Austria
cm@christinamerl.com

Abstract. Societies everywhere are currently undergoing deep transformation, including the transition from traditional organizational structures to network structures where connected autonomous individuals are challenged to solve complex problems co-creatively and responsibly. Phenomena like the global health crisis, climate change, and the impact of digitalization are forcing governments, organizations, institutions of higher learning, and schools to re-think learning, teaching, and assessment practices in a way that they engage working learners, students, and pupils and prepare them for the future. The present paper aims to showcase a learner-centered, multi-method teaching approach, which is grounded in the principles of communities of practice, anchored in hyper-curricula for 21st century skills acquisition, and implemented in a hybrid hyper-structure for learning. This multi-method teaching approach, referred to as 2CG® method, aims to connect learners across hierarchical, cultural, and disciplinary boundaries and helps them unlock their creative potential by means of customized artistic impulses. In this paper, we look at how the 2CG® teaching approach has been implemented in Lab 21, a hybrid hyper-structure for learning that immerses learners in practice-based learning experiences so that they can develop 21st century skills while exploring new pathways of action. Practice examples demonstrate how communication skills programs, coding and media literacy programs, innovation and wellbeing programs in organizations, institutions of higher learning, and schools can serve as an effective context for this new way of learning, which lays out a human vision for education. Finally, the paper points to current limitations of the approach and makes suggestions for future research.

Keywords: 21st century skills · Community of practice · Hyper-structure for learning · 2CG® multi-method approach · Artistic impulses

1 Fostering 21st Century Skill in Learners

1.1 Learning and Teaching in the 21st Century

Societies everywhere are currently undergoing deep transformation [1], including the transition from traditional organizational structures to network structures where connected autonomous individuals are challenged to solve complex problems co-creatively and responsibly. This development challenges governments, organizations, institutions

D. Guralnick et al. (Eds.): TLIC 2021, LNNS 349, pp. 199–211, 2022.
https://doi.org/10.1007/978-3-030-90677-1_20

of higher learning, schools, and individuals in unknown ways. Previous experiences cannot help them predict the future or find fast and sustainable answers to complex questions anymore [2]. New skills are needed to deal with complex phenomena like the current global health crisis, the effects of climate change, and the impact of digitalization. Educators, working learners (leaders as well as team members), students, and pupils have to become constant learners and rethink the way they teach, learn, lead, and collaborate in order to create sustainable, democratic societies and build inclusive and co-creative ecosystems of work.

This paper aims to present a hybrid, learner-centered, multi-method teaching approach that focuses on 21st century skills acquisition and is applied in a hyper-structure for learning. The approach lays out a humanist vision for education and emphasizes the strong need for human intelligence in times where the focus clearly is on the advancement of technology and artificial intelligence (AI) [2]. The learning model presented in this paper is grounded in the principles of communities of practice (CoP) and based on the assumption that 21st century skills can help learners build good personal and professional relationships; collaborate across hierarchies, disciplines, and cultures; create a sense of belonging; know their purpose; and do meaningful work. The overall goal is to raise multidisciplinary thinkers, problem-solvers, innovators, and caretakers who can overcome the big challenges of the 21st century.

1.2 Definitions of 21st Century Skills

Twenty-first century skills comprise learning skills (the 4 C's), literacy skills (IMT), life skills (FLIPS) [3], and 21st-century digital skills (see Fig. 1) (see Table 1), whereby digital literacy is seen as a mindset that enables learners to perform intuitively in digital environments, and to both easily and effectively access a wide range of knowledge embedded in such environments [4].

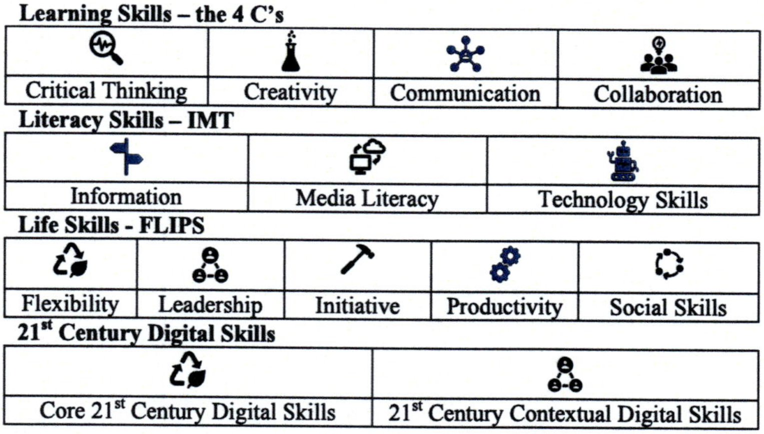

Fig. 1. 21st century skills framework.

Table 1. 21st century skills framework.

Learning Skills or the 4 C's	
Critical thinking, creativity, communication, and collaboration.	To acquire and build the 4 C's, learners need to train the mental processes that are required for connecting, collaborating, and complex problem solving, and which they will need to adapt and to improve upon modern work environments.
Literacy Skills or IMT	
Information management, media literacy, and technology skills	Learners need to be able to discern facts, to validate information, and to transform data into effective stories. Emphasis needs to be put on determining trustworthy sources and to separate factual from misinformation that floods the Internet and print publications.
Life Skills or FLIPS.	
Flexibility, leadership, initiative, productivity, and social skills	These are needed to effectively build relationships, find purpose, adapt to constant change, and build inclusive, kind, sustainable and democratic societies.
21st Century Digital Skills	
Core 21st-century digital skills and contextual 21st-century digital skills	Seven core skills (technical, information, management, communication, collaboration, creativity, and critical thinking and problem solving) and 5 contextual skills (ethical awareness, cultural awareness, flexibility, self-direction, and lifelong learning).

2 Acquiring 21st Century Skills in a Hyper-structure for Learning

2.1 2CG® - A Multi-method Teaching Approach

The 2CG® multi-method approach [5] is an experiential, learner-centered teaching approach that hones human skills with a focus on 21st century skills (see Fig. 1) (see Table 1). 2CG stands for content- and context-specific generic competency coaching and targets working learners—leaders as well as team members—students and pupils who need to be prepared for the future. The approach is grounded in the principles of CoP [6] (see Table 2); it connects learners across hierarchies, and cultural and disciplinary boundaries, and immerses them in context-specific, practice-based learning experiences. More specifically, the 2CG® method combines context-specific content with customized creative impulses from poetry, puppetry, literature, theater, film, painting, and music. The customized artistic impulses connect learners with their emotions, which can help them better understand complex issues [7], disrupt unhelpful thinking patterns and start thinking out of the box. It should be pointed out here that the 2CG® teaching approach advocates for a mix of methods, and also for the freedom to create methods, rather than simply using pre-existing and fixed procedures [8]. The approach is most effective when embedded in a hyper-curriculum as outlined in reference 9 and implemented in a hyper-structure for learning as presented in this paper. Overall, the teaching approach puts the needs of learners center stage and enables them to connect with their intuition, unlock their creativity and explore new pathways of thinking and doing [10].

Table 2. The 2CG® teaching approach is grounded on the fundamental pillars of CoP.

CoP Pillar 1. Intrinsic motivation is crucial for learning.

Intrinsic Motivation in a CoP-based Hyper-Structure for Learning	Learners are enabled to discover their passion and shared interest; they define their shared practice and find out what their purpose is. They need to define for themselves: What's the difference we want to make?

CoP Pillar 2. Learners engage in a shared practice.

Shared Practice in a CoP-based Hyper-Structure for Learning	Learners continually reflect how their learning activities relate to their professional, real-life context, and practice. Deep learning can happen when the learning is anchored in a shared practice, a shared interest, and a shared purpose.

CoP Pillar 3. All learning is social.

Sense of Belonging in a CoP-based Hyper-Structure for Learning	Learning is a social process, both in real life and in virtual settings. Social cohesion, trust, and a sense of belonging are the foundation of fruitful learning and collaboration. For deep learning to take place, learners need to define how they want to work and communicate with each other.

Pillar 4. Deep learning follows the principle of impulse – action – reflection.

Action Learning in a CoP-based Hyper-Structure for Learning	Social constructivist learning processes are based on the action learning principles of 'impulse – action – reflection' and incorporate the cognitive, emotional, and psychomotoric dimensions of learning.

2.2 Lab 21 – A Hyper-structure for Learning

Since it is no longer sufficient for learner-centered educational design to be personalized, functional, and to respond to short-lived market demands as pointed out in [11], the learning model presented in this paper holds that effective learning and future skills acquisition takes place in a practice-based hyper-structure for learning, herein referred to as Lab 21. Lab 21 is an immersive, practice-based, hybrid learning environment—a real, virtual, or hybrid space that can be linked to any relevant practice in any professional context. Lab 21 users experience a sense of belonging—an atmosphere of mutual trust makes them feel safe so that they can share their thoughts and ideas openly. They receive context-specific content and customized artistic impulses with the 2CG® method, which enables them to disrupt old thinking patterns and develop future skills while exploring new paths of thinking and doing.

2.3 Three-Dimensional Learning in Lab 21

One key priority of the Lab 21 hyper-structure for learning is to create a safe, non-judgmental, respectful culture of 'learning how to learn' as discussed in [12]. This implies connecting learners across hierarchical, cultural, and organizational boundaries; nurturing them with inspirational input; and encouraging them to take risks, bring their ideas to the table, collaborate, and act co-creatively and responsibly.

Essentially, Lab 21 learning experiences comprise three dimensions or overlapping layers: 1. the practice layer (shared interest, shared undertaking, artefacts, outputs and outcomes); 2. the social collaborative dimension (peer learning, multiple learning loops, inputs); and 3. the reflective dimension of learning (evaluation and assessment). Through these three layers or dimensions of learning, learners become part of a peer coaching culture where they can trust and rely upon each other and connect to what they do. Their teacher or facilitator guides them in exploring creative possibilities that allow them to see connections and relationships in the most unlikely of places within today's data-rich environments [13]. What is more, during their learning journey, the learners themselves become aware of what skills they already have and what skills they need to further develop in order to become fit for the future or stay relevant and connected in the workplace.

2.4 Interactive, Iterative Learning Sequences in Lab 21

Lab 21 users experience regular interactive learning sequences (see Fig. 2) that consider both the learning process (or learning journey) and the learning output (impact or outcome) by following the action learning principles of 'impulse – action – reflection' [14]. These principles incorporate the cognitive, emotional and psychomotoric dimensions of learning, which can help balance social inequalities and different cognitive learning levels [15]. Step by step, learners build awareness and identify and analyse the problems they need to solve. Through multiple learning loops as outlined in [16], they receive tailored feedback from their peers, their instructor or facilitator, and from experts. This is how they build 21st century skills 'on the go,' while being immersed in their practice and producing concrete outputs or artefacts, such as prototypes, concepts, future scenarios, stories, manifestos, etc. Developing 21st century digital skills as outlined in [4] (see Fig. 1, Table 1) will help learners understand how they can use technology as a vehicle for communication, collaboration, and co-creation.

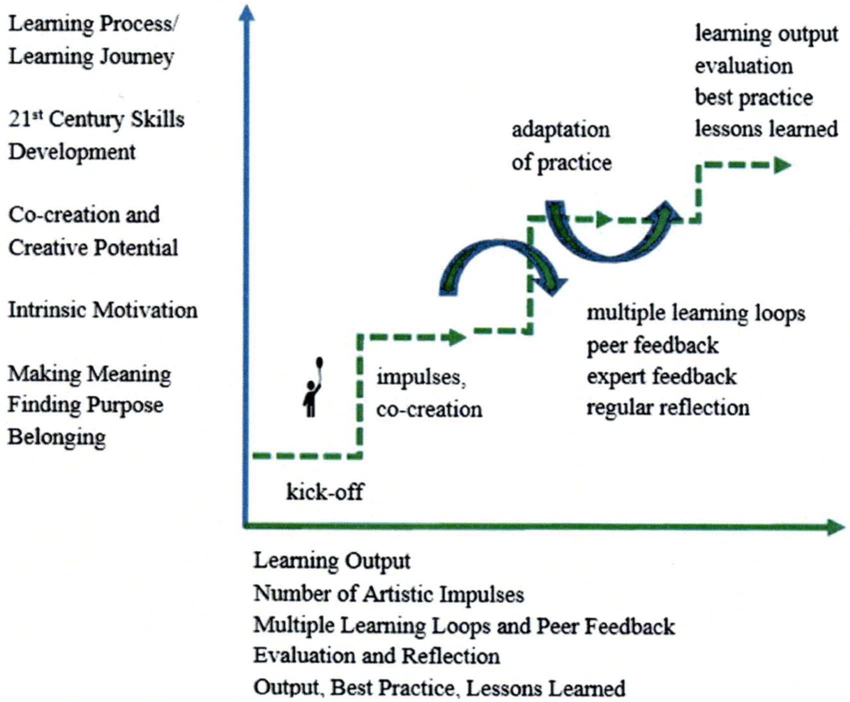

Fig. 2. Interactive, iterative learning sequences in Lab 21, based on principles of action learning and multiple learning loops.

2.5 Learning as a Rich Human Experience in Times of AI

In times where the focus clearly is on the advancement of technology and AI, it is important to meet learners' increased demand for transparency, trust, stability, and loyalty and to help them understand the role and capabilities of humans in the era of augmented intelligence where people will be increasingly working alongside machines.

Accordingly, the teaching model presented in this paper holds that hyper-structures for learning, such as Lab 21, need to become cultures of wellbeing as outlined in reference 17, where learners are nourished and supported in developing their future skills, and where they perceive learning as a delightful and rich human experience. This requires educators to take the role of facilitators or activators; they need to support learners in building a healthy relationship with their own narrative, an optimistic mindset as laid out in reference 18, and a positive self-belief.

3 Lab 21 and 2CG® in Practice

3.1 Target Groups and Context

The 2CG® mixed-method approach has been tried and tested in multiple settings with different target groups: 1. part-time bachelor programs for industrial engineering, business management and systems engineering; 2. full-time bachelor programs for cross-cultural business management and intercultural communication; 3. full-time bachelor study programs for media management, marketing and communications consulting; 4. technical colleges for programming and media literacy; 5. teacher training programs; and 6. new work and digitalization programs, health and wellbeing initiatives, and innovation and continuing education programs in public and private sector organizations.

Language and communication skills programs, coding and media literacy programs, as well as innovation and wellbeing programs in organizations, institutions of higher learning, and schools have served as an effective context for 21st century skills acquisition. The practice examples given below draw from face-to-face, blended learning, mobile learning and live online learning insights and demonstrate how the Lab 21 hyper-structure for learning has been implemented and how the 2CG® teaching approach has been applied through hyper-curricular activities. The overall goal of the practice examples has been to prepare working learners and students for the future and enable them to effect change in the real world. The learning design of the individual learning projects has been tailored to the needs of learners in their specific context.

2CG® Practice 1: Cross-Organizational, Cross-Disciplinary Network of Learners

Method and Target Group: Applied Strollology + Sequential Learning Experiences for City Dwellers of Different Cultural Background	We perceive a cityscape in sequences, and we are biased through infrastructure, advertisements, images, videos, and our own personal experiences. In other words: we see what we have learned to see, create an overall impression of the city, and adopt helpful and less helpful behaviors.
Learning Goal and Practice: *Developing the 4C's (communication, critical thinking, creativity, collaboration) and Feeling at Home in the City.* *Blended Format*	City strolls, including café and museum visits, were used as an archaic instrument to perceive and experience the city in sequences. Participants should learn to perceive their city with all their senses — vertically, in a linear way, and fictitiously — and understand its history and culture, feel at home, and adopt constructive communication behaviors. They were guided by their facilitator, experts, and artists.
Learning Impact and Artefacts	Curiosity, new ways of seeing, new perspectives, and new sensory experiences have changed the way participants perceive and move in their city. Sharing their own stories, feelings and learning more about the city's history and traditions has allowed for deep learning well as new communication behaviors. Artefacts produced included film material, poems, and collages that represent how they wish to live in their city.
Developed 21st Century Skills	Attention to detail; analytical skills; awareness of factors that have an impact on wellbeing; change of perspective; understanding complex situations; connected thinking; creativity: reflective skills and building an optimistic mindset.

2CG® Practice 2: Teacher Training College

Method and Target Group: Photo Journalism as part of a Language Teacher Training Program	Photo stories were used to help participants gain unexpected perspectives, create a multiple perception of reality, and get access to past experiences and future hopes.
Learning Goal and Practice: *Teaching 21st Century Skills in Secondary School* *Face-to-Face Format*	Language teachers should experience how they can integrate the method of photo stories into their teaching program. Visualising different realities, perspectives, and viewpoints of one and the same space helped them to understand how they can immerse their students in a shared practice and support them in developing future skills.
Learning Impact and Artefacts	Participants were astonished at their peers' photo stories and they appreciated the diverse layers of one and the same reality. They learned how photo stories can help learners to express their thoughts, broaden their horizons, and understand that there is not one true story.
Developed 21st Century Skills	Understanding how they can teach 21st century skills, including awareness building, active listening, media literacy, creativity, storytelling, empathy, etc.

2CG® Practice 3: Public Sector Organization

Method and Target Group: Strategy Workshop with Impulses from Poetry, Puppetry, and Performance Art for Public Officials *Face to Face and Live Online Format*	Three half-day workshops with customized content and impulses from poetry, puppetry, and performance art helped learners to disrupt their thinking patterns, think out of the box, and explore new pathways of action.
Learning Goal and Practice: *Creating Healthy Ecosystems of Work*	Department leaders, team leaders, and team members of a public sector organization should identify what wellbeing means for them in their increasingly digital ecosystems of work; what they would need to create a culture of wellbeing; what factors have an impact on their wellbeing; and how this relates to productivity, creativity, and innovation.
Learning Impact and Artefacts	Participants identified which problems they needed to solve and developed a manifesto for wellbeing at work which served as a basis for further action steps.
Developed 21st Century Skills	Awareness, critical thinking, analytical thinking, connected thinking, empathy, change of perspective, social intelligence, collaboration, creativity, initiative, communication skills, etc.

2CG® Practice 4: Study Program for Learning Workers

Method and Target Group: Case Clinic and Values-Based Communication for Working Learners of Different Cultural and Ethical Background *Live Online Format*	The case clinic and values-based communication method was implemented in the framework of a part-time business management, industrial engineering, and systems engineering bachelor program. Learners were provided with a real-life case where trust had been misused in a team structure.
Learning Goal and Practice: Exploring and Understanding the Impact of Communication Behaviors on Building Trust	Working learners should learn to put themselves into the shoes of others, develop empathy, think strategically, give peer advice, and consider the role of communication behaviors in team-building processes.
Learning Impact and Artefacts	In a live online half-day workshop, learners developed concrete ideas and an understanding of the role trust plays in team processes and collaboration. They defined core values to which their organizational and institutional structures need to stick to. Some divergent thinking between participants of European and Asian background drove participants to further investigate the matter and find ways to overcome this hurdle as it might impact the collaboration, innovation, and productivity of their community.
Developed 21st Century Skills	Change of perspective, different viewpoints, empathy, strategic thinking, collaboration skills, social intelligence, communication skills, critical thinking, etc.

4 Discussion

4.1 Main Findings

The practice examples presented in this paper—and it should be pointed out here that only a small selection of numerous practice examples as carried out in multiple contexts over the past 13 years has been shared—suggest that effective 21st century skills acquisition (see Fig. 1, Table 1) takes place in a hybrid hyper-structure for learning that is grounded in the principles of CoP and linked to a professional practice. The present paper has introduced Lab 21 as a hybrid hyper-structure for learning that can be linked to any professional practice in public and private sector organizations, institutions of higher learning and schools. The practice examples indicate that high learner engagement and impactful learning outcomes can be obtained with the 2CG® multi-method teaching approach that combines context-specific content with customized artistic impulses from literature, poetry, puppetry, theater, film, photography and music. The approach is embedded in hyper-curricular activities and has been implemented in the context of language and communication programs, innovation and wellbeing programs, coding and media literacy programs in organizations, institutions of higher learning, and schools.

Overall, it can be argued from the practice examples that effective 21st century skills acquisition, high learner engagement and impactful learning outcomes can be obtained in a safe space for learning and sharing where learners are immersed in a relevant practice and enabled to disrupt old thinking patterns and explore new pathways of action with the help of creative, artistic impulses (see Fig. 2). The practice examples further seem to confirm that simple 'onlinification' of traditional teaching methods that rely on textbook examples and exercises is not sufficient as it tends to result in low learner engagement, as pointed out in reference 19.

The learning model presented in this paper clearly provides a human vision for education; it hones human skills with a focus on 21st century skills where technology is used to serve the purpose of dealing with complex issues co-creatively and responsibly.

4.2 Limitations: Quantifying Transformational Change

The overall goal of the 2CG® multi-method teaching approach as applied in the Lab 21 hyper-structure for learning is to prepare learners for the future and enable them to effect change in the real world. This change should manifest itself in the learning outcomes, in the behavior of learners, and in the organizational structures. Admittedly, quantifying transformational change in (learning) behaviors and organizational structures has not yet been sufficiently achieved. So far, transformation and change processes have been expressed through learner feedback, including new insights, changed worldviews, and positive stories about improved practices. For example, learners report that they have incorporated new feedback structures into their team culture as a result of their learning insights; they tell that they have identified what needs to be changed in their context; they report that they have adopted new practices in their team and discovered blind spots; and they have adopted mindfulness practices and their insights and thoughts have created an internal discourse or triggered further conversations with team members. However, there are not yet any exact figures that reflect the learning outcomes and transformations.

4.3 Summary and Suggested Future Research Directions

Numerous Lab 21/2CG® case studies and pilot projects implemented in institutions of higher education, schools, and public and private sector organizations during the past 13 years have validated the following assumptions:

1. Effective 21st century skills acquisition needs to be linked to a professional practice and takes place in the context of hyper-curricular activities;
2. Language and communication skills programs, innovation and wellbeing programs, and coding and media literacy programs can serve as an effective framework for 21st century skills acquisition;
3. Lab 21, a hybrid hyper-structure for learning that is grounded in the principles of CoP, can provide a safe space for learning, sharing, and innovating;
4. Educators and trainers need to act as facilitators and activators within such a hyper-structure for learning where trust, peer learning, meaning-making, identity-building, sense of belonging, collaborative team spirit, and the continuous improvement of the shared practice are nurtured;
5. Deep learning—a high learning impact, positive learning outcomes, and learner satisfaction—can be obtained if learners get the possibility to emotionally connect with the topic at hand; the 2CG® multi-method teaching approach, which combines context-specific content and customized artistic impulses, can help learners to effectively acquire 21st century skills in the context of a real-life practice;
6. Learning needs to become a way of wellbeing; complexity and chaos need to be embraced as they allow for creativity and innovation; cultivating an optimistic mindset, making meaning, belonging, and constant exploring is part of the learning;
7. Project-based learning can be a fruitful way to develop 21st century skills as learners get the opportunity to engage in a real-life practice and deal with complex issues in a co-creative, collaborative, responsible way, thereby making progress, getting feedback, reflecting individually and with their peers;
8. As it is impossible to change the whole educational system at once, small hyper-curricular activities should aim at giving learners the opportunity to work on projects that they can control and implement; further applied research is needed to improve and standardize hyper-structures of learning;
9. Effective ways to measure 21st century skills acquisition and behavioral change, including qualitative and quantitative indicators and parameters, need to be further explored and developed with the help of comparative studies in multiple contexts;
10. In the 21st century, everyone needs to take the role of learner, just like everyone needs to experience and become aware that learning is a new way of being which can happen everywhere—in physical and digital environments; pilot projects and multidisciplinary research are needed to further explore and standardize the educational practice presented in this paper.

5 Concluding Remarks

How we design our learning environments—or our ecosystems of work—will define how we collaborate, how we support each other, and how we learn from each other. In most organizations, institutions of higher learning and schools, learning is still separated from the practice. More Lab 21 pilot projects and fieldwork are needed to further develop the 2CG® multi-method teaching approach and to find adequate ways to quantify learning outcomes and transformations of individual and organizational behaviors and structures. Effective 21st century skills acquisition that prepares learners to effect change in the real world and to contribute to sustainable and democratic societies needs to be embedded in hyper-curricular activities and anchored in hyper-structures for learning.

Acknowledgments. The author of this paper would like to express her thanks to the stakeholders in organizations, institutions of higher education, and schools who have supported the new way of teaching and learning. A special thanks goes to all the (working) learners for their ongoing commitment and invaluable contributions that have delivered important educational insights and many joyful moments of learning, sharing and innovating.

References

1. UNESCO. https://unevoc.unesco.org/e-forum/RethinkingEducation.pdf. Accessed 13 Mar 2021
2. Schumacher, I.: COVID-19 and Its Implications for Environmental Economics, IPAG Business School Paris, France, Springer (2020)
3. World Economic Forum: The skills needed in the 21st century, New Vision for Education (2015)
4. Van Laar, E., van Deursen, A., van Dijk, J., de Haan, J., The relation between 21st-century skills and digital skills: a systematic literature review. Elsevier Ltd. (2017)
5. Merl, C.: Fostering 21st century skills in engineering and business management students, ICL 2018: The Challenges of the Digital Transformation in Education, pp. 145–156 (2019)
6. Wenger-Trayner, E., Wenger-Trayner, B.: Learning to Make a Difference: Value Creation in Communities of Practice. Cambridge University Press, Cambridge (2020)
7. Sowa, H.: Gemeinsam Vorstellen lernen. Theorie und Didaktik der Kooperativen Vorstellungsbildung. In: Glas, A., Heinen, U., Krautz, J., Miller, M., Sowa, H., Uhlig, B. (eds.) München: kopaed, pp. 65–69 (2015)
8. McNiff, S.: Philosophical and practical foundations of artistic inquiry: creating paradigms, methods, and presentations based in art. In: Leavy, P. (ed.) Handbook of Arts-based Research, pp. 22–36. Guilford, New York (2017)
9. Kazanjian, C.J.: Hyper-curriculum: transcending borders of standardization in the cosmopolitan classroom. Int. J. Curriculu. Instruct. ERIC Number: EJ1207327 (2016)
10. Poetry in Business, Wie das Heute im Morgen entsteht (2018). https://issuu.com/christinamerl/docs/poetry_in_business_by_christina_mer. Accessed 14 Mar 2021
11. Homepage of Medium. https://medium.com/age-of-awareness/why-personalization-in-education-misses-the-point-aa1ac84048ed. Accessed 14 Mar 2021
12. Hess, E.: Hyper learning: how to adapt to the speed of change, Edward D. Hess (2020)
13. Weinberger, D.: Too Big to Know: Rethinking Knowledge Now that the Facts aren't the Facts, Experts are Everywhere, and the Smartest Person in the Room is the Room. Basic Books, New York (2011)
14. Lewin, K.: Action research and minority problems. J. Soc. Issues **2**(4), 34–36 (1946)
15. Laloux, F.: Reinventing organizations, a guide to creating organizations, Nelson Parker (2014)
16. Shelley, A.: Knowledge Succession, e-book, Business Expert Press (2015)

17. Healthy Minds Innovations. https://hminnovations.org/. Accessed 7 Mar 2021
18. Pakhalé, S.: Culture of Creation, Alberto Alessi E.a., Nai010 Publishers (2020)
19. The Conversation: https://theconversation.com/coronavirus-universities-are-shifting-classes-online-but-its-not-as-easy-as-it-sounds-133030. Accessed 14 Mar 2021

The Importance of Language and Communication Skills in the IT Industry in Bosnia and Herzegovina

DŽalila Muharemagić(✉)

University of Bihać, 77101 Bihać, Bosnia and Herzegovina
dzalila.muharemagic@unbi.ba

Abstract. Bosnia and Herzegovina belongs to a group of developing countries with high unemployment rate. In the recent period, IT industry in Bosnia and Herzegovina indicates an enormous growth rate with 100% employability of all graduated IT professionals from higher education institutions. The educational sector, on the other hand, does not respond to the demands of the IT industry, neither with the number of educated graduates nor with the adequate skills. Research conducted by UNDP and Bit Alliance in B&H indicates the need for improvement of necessary skills, which include both the language, as well as communication skills. Studies conducted within certain universities in B&H indicate that the specific language classes offered as English for Specific Purposes classes do not include most of the requirements of this methodology of language teaching. The study aims at investigating the frequency of foreign languages used for work purposes at some of the major IT companies in Bosnia and Herzegovina, the importance of language and communication skills for hiring a candidate, and the satisfaction of both the employees (former graduates) and employers with the language skills obtained during the university studies. The purpose of the study is to show the need for changes in curricular policies within universities related to the language and communication skills of future IT graduates and as a foundation for more thorough study of the specific language needs of IT companies in Bosnia and Herzegovina.

Keywords: IT industry · Language and communications skills · English for specific purposes

1 Introduction

Language and communication can be seen as an essential part of any human activity, meaning that the education in relevant professional skills should be complemented by development of language and communication skills. Although the statement sounds too obvious to be discussed, numerous studies [1–5] conducted in the world's largest economies indicate foreign language and communication skills gap between industry needs and the offerings of the educational system. In this sense, 40% of surveyed employers in EU [1] disagree with the statement that the national educational systems are providing graduates with sufficient foreign language skills necessary for the job. In France

D. Guralnick et al. (Eds.): TLIC 2021, LNNS 349, pp. 212–223, 2022.
https://doi.org/10.1007/978-3-030-90677-1_21

[2], 60% of employers face difficulties in employing candidates with adequate language skills due to inadequacies of the school and university education. Similarly, a study of the Confederation of British Industry [3] indicates that only 36% of employers are satisfied with their employees' language skills. Situation in United States shows similar trends [4], with 34% of employers reporting a foreign language skills gap between their business needs and what employees can offer.

Bosnia and Herzegovina (B&H) is not an exception to these trends. Being a post-war developing country with complex state hierarchy impacting both the economy and the educational system, a shortage of skilled workforce is inevitable. A synthesis report [6] of European Commission on Western Balkans higher education provision and labor market needs indicates a high level of gaps in foreign language skills — among others, inadequacy of a higher education system to provide graduates with adequate skills for easy transition to the labor market and the need for change in terms of both the content of the curriculum, as well as the teaching methodology in higher education. The same implications have been identified from several studies [7–9], specifically for the foreign language classes at some of the B&H universities through the perspective of English for Specific Purposes (ESP). Consequently or not, unemployment rates are steadily high (15.7%) [10], especially among recent graduates (37.1%) [6], causing brain drain and slow economic development. However, one industry branch has been indicating steady rise with 210% of employment growth since 2012 [11]. IT industry in B&H is showing positive figures in all aspects — employment growth is 10 times higher than in other industries, salary rate is 71.77% higher than the country's average, and it shows 201% to 1419% of revenue growth in the last 5 years. It would be expected that, with such figures, the government and educational sector are eager to fulfill the requirements of the industry in order to provide it with an adequate skilled work force. Although Council of Ministers of B&H adopted a policy on development of information society in B&H 2017–2021, and it was declared as strategic or industry of significant importance in numerous strategic documents throughout the country, several studies [6, 11] indicate a mismatch in skills provided by the educational sector and the needs of IT industry. It is stated that 41.9% of open job positions in IT industry in B&H remain vacant due to the lack of IT graduates in the country. The same studies show urgent need for modernization of curricula, promotion of student-centered teaching methodologies, and greater cooperation of higher education and industry in determination of labor market needs. Among the inadequate response to labor market needs are the insufficient professional skills, as well as the soft skills, including foreign language and communication.

Comparing the abovementioned situation with the premise from the beginning of the introduction, the following research questions were formulated for the purpose of this paper: what is the frequency of foreign language use in B&H IT companies; how important are the foreign language and communication skills for competitiveness of the companies; how important are these skills in the hiring process; and what is the level of satisfaction of both the employers and employees with language skills obtained during the formal education? The quantitative study involved 17 participants from 14 IT companies in B&H. By gathering data directly from the labor market, it is expected that the paper would contribute to raising the awareness of academia and policy makers on an urgent need for an overarching curricular and teaching methodology reforms in higher education of IT professionals in B&H.

1.1 Theoretical Background

Defining Foreign Language and Communication Skills. Foreign language and communication skills in terms of this paper are considered from the perspective of Canale and Swain's [12] definition of *communicative language competence*, since it focuses on grammatical, sociolinguistic and strategic skills, knowledge, and abilities. The definition can be seen as integrative in preparing second language learners to

"exploit – initially through aspects of sociolinguistic competence and strategic competence acquired through experience in communicative use of the first or dominant language – those grammatical features of the second language that are selected on the basis of, among other criteria, their grammatical and cognitive complexity, transparency with respect to communicative function, probability of use by native speakers, generalizability to different communicative functions and contexts, and relevance to the learners' communicative needs in the second language."

The grammatical aspect of the definition includes all the relevant linguistic components of the language, such as the

"knowledge of lexical items and of rules of morphology, syntax, sentence-grammar semantics, and phonology."

The sociocultural aspect on the other hand involves

"sociocultural rules of use and rules of discourse. (…) The primary focus of these rules is on the extent to which certain propositions and communicative functions are appropriate within given sociocultural context depending on contextual factors such as topic, role of participants, setting, and norms of interaction."

Finally, the strategic competence relates both

"to grammatical (e.g. how to paraphrase grammatical forms that one has not mastered or cannot recall momentarily) and those that relate more to sociolinguistic competence (e.g. various role-playing strategies, how to address strangers when unsure of their social status)."

Such definition covers several aspects of foreign language and communication skills education preparing future professionals with not only linguistic aspects of the language, but also focuses on contexts of language use, learners' needs, and use of authentic materials.

In line with such definition are also the findings of the Languages for Specific Purposes (LSP), which focus on the purpose of the language learning, rather than the language itself. This makes the LSP methodology specifically appropriate for education related to language use in the workplace, since such methodology uses the following criteria for development of language courses [13]: restriction to necessary skills, selection of language items which serve the purpose, themes and topics relevant for the learner's needs, and communicative needs reflecting the purpose of using the language.

The communication skills considered within the paper are taken as described by Reimer [14], including oral, listening, visual (non-verbal), interdisciplinary, and intercultural elements.

Importance of Foreign Language and Communication Skills for International Businesses - English for Specific Purposes Courses as Examples of Good Training Practices. As stated at the introduction, several studies [1–5] have been conducted on the topic of foreign language and communication skills and their relation to the labor market. The Making Languages our Business report from the United States labor market [4] indicates that 90% of U.S. employers express a high demand for foreign language skills. The same research shows that one in four businesses surveyed has reported loss of opportunity or a business partner due to the lack of foreign language capacities, while an ever-growing number of employers is relying on employees with language skills other than English to reach their business goals. The increased demand for foreign languages has been reported among EU employers, as well [15], indicating that a "*clear majority of employers believe that foreign languages add value to the company by supporting growth and by helping to build more effective relationships with customers.*" Another report from EU conducted in 2015 [16] shows that knowing one or more foreign languages has a strong connection with employability, with English and Russian having been reported as the most important languages. In terms of importance of English language, a survey from 2016 [17] conducted in countries where English is not an official language shows that 95% of surveyed employers consider that English skills are important. With regards to the IT industry, 73% of employers from the countries taking part in the research reported that English language is important for their organization.

In order to respond to the needs of the labor market in terms of foreign language skills, some higher education institutions have developed good examples of language courses. This mostly refers to ESP courses in higher education. A study [18] of ESP courses in eight European countries showed that most of the universities in these countries offer courses in LSP. Some good examples are universities where special departments or centers have been developed in order to provide organization of specific language courses for faculties and companies (Germany, Finland, and Sweden to some extent). Such examples are also present in Austria with technical universities offering education of "humanist engineers" [19] where department for Humanities serves solely to prepare and organize soft skills courses for engineering students incorporating practice of such skills into the ESP courses. Although not declared as ESP course, a good example of providing engineering students with 21st century can be found at another Austrian university [20] where several approaches of blended learning concepts have been introduced to incorporate professional skills into the curricula of engineering students. A response to labor market needs in terms of ESP courses can be found in Japanese universities, as well, where companies such as Nissan, Mazda, and Nippon have indicated a need for globalized workforce that could speak English. Positive example of response to these needs [21] is a state-of-the-art ESP course within the Center for English Language Education. The program provides future engineers with skills aimed at the constantly changing workplace — skills focusing on learning processes, team work, project management, discussion, negotiation, and debate — incorporating language learning in such courses.

In B&H, foreign language courses are traditionally offered at most study programs within universities. The importance of English language within higher education has grown significantly in the last decade, partially due to the increased number of student exchange programs, as well as due to the employment requirements [22]. Although many universities are offering ESP courses for different disciplines, studies [7, 8, 10, 23] have shown that these courses do not comply with ESP methodology and require improvement in terms of course content, teaching methodology, and approach to learners. The number of years studying English language also varies, with one or two semesters of obligatory language subjects as the most frequent offer. Information on cooperation of the educational sector and industry in defining language and communication skills is scarce, which offers potential for future research projects and initiatives.

2 The Importance of Language and Communication Skills for IT Industry in Bosnia and Herzegovina

2.1 Research

Methodology. Considering the research problem of the study, a quantitative study [24] was selected as the most appropriate research method in order to obtain objective data generalizable to the wider context. In line with the aim of covering the wider context of IT companies in B&H as the target population, purposive sampling was directed towards companies allied through the Bit Alliance association of IT companies in B&H, as well as companies which are not members of the alliance. No distinction in selection of companies was made in terms of number of employees, year of establishment, nor the type of ownership (domestic/international).

Instrument. The quantitative research was conducted by means of structured questionnaires measuring factual and attitudinal information on the respondents — in this case, managers and IT staff of the relevant companies. The questionnaires were formulated using online tools — more precisely, Google Forms — and the link was sent out to the companies via Bit Alliance newsletters, via contact e-mails available at the companies' web sites, as well as the private contacts of the author. The questionnaires were written in Bosnian language and consisted of total of 28 questions divided into four categories: general information on the respondents (company/sector, job title, number of years of work experience, academic degree, academic title, age and gender); information on the frequency of use and importance of foreign language and communication skills in the workplace (frequency of use, languages/communication skills used, importance for the job title, importance for competitiveness of the company, purpose of use); relevance of language and communication skills for hiring a candidate (level of relevance, how the skills are proven/tested, priority in employment); and information related to language and communications skills trainings offered through formal education (number of years learning foreign languages, languages/communication skills provided, usefulness for current job position, content of the language training provided, need for further informal training, need for improvement, language/communication skills needed). Depending on the purpose of the question (factual or attitudinal information), the questionnaire included short-response open-ended questions, as well as single and multiple-choice questions.

Participants and Data Collection. Data were collected in the period from January to beginning of March 2021, with 14 companies and total of 17 respondents taking part in the research. According to the data obtained and available on official websites, the companies are located in different parts of B&H including major cities such as Sarajevo, Banja Luka, and Mostar. The companies cover several areas of IT industry including software development, IT consulting and services, IT security solutions, and IT education. The collected responses were partially analyzed using SPSS software.

2.2 Results and Findings

The analysis of generated results in this paper is presented through four sections in line with the categories of questions listed in the methodology section.

Demographic Information. According to the data obtained in the general information section of the questionnaire, the respondents can be grouped according to the following job titles within the company: CEO, Founder, Chief Communications Officer, Programmer, Technical Lead, Educator, Professor, Project Manager, Content Writer and Mental Health Coach, and Trainee. Most respondents (over 80%) have 1–5 years of experience in the current job position. Regarding the degree obtained, 47.1% have a Master's degree, 41.2% Bachelor's degree, and 11.8% doctoral degree. According to the responses on academic title, the following titles can be identified: Bachelor of Information Technologies, Master of Information Technologies, Bachelor of Mathematics and Informatics, Master of Economy, Master of Psychology, Master of English Language Teaching, Android-IOS Developer, and Assistant Professor. With regards to the sex and age variables, 70.6% are male respondents, while 41.2% of respondents are of the age above 40, with 29.4% for both the age range between 26 and 35 and between 36 and 40.

Information on the Frequency of use and Importance of Foreign Language and Communication Skills in the Workplace. The first question in this section of the questionnaire was related to the frequency of foreign language use in the current job position, where 94.1% of respondents reported to use foreign language every day, with the same percentage for English as the foreign language used most frequently (Fig. 1).

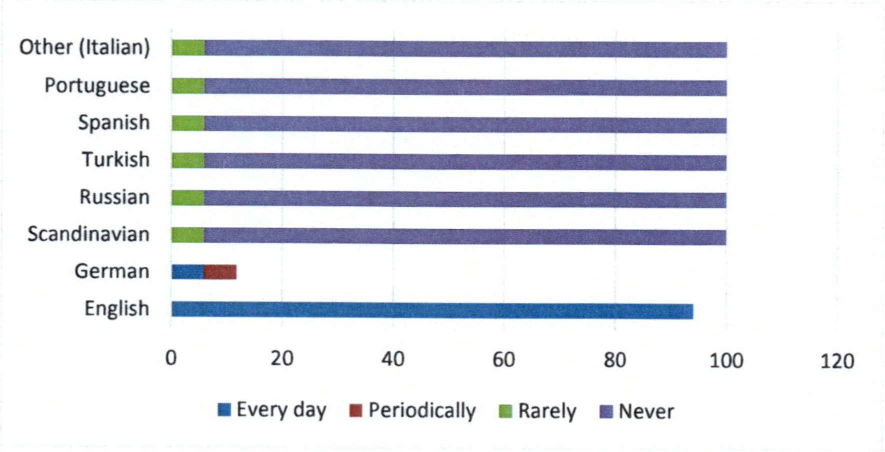

Fig. 1. Frequency of foreign languages use per language (%).

For the questions related to the importance of foreign languages, 88.2% consider the foreign languages to be highly important for the current job position and 11.8% important, while 94.1% consider it highly important for the competitiveness of the company and 5.4% important. With regards to the purpose of foreign language use in the workplace, 88.2% of respondents use foreign languages for communication with clients (Fig. 2).

Fig. 2. Purpose of foreign language use in workplace (%).

When referring to communication skills, 76.5% of respondents considers communication skills to be highly important for the current job position, 17.6% important, and

5.9% does not consider it important or unimportant. For the question on type of communication skills used in the workplace, Verbal communication, Written Communication, and Listening are used every day in most of the cases (Fig. 3).

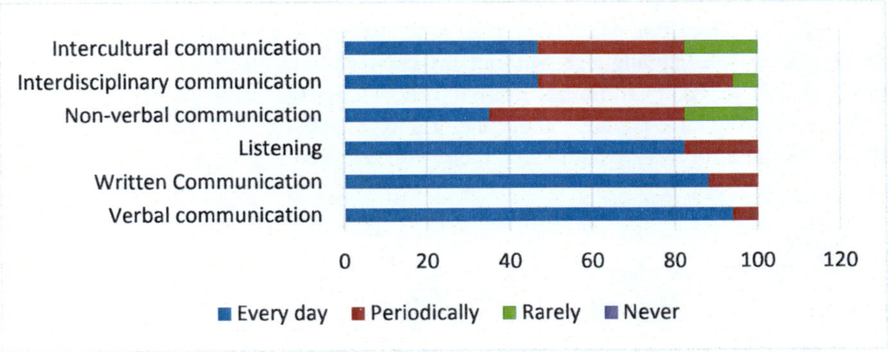

Fig. 3. Frequency of communication skills used in the workplace (%).

Relevance of Language and Communication Skills for Hiring a Candidate. When employing a candidate, 76.5% reported that foreign language skills are a compulsory skill, 17.6% reported that knowledge of foreign languages is desirable but not obligatory, and 5.9% reported it is not required at all. As a proof of foreign languages skills, 73.3% of respondents reported that the company organizes in-house testing (Fig. 4).

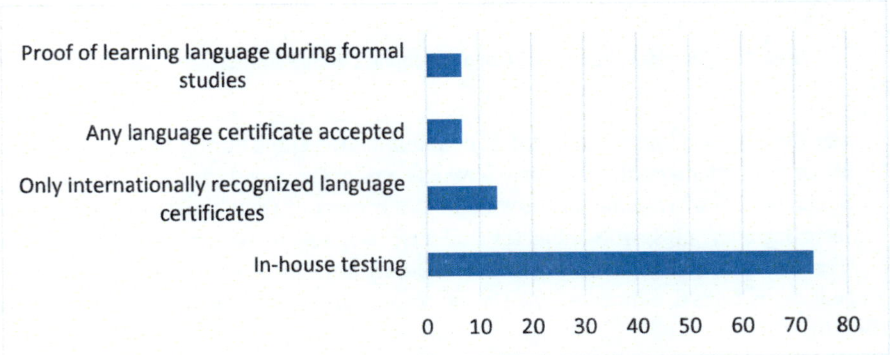

Fig. 4. How candidates are proving language skills (%).

In total, 88.2% of respondents confirmed that the foreign language skills are considered as an advantage when choosing between candidates with same professional qualifications.

Information Related to Language and Communications Skills Trainings Offered Through Formal Education. Most of the respondents studied foreign languages between 5 and 10 years (64.7%) and more than 10 years or living in a foreign country was reported by 35.3%. English was reported in 64.7% cases, both English and German in 29.4%, and only German in 5.9%. With regards to the usefulness of the acquired knowledge on foreign languages during formal education, 52.9% of respondents considers it to be highly useful. For the question on the content of the foreign language classes during formal education, 41.2% of the respondents reported having grammar and communication in foreign language, without any connection to the content language of the profession. In terms of additional informal foreign language trainings, 52.9% of respondents reported to have attended additional trainings, and 100% of these confirmed that the additional trainings have provided them with relevant knowledge (Fig. 5).

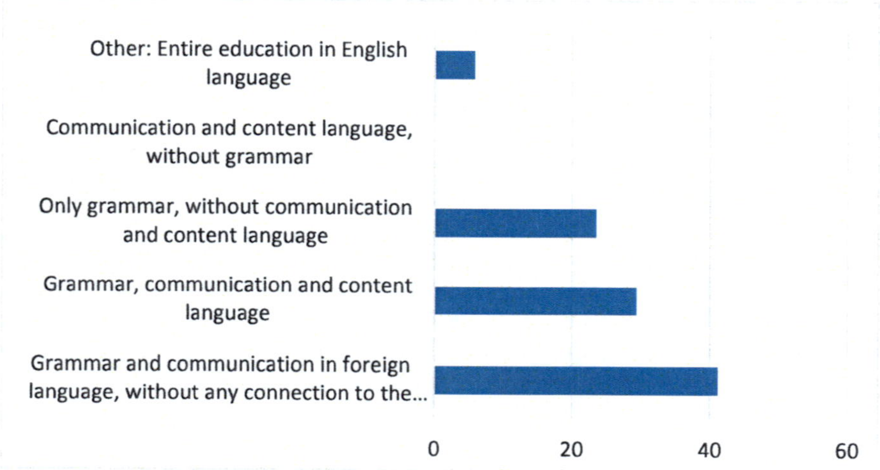

Fig. 5. Content of the foreign language courses during formal education (%).

For the question on development of communication skills during formal education, 52.9% of respondents confirmed not to have any type of training during formal education. However, 47.1% of respondents confirmed some form of training, including separate subjects during studies, workshops for students, assignments within other subjects, etc.

For the question regarding whether changes are necessary in terms of foreign languages in formal education, 94.1% of respondents provided a positive response (Fig. 6).

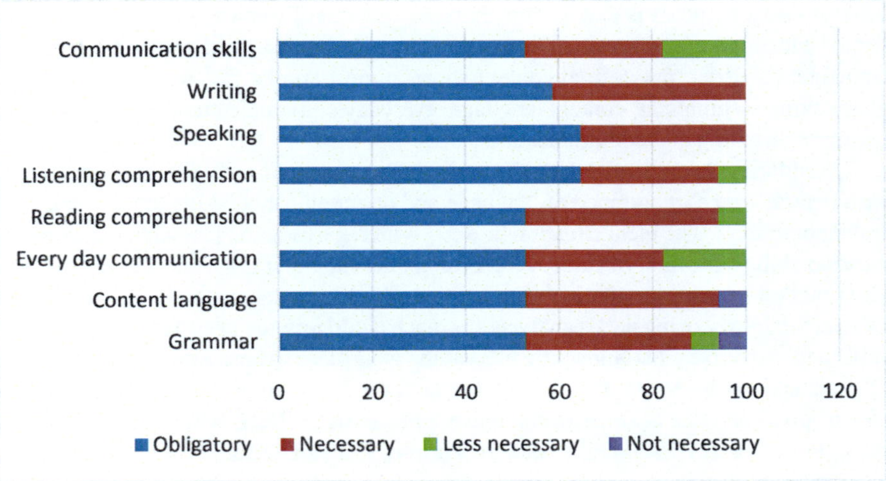

Fig. 6. Desired content of the foreign language courses in formal education (%).

3 Final Remarks and Recommendations

The paper starts with premise that the language and communication skills can be seen as an integral part of any human activity, confronting it to the situation on the global level where the offer and demand in terms of language and communication skills education are in mismatch. The employers worldwide are reporting high demand for foreign languages and communication skills in reaching their companies' international goals. On the other hand, national educational systems are falling behind in providing graduates with adequate skills. As B&H is dealing with slow economy growth and high unemployment rates in all sectors except the IT sector, searching to respond to educational demands of this industry seems to be utterly important. Through the quantitative survey of 14 IT companies in B&H, the study aimed to contribute to raising awareness of the industry's needs in terms of foreign language and communication skills. Based on the acquired data, the following conclusions can be made:

Foreign language and communication skills are used on an everyday basis in most of the companies, with English being the language number one in communication with clients, development of new products and services, and professional development of employees. The skills have also been confirmed as highly important for the competitiveness of the companies. The companies reported language skills to be mostly compulsory in the employment process, with advantage for candidates with good language skills. This information is of great value, considering the unemployment rates in the country and struggle of the recent graduates to enter the labor market due to the lack of experience.

Verbal communication, written communication, and listening are reported as the most frequently used communication skills in the workplace, while according to the responses, the educational system is not providing education for these skills in adequate manner. The system also lags in terms of providing appropriate content of the foreign language classes, as most of the respondents believe changes are necessary in this area.

What is also significant is that the mostly reported form of foreign languages courses in formal education offers grammar and communication content, without the professional language. However, the desired content shows almost similar requirements of all language skills — grammar, content language, everyday communication, reading, writing, speaking, listening and communication skills.

Considering the collected data, the findings show compliance with the global trends with regards to the importance of language and communication skills to the workplace and labor market. The ideal situation in B&H with regards to the language and communication skills education of future IT professionals should at first start with identifying labor market needs through close cooperation of educational sector and industry. The LSP methodology using needs analysis approach could provide basis for the insight into skills and discourse most frequently used at the workplace, where active involvement of IT companies is essential. Besides, higher education institutions should follow examples of good practice at some of the universities outside B&H, which have efficiently incorporated professional skills such as learning strategies, team work, critical thinking, problem-solving, and project management into language courses. A shift has to be made from traditional general language courses to more target-need-oriented courses. Formation of centers for language education which would offer custom-made courses in line with labor market needs spreading over entire period of study would enable steady and up-to-date development of language and communication skills of future IT professionals.

Finally, it could be stated that the both the importance of IT sector for B&H economy and importance of language and communication skills for the industry are evident and should receive urgent attention from government and academia.

References

1. Beadle, S., Humburg, M., Smith, R., Vale, P.: Study on Foreign Language Proficiency and Employability Final Report. European Commission, Brussels (2015)
2. Analysis of the foreign language skills requirements of French employers. https://www.france-education-international.fr/sites/default/files/atoms/files/lemp-summary-survey-report.pdf. Accessed 13 Mar 2021
3. Tinsley, T., Board, K.: Languages for the Future. British Council (2014)
4. IPSOS Public Affairs: Making Languages Our Business. American Council on the Teaching of Foreign Languages (2019). https://www.leadwithlanguages.org/wp-content/uploads/MakingLanguagesOurBusiness_FullReport.pdf. Accessed 13 Mar 2021
5. Green, C., Evans, S.: A survey of language use in the professional workplace in Hong Kong. Hong Kong J. Appl. Linguist. **5**(2), 60–82 (2000)
6. Bartlett, W., Uvalić, M., Durazzi, N., Monastiriotis, V., Sene, T.: Higher Education Provision and Labour Market Needs in the Western Balkans Synthesis Report. European Union (2016)
7. Babić, Ž: Exploring ESP syllabi some implications for further pedagogical considerations. Zbornik Inst. Pedagoška Istraživanja **48**(1), 164–179 (2016)
8. Dabić, T.: Analiza potreba u okviru engleskog jezika za posebne namjene. Univerzitet u Istočnom Sarajevu: Radovi Filozofskog fakulteta, Filološke nauke **13**, 443–455 (2011)
9. Muharemagić, DŽ.: Specific features of ESP classes in developed and developing countries a multiple case study in Austria and B&H. In: English for Specific Purposes a Multidimensional Challenge Conference Proceedings, pp. 45–53. University of Applied Sciences Technikum, Wien (2018)

10. International Labour Organization. https://www.ilo.org/budapest/countries-covered/bosnia-herzegovina/WCMS_471903/lang--en/index.htm. Accessed 13 Mar 2021
11. Ekonomski Institut Sarajevo. Analiza potrebnih vještina za softversku industriju u Bosni i Hercegovini (2019). https://www.ba.undp.org/content/dam/bosnia_and_herzegovina/docs/Analiza_potrebnih_vjestina_za_softversku_industriju_u_Bosni_i_Hercegovini_2019.pdf. Accessed 13 Mar 2021
12. Canale, M.: Theoretical bases of communicative approaches to second language teaching and testing. Appl. Linguis. **1**(1), 1–47 (1980)
13. Strevens, P.: Special purpose language learning: a perspective. Lang. Teach. Linguist. **10**(3), 145–163 (1977)
14. Reimer, J.: Communication skills for the 21st century engineer. Glob. J. Eng. Educ. **11**(1), 89–100 (2007)
15. European Strategic Framework for Education and Training. https://ec.europa.eu/assets/eac/languages/policy/strategic-framework/documents/languages-for-jobs-report_en.pdf. Accessed 13 Mar 2021
16. European Commission Joint Research Centre: Languages and Employability (2015). https://ec.europa.eu/jrc/en/publication/eur-scientific-and-technical-research-reports/languages-and-employability. Accessed 13 Mar 2021
17. Cambridge English Language Assessment: English at Work Global Analysis of Language and Skills in the Workplace (2016). https://www.cambridgeenglish.org/images/english-at-work-full-report.pdf. Accessed 13 Mar 2021
18. Fortanet-Gómez, I., Räisänen, C.A. (Eds.). ESP in European Higher Education: Integrating Language and content. John Benjamins B.V., Amsterdam (2008)
19. Maderdonner, O.: The Humanist Engineer. In: Proceedings of the 6th Austrian UAS Language Instructors' Conference, Vienna. UAS Technikum Wien, Vienna (2010)
20. Schefer-Wenzl, S., Miladinovic, I.: Developing 21st century skills in engineering studies with E-learning. Int. Conf. E-Learn. Workplace **2020**, 1–3 (2020)
21. Anthony, L.: The Waseda University CELESE program: a large-scale, centralised ESP program for scientists and engineers. In: Proceedings of the 40th Annual Meeting of the English Language Education Society of Japan, pp. 39–46. Waseda University, Tokyo (2010)
22. Delibegović-Džanic, N., Imamović, A.: The Status of English in Bosnia and Herzegovina: past and present. In: The Status of English in Bosnia and Herzegovina Buckingham, L. (Ed.). Multilingual Matters (2016). https://doi.org/10.21832/9781783095971
23. Obralić, N., Akbarov, A.: A case study: the significance of the ESP in-session course at International Burch University. J. Foreign Lang. Teach. Appl. Linguist. **1**(1), 191–205 (2014). https://doi.org/10.14706/JFLTAL141113
24. Dornyei, Z.: Research Methods in Applied Linguistics. Oxford University Press, Oxford (2007)

Taking Project-Based Learning Online

Gary Natriello[1](✉) and Hui Soo Chae[2]

[1] Teachers College, Columbia University, New York, NY 10027, USA
gjn6@tc.columbia.edu
[2] New York University, New York, NY 10003, USA
huisoo.chae@gmail.com

Abstract. We present a case study of a university course using project-based learning, collaborative small teams, and the Scrum project management framework. The course was moved entirely online in Fall 2020 due to COVID-19. Students in the course were unable to come to campus and so were distributed widely across time zones with concentrations in North America and Asia. Students had access to a set of tools arranged specifically for the class, including: a WordPress website, the LearnDash learning platform to organize resources and activities, Slack for class and for individual teams to facilitate communications, and Asana for individual teams to track project progress. Students also brought other tools of their choice (e.g., Google docs, WeChat, Notion). None of the tools were institutionally based, with the exception of an institutional Zoom account that was used for six whole class sessions as well as by individual teams for their collaboration meetings. Students were assessed based on team projects, as well as individual short papers. Additional data from Slack and Asana allowed the instructors to follow the progress of each team throughout the semester and to offer guidance along the way. We review the course activity data, examine areas for improvement, and pose suggestions for improving the learning experience.

Keywords: Project-based · Online learning · Case

1 Introduction

In this paper, we discuss our experience offering a course in an online format as a result of the COVID-19 pandemic and the shift to online learning during the Fall 2020 semester. Ironically, the course was on networked and online learning offered for masters and doctoral students. Although the course had been offered in both online and on campus formats over the years, for the past several offerings, it had been offered on campus to take advantage of a robust teaching environment that supported a variety of technologies and allowed students to assemble and re-assemble in multiple team configurations. In its most recent offering the course had evolved into a fully project-based [3] and small team oriented experience [2]. We had found these arrangements conducive to providing a demanding, high engagement learning experience, and our goal as we planned to move the course online was to create an experience equal or superior to the on campus experience.

D. Guralnick et al. (Eds.): TLIC 2021, LNNS 349, pp. 224–236, 2022.
https://doi.org/10.1007/978-3-030-90677-1_22

1.1 The Challenge

The Fall 2020 semester presented several unique challenges that we needed to address as we re-designed the course. First, as a result of the pandemic, the 21 students enrolled in the class were spread over multiple time zones, with concentrations of students in North America and Asia. This meant that the time difference for many students in Asia was 12 or more hours from eastern time. A few students were in other time zones in the eastern hemisphere. One of our immediate design challenges was to provide learning experiences that were not limited to times when it was difficult for students to participate, a challenge made more difficult by students who were working full-time, some in North America and others in Asia.

Second, for many of our students, Fall 2020 was their first semester of graduate studies with no prior experience on the campus as a member of the college and university community. This meant that the students had no attachment to the university or to their peers. Our challenge as course instructors was to create the social connections that make graduate and professional school more than the sum of classes even as many of our students were living with parents and/or their children in family environments.

2 Course Design

2.1 Course Site

With the unique challenges of Fall 2020 in mind, we designed the online learning experience from the ground up, looking for hidden opportunities to make the most of the times. We began by creating a custom WordPress website at latrlab.com that allowed us the flexibility to expose students to current issues in the area as well as an opportunity to learn about the instructors in a context less limited than a typical course platform setting. At the site, we posted items related to distance learning and related topics such as self-directed learning. In addition, the site featured books of interest, music we were listening to, and recipes for dishes one of us was cooking while homebound. Although we could not invite students to share a meal as we might do on campus, we tried to share interests beyond the course in the hopes of creating some informal connections.

2.2 LMS

To organize course materials for the semester we used the LearnDash WordPress plugin which provided LMS functionality. To set a tone for the semester designed to evoke a sense of student agency and self-direction, we changed the titles of the levels of content in LearnDash from courses, lessons, and topics to environments, experiences, and resources. These three levels of content provided sufficient options for organizing the content of the class. At the homepage level, we re-branded the area holding the courses (now termed "environments") as "journeys" to create a sense of adventure.

The environment level provided information on the activities for the entire semester. The semester was organized around four projects that were presented at the experience level. Readings, media, and other materials needed for the semester were provided at the resource level, organized under each project.

2.3 Small Learning Groups

A key feature of the course was the organization of students into small learning groups or teams. Our practice when teaching on campus had been to create new teams for each project so that students would have an opportunity to work with everyone in the class. With the move online, we decided to form teams at the beginning of the semester and keep them stable throughout all four projects. We believed that the online experience would not allow students to become familiar with their peers as quickly as they might on campus, so we hoped that working closely with a small team on successive projects would lead to a deeper collaborative experience, even at a distance. In designing the team experience, we settled on teams of 4 or 5 members based on the consensus of researchers in this area [1]. For the most part, students were assigned to teams with other members in their time zone to make coordination easier.

With the team membership stable throughout the four projects for the semester, we assigned students to roles that rotated with each project. The four roles were intended to promote the sharing of responsibilities and equalize the workload. They were also designed to create some internal monitoring of the work to ensure that it was done thoroughly. The product owner and Scrum master roles were taken from the standard Scrum process. The resource monitor and deliverable agent roles were specific to the projects.

Product Owner

- Clearly expresses product backlog items
- Orders the items in the product backlog to achieve project goals
- Ensures that the product backlog is visible, transparent, and clear to all
- Shows what the team will work on next
- Ensures that the team understands items in the product backlog

Scrum Master

- Facilitates meetings, conversations, and improvements
- Keeps the team focused
- Leads without authority and puts the team first
- Reinforces agile principles throughout the project

Resource Monitor

- Coordinates coverage of resources
- Manages resources summaries or other artifacts
- Checks to see that resources are utilized in project deliverables

Deliverable Agent

- Coordinates the preparation of project deliverables
- Prepares Files for Upload
- Uploads Team deliverables

2.4 Tools to Support Interaction – Zoom, Slack, Asana

Because we intended for the course to be highly interactive, and we wanted that interaction to be robust both among instructors and students and among the students themselves, we arranged for several tools to facilitate the interactions and collaboration. We made use of the college Zoom software for synchronous sessions of the entire class. We limited these sessions to a total of six, one each at the beginning and end of the semester, and one to introduce each of the four projects that also served as the session for sharing student work on the just concluded project. Because of the diverse time zones, we scheduled these sessions either in the evenings or early mornings, eastern time. This allowed students in Asia to participate in the sessions even with the 12-h time difference. Although at times student work obligations prevented them from participating in the whole-class synchronous sessions, most students were able to join these sessions. The synchronous sessions were not mandatory, and they were not recorded.

Zoom was available for all members of the class, and student teams were encouraged to use it for their synchronous meetings. In addition, they were encouraged to invite members of the instructional team to join these team meetings whenever their help or advice was needed. The teams held such meetings routinely and invited instructors to join at least one team meeting for each project.

To facilitate text communication for the entire class, a class Slack was activated. This allowed instant communications, as well as the sharing of information, resources, and other documents. In addition to whole-class communications, each team had its own slack channel to facilitate communications at the team level.

With the course heavily dependent on teamwork, we decided to provide scaffolding for the team activities in the form of the Scrum project framework [5]. We had used elements of Scrum in earlier classes, but we had never used it in distance mode, so this was even more of an experiment than other features of the course design. To support the use of Scrum by the teams, we provided team workspaces in Asana that allowed each team to manage their projects.

2.5 Assessment

Our approach to assessment was designed to advance our goal of helping students to become competent evaluators of their own progress and performance. We also aimed to have all work in the course be authentic, that is, we wanted the work to be like the work that students would encounter in work situations. We wished to avoid the kind of work only found in class assignments.

Accordingly, we used a combination of individual and group evaluations. The four small group or team projects were counted for 60% of the final grade. Group projects each had a clear deliverable due at the end of the project. Group projects were assessed based on achievement of project goals, use of resources to inform the project designs, and documentation of project work.

Along with the team product, for each project, individual students were required to prepare a reflection on their learning in the project. These reflections were counted for 20% of the final grade and were assessed on the analytic treatment of course materials and connections made between coursework and individual learning goals. This assignment was intended to provide students with experience reflecting on their learning to establish what we hoped would become a personal habit [4].

Participation was counted for 20% of the final grade. Participation was assessed based on contributions to class activities and discussions and support of other students through helpful commentary on their work. Such support might be for other students individually and for teams. To create a strong climate for collaboration, we emphasized that each student had an obligation to make sure that all of their colleagues in the class did well.

3 The Course in Operation

3.1 Beginning

With many students new to distance learning and many also new to graduate school, we anticipated that there would be rough spots getting everyone oriented at the start. To minimize the problems, we spent a good deal of time communicating with enrolled students prior to the start of the semester. Our communications included gathering some basic information from the students, as well as coaching them to register accounts on the systems we planned to use. In some cases, this outreach worked well, but in other cases, it proved more difficult. In addition to the typical problems that might affect any semester, in this semester, we had some issues regarding network connectivity — particularly for students in Asia. We also had a good deal of enrollment churn as students sought a course load that would work for their schedules. This made initial scheduling of class sessions uncertain as we added and lost students in different time zones.

Although we devoted a good deal of attention to orienting students, we may have underestimated the challenge our unique design posed for some students. After all, we were introducing the project method, small teams, and the agile framework for the first time. Nonetheless, after a few weeks, everyone developed an understanding of how the course would operate. The team organization helped in this regard as team members assisted their teammates. Fortunately, our project sequence included consideration of start-up issues.

3.2 The Four Projects

The projects were the center of the learning experience. We developed four design projects that engaged the teams in designing a solution to the challenge presented. The four projects were intended to move students through the course material as their teams worked to understand the project, review the resources provided, draw on their own experiences and additional resources they brought to the task, and develop a solution. The projects were sequenced so that earlier projects provided important experiences that could be applied to later projects. The four projects were summarized as follows:

Project 1 – Design a Theory of Learning

Introduction - This project introduces the course and its participants, the sociological perspective that will be used to frame issues, the concept of learning as it will be used throughout the semester, and the contemporary context.
Design Task - For this project, your team will design a theory of learning to serve as a foundation for all of your work in the course.
Deliverables - Create a one-page graphic exhibit illustrating the team's learning theory. Create a one minute video discussing the exhibit.

Project 2 – Design an Online Version of an Existing Degree Program

Introduction - This project covers social forms developed for online learning through borrowing from those developed prior to the information age. These include courses, schools, work-related venues such as corporations and professional associations, and libraries.
Design Task - For this project, your team will design an online learning experience by drawing on existing forms of education developed prior to the information age. Your team has been hired by the director of the Learning Analytics Program at Teachers College to create an online version of the program. Your goal is to create a program that offers students in the program a superior learning experience, better than the campus version of the program.
Deliverables - Create a new program description and related materials for prospective students to inform them about the program. Identify one thing that you would like to teach or share to illustrate your online version of the program and create an online element (e.g., a lesson, module, experience) for the one thing you would like to teach or share.

Project 3 – Design an Online Learning Community for a School

Introduction - This project examines those social forms that are emerging with the transition to the post-industrial or information age. In most cases, these forms could not emerge in their current form in earlier eras.
Design Task - Your team has been hired by the Executive Director of the Center for Academic Excellence and Support at the NYU School of Professional Studies to create an online learning community that will become the central hub for the school, encompassing faculty, staff, students, and partners in the professions represented at the school.

Deliverables - Submit a list of resources on emerging forms that your team reviewed. Create a data plan identifying the data elements and anticipated analyses that would allow you to assess engagement and participation. Develop a webinar script and slides to introduce students to the online learning community along with feedback from a trial run of the webinar with another team.

Project 4 – Design a Global Online Learning Network

Introduction - This project asks you to consider issues that arise when online learning activities span existing learning organizations.
Design Task - Your team has been hired by a newly formed global network of universities to create an online learning ecosystem to support the work of the network and the leaders of the member universities in the network.
Deliverables - Create a 5- to 10-page design document that describes the details of your proposed online learning ecosystem. Create a plan for gathering data that will allow you to determine the effectiveness of your design. Create a Pecha Kucha style (20 slides/20 s per slide for a total of 6 min, 40 s) presentation on your proposal for the global learning network.

The first project was designed to create a foundation for the work of the teams by allowing them to get to know one another while discussing their own approach to learning. This seemingly simple project was really a vehicle for teams to practice teamwork and experience the agile framework for the first time. Teams were able to draw on their product from the first project, their learning theory, for the remainder of the semester.

With each subsequent project, the difficulty and complexity of the work increased as the teams moved through program level, organizational level, and multi-organizational level initiatives. As the scope increased, teams were able to draw on earlier projects to inform work on the later projects. The deliverables for the projects were varied to give students exposure to different ways of summarizing and presenting their work.

3.3 The Teams at Work

As the teams worked on the projects, we discovered that, in addition to the tools and applications we had provided, students brought other tools into the mix. These included WeChat and WhatsApp for communications, Google Docs for collaboration, and Google Sites, Google Slides, and PowerPoint and Keynote for presentations. We accommodated these tools as much as possible while reminding students that they needed to record their ongoing work and their final products in formats that could be shared with the instructional team and with other students who might not have ready access to individual tools. We reminded students to document their ongoing work by connecting it to the applications we had provided for everyone. So, for example, we asked them to capture interactions in Slack and to connect work products to Asana. The teams were reasonably, but not perfectly, responsive to this request.

Our efforts to have the teams use the Scrum framework as a way to structure their work process were less successful than we had hoped. We had intended for the teams to use the Scrum roles and rituals throughout the semester and to produce several iterations for each project. We relied on those team members in the product owner and Scrum master roles to organize and reinforce the work arrangements, but several factors seemed to reduce their effectiveness. First, the students did not grasp fully the responsibilities of the roles. This was compounded by the fact that we were asking them to play directing roles with peers they just met. Embracing and faithfully enacting Scrum is challenging even for established teams, so we might have anticipated the difficulty in our context. Second, the first project was relatively straightforward, and the teams were small and required little in the way of structure. When a team of four or five sizes up a project and concludes it can be done with just some good conversations, it is difficult to motivate them to become engaged in more extensive role play, despite our telling them that they should practice the Scrum framework on a simple project because they will need it as the projects become more complex. Third, working at a distance imposed some additional barriers to implementing the method that seemed to require a daily communication even if brief. Fourth, the time frame for the projects of 3 or 4 weeks, although not too short for teams working full-time, may have been too short for our students carrying a full load of classes and, in some cases, holding full-time jobs in addition to our class.

3.4 Data on Team Work

We gathered data on the work of the teams in multiple ways. For each of the four projects in the course, students were asked to submit a self-directed learning reflection that included a section on their experience working on their team. In addition, both Slack and Asana allowed us to gather data on the work habits of the teams over the entire semester.

Figure 1 displays the cumulative number of total messages recorded throughout the semester. From about the sixth of September, the start of the semester, until the fifteenth of December, the end of the semester, there is a fairly constant upward slope, indicating

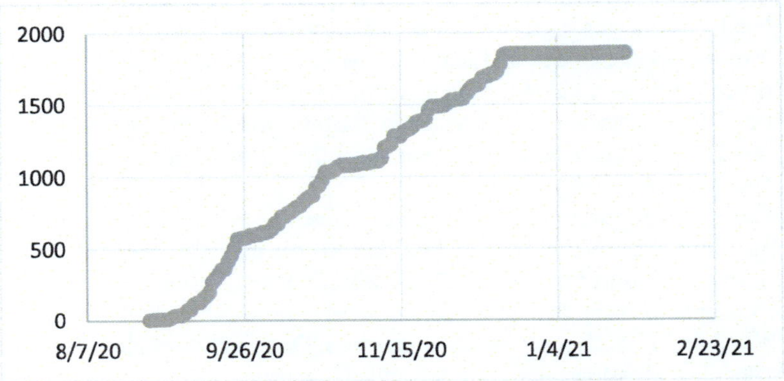

Fig. 1. Cumulative slack messages posted by date.

a steady accumulation of messages among team members. Over the semester the teams registered 1856 messages in Slack.

Figure 2 provides a breakdown of the Slack messages. Teams 1, 3, and 4 made extensive use of Slack for communications, accumulating hundreds of messages over the semester. Teams 2 and 5 rarely used Slack. The disparity in team use of Slack appears to be the result of Teams 2 and 5, those teams with all members in Asia, relying on WeChat, the most popular messaging application in China. As noted earlier, although we provided Slack and encouraged teams to use it, we did not prohibit the use of other applications for communications.

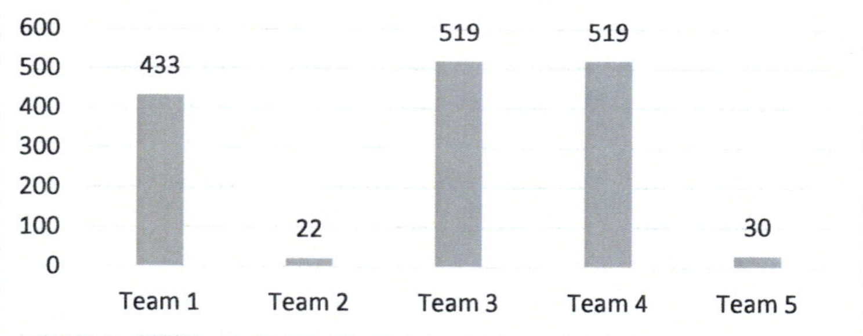

Fig. 2. Total slack messages by team.

We asked the teams to use Asana to define their tasks for each project and to track their progress. To understand a bit about how the teams approached their work on the four projects, we used Asana to record the number of tasks and subtasks defined for each of the projects by each of the five teams. Figure 3 displays the number of tasks defined and tracked by each team for each project.

Never having used Asana for teams working remotely, we had little idea of how the teams would actually use it to guide and document their project work beyond a few initial hunches. For example, because the projects grow in complexity over the semester, we anticipated that the teamwork plans would reflect this growing complexity by developing more detailed plans with a greater number of tasks and subtasks. As Fig. 3 shows, our hunch was not borne out. Teams 1 and 5 appear to have made less use of Asana as the projects became more complex. In fact, Team 1 made no use of Asana for the fourth project, and Team 5 made no use of Asana for the third and fourth projects. The situation for Teams 2, 3, and 4 was more varied and these three teams defined and tracked tasks for all four projects.

We also encouraged teams to schedule team meetings with the instructional team at any point in the term. We told the students to consider the instructors their consultants who were there to ensure their success. These team-initiated sessions were an opportunity for the students to share their progress and receive feedback and advice. The student teams scheduled 11 of these consultative sessions. One team scheduled 4 sessions, one team scheduled 3 sessions, one team scheduled 2 sessions, and two teams scheduled a single session.

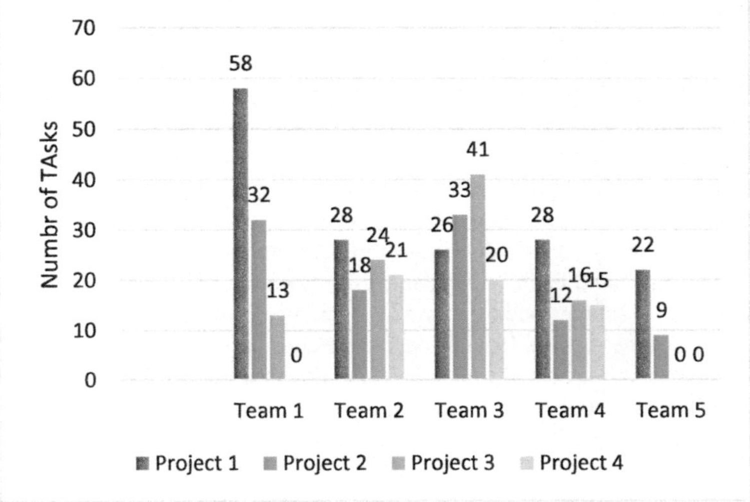

Fig. 3. Number of Asana tasks by team and project.

4 Avenues for Improvement/Refinement – What We Learned During the Semester

We approached the Fall 2020 semester as an experiment in moving our project-based course online. We made many adjustments in an attempt to optimize the course for the online format. The result was a fairly successful offering as judged by the positive comments in the student learning reflections submitted throughout the semester and by the student course evaluations collected at the end of the semester, which were equal to or better than the evaluations from prior offerings of the course on campus. We learned a great deal about ways in which the course and its various features operates online, and we identified a number of avenues for further investigation and possible improvements.

4.1 Instructor Workload

Although the work of the instructor in a project-based course tends to be more extensive than in a more traditional lecture course, we found the work load of an online project-based course even more demanding. To keep each of the teams engaged and actively learning, we found it necessary to do a great deal of coaching. This coaching took place through extensive communications via Slack and during the meetings called by the teams with the instructors. In our particular case, the demands of online delivery were compounded by having students in different time zones, which resulted in communications about the course going on at all hours. We recognize that some of the additional time required was the result of our approach that placed the instructors in the role of consultants providing advice and assistance throughout the four projects. We expect that some of the adjustments in the course that we discuss below will relieve some of the workload pressure for instructors.

4.2 Start-Up Issues

We encountered issues at the beginning of the term that delayed the start of work on the first project. Although we provided instructions to students explaining how to register for the applications that we would use for the term, some students seemed to struggle to follow them. As a result, it took longer than ideal for the enrollment in the course to be clear, and it required a good deal of effort to make sure that everyone was connected to the course resources. For future offerings of the course, we are considering a more structured set of instructions delivered both in text and at a synchronous whole class session. The goal would be for all students to demonstrate their connection to required applications and resources and to provide biographical information for a course roster before the end of the session. In addition, we are considering a short video to introduce the course format and explain the need for students to engage quickly. While these steps should improve the start-up experience, we intend to keep the initial project relatively simple as a means of familiarizing students with the process by experiencing it.

4.3 Re-thinking the Tool Set

Although the tools selected for the class worked well, there may be some advantages to reducing the number of different tools that students needed to learn to use. Students made good use of Slack for communications, but the teams had a more difficult time integrating Asana into their work. This may be the result of tool overload or it may be connected to the difficulty of using Scrum as a framework for their work.

4.4 Re-positioning Scrum as a Framework for Project Work

The teams varied in their ability to employ the Scrum framework to shape their work on the projects. This appears to be the result of at least three factors. First, there is the known difficulty of mastering the elements of the framework. Second, the fact that our teams were coming together for the first time and working remotely made the challenge greater. Third, because our students were not working full-time on our class and had other calls on their time in the form of other classes, jobs, and home responsibilities, they had less time to devote to mastering the elements of Scrum. We are considering several ways to address this issue. On the one hand, we think we can develop a modified limited Scrum approach more suited to the constraints of a class. On the other hand, we can develop more detailed instructional materials and scripts to make learning the framework easier for distance students. Finally, we may restructure the four projects to make them more iterative to take advantage of that aspect of Scrum.

4.5 Enhancing the Structure/Protecting the Innovation Space

When this course was designed for on campus delivery, we minimized the structure to provide sufficient space and opportunity for students to be imaginative and innovative in how they approached the work for the term. This worked well because we were readily available for extensive discussion and coaching at each point. In fact, creating conditions where students needed to ask questions of the instructional team was part of our pedagogical approach, since we thought the student-generated questions would lead to greater learning. When we re-designed the course for students at a distance, we recognized that more structure might be necessary, and we enhanced the structure accordingly in the form of more explicit directions for both procedural and content elements of the course. The level of structure worked well for some students, but others were confused. For those in the latter group who asked questions, the level of structure was sufficient when combined with the resulting discussions with the instructional team. However for those students who were confused and failed to ask questions, the level of structure proved problematic. For future offerings of the course, we are considering the addition of some "learn more" sections where students can access more structure or more detailed guidance immediately online.

4.6 Making Greater Use of Real-Time Performance Data

Since we hoped to provide students with an experience doing authentic work, our message to them was to pay as much attention to how they did the work in the class as they did to their final products. We wanted students to work collaboratively and steadily throughout the semester and to build expertise and competence as they moved from project to project. Although we provided a good deal of informal feedback throughout the term, we refrained from assigning grades until the end of the class. Most students accepted this approach, but they remained somewhat more reliant on instructor assessments than we had hoped. We aimed to help them to become evaluators of their own performance and the performance of their teams. As we look to the next offering of the course, we plan to make greater use of the data generated by applications and tools in use to provide students with feedback on their performance in real time. For example, we might make summaries of teamwork reflected in Slack and Asana accessible to all on a dashboard. This initial online offering provided opportunities to examine the data generated that might be shared throughout the term.

5 Summary

Despite the challenges, taking our project-based course online proved very rewarding both for us and for our students. We achieved our goal of helping students master the subject matter at a distance while also developing new collaboration and design skills. All members of the class made it successfully through the semester despite the pandemic and the necessary arrangements of living and working at a distance. All of the teams succeeded in completing the four projects for the semester. And the instructional team lived to tell this story!

References

1. Hare, A.: Chapter 10: Group size. In: Hare, A. (ed.). Handbook of Small Group Research, 2nd edn, pp. 214–231. Free Press, New York (1976)
2. Jaques, D., Salmon, G.: Learning in Groups: A Handbook for Face-to-Face and Online Environments. Routledge, New York (2007)

3. Krajcik, J., Shin, N.: Project-based learning. In: Sawyer, R. (ed.) The Cambridge Handbook of the Learning Sciences, 2nd edn., pp. 275–297. Cambridge University Press, New York (2014)
4. Loo, R., Thorpe, K.: Using reflective learning journals to improve individual and team performance. Team Perform. Manag. **8**(5/6), 134–139 (2002)
5. Pope-Ruark, R.: We Scrum every day: using Scrum project management framework for group projects. Coll. Teach. **60**(4), 164–169 (2012)

Instructional Design in a Digital Age: A New Model Is Taking Hold

Lisa M. D. Owens[1] and Crystal Kadakia[2](✉)

[1] Learning Cluster Design Group, Cape Coral, FL, USA
LMDOwens@tdStrategies.net
[2] Learning Cluster Design Group, Atlanta, GA, USA
Crystal@LearningClusterDesign.com
http://www.learningclusterdesign.com

Abstract. For over 50 years, instructional design and development models for training have centered around the goal of designing one deliverable for a given learning challenge. Though learning is becoming more bite-sized, these bits are often considered accessories or in support of a formal program like a course or class. The Learning Cluster Design model, in use since 2016, provides an approach that re-examines traditional assumptions and evolves learning and development (L&D) for today's digitally enabled culture. The model centers around a new goal: delivering Learning Clusters, a set of learning assets across social, formal, and immediate learning touchpoints and moments of learning need, tailored to diverse learner personas to impact on-the-job performance for a given learning challenge. In this paper, we will share how L&D practitioners and professionals are using the model, what is working, and what is challenging. Overall, our study showed a 77% Net Promoter Score for the LCD model, indicating a high degree of loyalty and positive referral, by a wide range of L&D professionals, representing a variety of roles, industries, and experience levels. We share three different case studies of the model: an internal L&D "solo" practitioner, a non-L&D use case, and an L&D practitioner layering in the LCD model. The study showed that over half of the respondents are making the LCD model a norm. We also learned about the use of the language, leadership sponsor support, and the value of the model. Ultimately, the LCD model is helping to drive continuous learning culture through a more relevant and strategic approach for users.

Keywords: Instructional design models · Learning design · Modern learner · Continuous learning · Learning in the flow of work

1 Introduction

In 2015, the authors were asked to develop a workshop to teach learning and development (L&D) professionals how to better design training to meet the needs of the emerging group labeled "modern learners." In researching the topic, the authors discovered that there is frustration with L&D, both within the industry and outside the industry — from L&D practitioners, from learners as the end users, and from the C-Suite who set

D. Guralnick et al. (Eds.): TLIC 2021, LNNS 349, pp. 237–249, 2022.
https://doi.org/10.1007/978-3-030-90677-1_23

their budget. Frustrations included lack of confidence of impact back on the job, lack of accessibility in the moment of learning need, and training seen as an extra to-do outside of work to do rather than an essential part of the work. In their former roles in corporate L&D, the authors had experienced these frustrations but had not realized how wide-spread the issues are.

The result of the authors' research and work is the Owens-Kadakia Learning Cluster Design (LCD) model, which addresses the frustrations with a new goal and, therefore, seeks to modernize not just training programs, but also the L&D Industry itself. The LCD model was first presented in 2016, and, more recently, described in the book Designing for Modern Learning: Beyond ADDIE and SAM [1]. This paper provides an overview of the model and reports the February 2021 research among current LCD model users to determine what is working, as well as where the model, or the teaching of the model, may need to evolve to assure greater positive impact for the L&D industry.

2 The Owens-Kadakia Learning Cluster Design Model

2.1 A New Goal for Learning and Development

L&D or Training organizations are charged with using their knowledge about how people learn to assure that the workforce is capable of doing their job. When viewed narrowly, this job description leads to L&D professionals being order-takers, with a goal of delivering one training program after the other, often based on management's view of what training topic will "fix" employees' work performance.

The industry's Instructional Design (ID) models reinforce this narrow view of L&D's goal. Today's popular ID models were primarily created in the 1960s and 1970s, and all of them focus on creating just one training course, curricula, or program for each given learning goal. Even the more recent blended-learning methods, which combine web-based or digital learning methods with traditional face-to-face methods, focus on formal, one-and-done programs, that usually must be completed in a sequence. This approach may have been sufficient in the past when the pace of business allowed for L&D to develop and deploy programs over months or years. Following the academic model for teaching, these one-and-done trainings were viewed as an efficient way to teach large numbers of employees on subjects that were relatively stable and unchanging.

As the authors reviewed the workplace context of today, many differences were apparent. The pace of change is exponential and CEOs worry that employees cannot keep up. Waiting for the next class or course is not feasible, except for static topics such as compliance. With today's digital resources, employees are finding their own learning content on a daily basis. In fact, research indicates that employees use L&D-provided training just once every three to four months [2]. Employees are learning when, where, and how they need to learn to do their job, albeit with reduced efficiency and with questionable resources, as they slog through search after internet search. Though unintentional, due to the "one-and-done" overall ethos and supporting processes, L&D fails to deliver sustained change on the job for business-critical talent development goals. L&D needs a new model to stay in the game, remain relevant, and be a major contributor to the business. The new model needs to move beyond andragogy and evolve to include heutagogy as a central design philosophy.

The Owens-Kadakia LCD model reveals that the key is to swap out the goal of creating training or a single deliverable with a new goal of creating Learning Clusters. With this new goal, L&D will surround learners with a set of strategically selected learning assets to change on-the-job behavior, to serve both during intentional and unplanned moments of learning need, in a way that addresses a business issue or "pain point" and accommodates learner-to-learner difference.

2.2 Four Principles Underlying the Model

The LCD Model is built upon four principles, as shared in this excerpt from the book, *Designing for Modern Learning,* below [1].

Go Beyond One-and-Done. L&D's new role is to deliver and facilitate access to multiple learning assets to build employee capability. It's no longer sufficient for L&D to design one learning asset — a class or course — to meet a business's or an employee's learning goal.

Design the Whole, Not the Parts. Multiple learning assets must be viewed and designed as part of an integrated whole (what the authors have coined the "learning cluster"), both as part of L&D's design and from the learner point of view. These multiple assets cannot be effective if they're created ad hoc, without consideration of one another and how they work together to achieve the goal.

Focus on Learner Needs. Our context as training designers and deliverers is no longer as important as the context of the learner. Whereas in the past, L&D had limited tools to deliver learning, today we can deliver learning when, where, and how the learner needs it. L&D must reflect a deeper understanding of the learner and the capability gap first.

Change On-the-Job Behavior. L&D can and should be held accountable for improving performance on the job, not just at the end of a training class, course, or program. Improved performance means that there will be a change in behavior through application on the job, rather than simply acquiring the knowledge and skill during the program.

2.3 The Five Actions that Comprise the Model

The LCD model (see Fig. 1) is composed of five Actions. These Actions honor L&D's traditional expertise, while building new strategic and design capabilities to meet modern learning needs. These five Actions tell L&D how to create a learning cluster, and they measure its effectiveness on several levels.

Here is a description of the five Actions, from *Designing for Modern Learning* [1]. The names of the Actions create the mnemonic CLUSTER.

C: Change On-the-Job Behavior. In the Change Action, set the goal for the overall learning cluster (called a strategic performance objective). This goal articulates the connection between learners' on-the-job performance and the desired business results. This is one of three early Actions leading to the Surround Action.

L: Learn Learner-to-Learner Differences. In the Learn Action, identify learner personas within the target learner group whose behavior change will have the greatest effect on the desired business impact. Persona definitions go beyond demographics and job type to explore contexts of when, where, and how each persona will most likely need to learn. This Action guides strategic choices for learning assets in the Surround Action and content for each learning asset during detailed learning asset design.

U: Upgrade Existing Assets. In the Upgrade Action, apply the nine elements of modern learning to quickly improve current programs. As a bonus, the ideas for new learning assets identified here will jump-start the work in the Surround Action.

S: Surround Learners with Meaningful Assets. In the Surround Action, combine the work and insights from the other Actions to intentionally select learning assets across all three learning touchpoints (social, formal, immediate) to build a learning cluster. The latest neuroscience and general observation show us that learners learn not just through formal means like the classroom, but through 24/7 in-the-moment (immediate) means and through other people (social). Rather than focusing on only the formal and "learning for the first time" moment of learning need, in this Action, the model guides L&D to thoughtfully consider possible learning asset ideas across all three touchpoints. Then, select, design, and deliver those learning asset ideas that clearly connect to learner persona needs, while assuring that each distinct learner persona has learning assets across all three touchpoints that serve their moments of need. By using this strategic approach, we build confidence that the chosen set of learning assets should deliver both the desired behavior change on the job and the related business results.

TER: Track Transformation of Everyone's Results. In the Track Action, identify those measures — qualitative and quantitative — that will indicate the impact of the overall learning cluster back on the job. Then, track these measures and turn the results into a story about the impact of learning. Use the results for further improvements.

Fig. 1. The learning cluster design model is composed of five actions [1].

The model is not linear. Instead, Instead, users choose a starting point based on their business climate and initiative needs. For example, some initiatives and climate may suggest a need for quick "modern" wins and start with the Upgrade Action. Other initiatives may have recently had poor feedback from users and the Learn Action may be used to investigate deeper. Each Action references the others in a way that can later influence or even change the content within other Actions as the project progresses. In this way, the model reflects and role-models the heutagogy philosophy, by allowing the L&D practitioner to choose how to apply the model and to serve as the central strategist, rather than being directed by the model.

By shifting L&D's goal of learning design from one-and-done training to Learning Clusters, and by empowering a comprehensive consideration of learning technologies and assets, the LCD model allows L&D to work across the spectrum of learning tools available not just today, but tomorrow. From artificial intelligence to a tried-and-true job aid, from peer buddy designs to sophisticated social learning networks, from books to immersive in-person or virtual reality experiences, the LCD model guides L&D professionals to gather the appropriate data, focus on the goal that makes a difference, and ultimately, make crucial design decisions across the spectrum of choices to include the optimal mix of assets learners want and need to close a capability gap.

Once assets are chosen for a given learning cluster, the model doesn't seek to overtask L&D with delivery. The model promotes crowdsourcing, repurposing, chunking, and re-using to provide efficient ways to create, manage, and deliver. In some cases, L&D is simply facilitating access to the resources (and with advances such as chatbots, AI, advanced social or crowd-based LMS technologies, this is becoming more and more the reality). In others, L&D is leveraging their full design and delivery skill set.

In the case studies and data shared next, the authors found that LCD users are indeed leveraging the model and finding it valuable for modern learning design.

3 The Research Study

A questionnaire was developed to gain insights from current users of the LCD model within a few weeks' time. To encourage participation from busy people, there were only seven multiple choice questions plus three demographic questions. To garner additional insights, each quantitative question was followed by an optional qualitative question, which essentially said, "and what else would you like us to know?"

The survey was anonymous, but users were welcome to leave their email or email the authors directly to get a copy of this report. Sixteen of the 22 respondents left a valid email address. Nineteen of the 22 respondents completed at least one open-ended question, and on average, these 19 provided 3.2 comments each, thereby providing deeper insights into what is working and what might need to be modified to encourage more global adoption of the model over time.

3.1 Qualitative Data

In this section, we will share some case studies and additional insights based on the open-ended question responses. Then, we paint a picture of what respondents say are the struggles or challenges they face as they move forward on implementation of the LCD approach.

Case Studies. Open-ended question responses provide insights into the use of the LCD model in respondents' work, with seven of the 22 answering four or more. Here are case studies from three particular respondents, sharing their stories based on a combination of their qualitative and quantitative responses.

Case Study 1: Doing L&D Solo at a Mid-size Corporation. As an experienced Jill-Of-All-Trades type of training manager at a mid-size corporation, she tends to work independently. She learned about the LCD model through a webinar and, subsequently, grew her capability through attending the workshop. She highly recommends the LCD model to others and has shared her thoughts on the model with others. She has applied the model itself and the principles behind it in her workplace. She views the model as helping her better meet the needs of her learners as she strives for a culture of continuous learning in a digital landscape.

She likes the LCD language but uses only terms that will help her during dialogs with other non-training colleagues, which means she avoids phrases like "learner persona" and "the 9 Elements of Modern Learning," but embraces language like Learning Cluster, "Five Moments of Learning Need," and Modern Learners. She says, "I'm still learning how to use the LCD model approach, so I anticipate using all the terms more frequently." She fully expects to use the model more throughout the year and even to use it for some company-wide initiatives, which she says will help her get more leadership, stakeholders, and sponsors onboard with the LCD approach.

Her LCD work project was to create "a Microsoft Teams learning cluster for people to learn how to use Teams." She says of the LCD approach, "It helps me 'sell' training because it fits into the flow of work and life, rather than a person having to take 60 min or 90 min at a time to learn something."

She values the Surround Action within the model, coupled with the Learn Action and its learner persona approach, as well as the Track Action. As a whole, the LCD model, its philosophy, and the Social-Formal-Immediate Learning Touchpoints concept are all consistent with her own design approach. "I always have more interaction (in my training design) and try to include a social aspect, among other ways of interacting and learning," she says.

This L&D professional has embraced the entire LCD approach and has plans to continue expanding this approach throughout her work and within her company.

Case Study 2: The Unofficial L&D Professional. This person is excited to apply the LCD model and is in a unique position to apply it in a low-key way. He is an experienced L&D professional currently in a non-L&D role at work, and he teaches high school students after hours. As he applies the model, he is finding it is beneficial in his L&D endeavors both at work and after work.

This L&Der read the book and a related blog or two. Even though he isn't in an L&D role, he has created a learning cluster at work and shared the approach with others. He is applying LCD principles at work and with high school students. He views the approach as helping him across all aspects of L&D work, with the biggest value coming from the LCD model itself, and the Surround Action and principles in particular. He says, "It helped me think of other avenues of learning that I could craft, to help me in my work and with my personal after-work students. Both of these groups will benefit from me applying the LCD model more, I feel. In particular, I think it can help me personalize the learning to my students in better ways, to help them get the results they are looking for in their studies."

At work, he says that he "incorporated the ideas into the project mid-way through. It worked well as a learning experience for the participants." Of his work with high school students, he says, "based on the LCD model, which does make a lot of sense to me, I am revamping the way I am teaching my students, providing more learning touchpoints for them in different ways to meet their moments of learning need when I'm not around to help them. … they are all different and need different things …."

He is sharing his LCD approach with others. "I am coaching my manager on these ideas and showing examples of how we can use them. So far, I feel it's going well," he says. Of the LCD language, he says, "Because some of this is new vocabulary, it's a work in progress to help others know what I mean if I use the terms — but repetition and persistence will pay off."

This L&D professional in a non-L&D role can see how the model applies to a wide variety of situations, and is appreciative of the model and how it helps him continue to improve in his L&D vocation. LCD appears to bring a sense of professional fulfillment and growth.

Case Study 3: Layering LCD into L&D Work. This experienced L&D professional is currently a consultant, who delivers training and designs it, with some work in digital design. To understand the model, he read the book and enhanced his understanding with other LCD learning assets, including webinars and blogs. He is at the early stages of creating a learning cluster and is layering on the LCD principles in his work, as well as sharing the LCD concepts with others in his network. He sees LCD as benefiting a wide range of L&D work and values each of the five Actions and the set of LCD principles. He uses all of the LCD language with at least half of the terms being a regular part of his L&D conversations.

He tells us that "In designing network training programs during 2020 and now 2021, I have consciously thought about the model and how I can use it to design and deliver the most impactful training possible." He goes on to describe his most recent project. "I just launched a semester-long program with the Business Law Scholars at my University. As part of the first webinar, I provided a short survey to understand more about their learning personas." This is a good step forward to assure that a helpful mix of learning assets are provided for these Law students.

This L&D Professional describes experience with senior leaders and the challenge to make change happen. He notes, "Given the opportunity to work with senior management prior to designing and delivering training, I would emphasize the Strategic Performance Objective." He sees the SPO as being consistent with his past work with the Kirkpatrick levels of evaluation and his viewpoint on the purpose and value of training. "I have long approached training with the end game in mind. If the time and effort and money to design and deliver training doesn't result in actual behavior change on the job, what is the point?" he says.

This consultant sees the model as valuable, describing it like this: "So the LCD model creates a very effective way to communicate how and why training should be designed and delivered in a way that will ensure results. For me, it is the path to creating effective L&D programs in the 21st century."

Additional Insights. Despite the small base size for this study, a few common threads stood out.

Professional Fulfillment. As we saw in Case Study 2, the LCD approach may create personal fulfillment or professional satisfaction. This is reinforced by this respondent's comment: "It has sparked new creativity and excitement in my work."

Favorable Attitude. In general, respondents are favorable toward the LCD approach, as is further demonstrated by these quotes from other respondents:

- "I appreciate the model, have begun using it in our team's work, and see us continuing to grow and develop with this being one of the fundamental models we use going forward."
- This is "the best tool to communicate MODERN learning, especially in the age of COVID."
- "I absolutely love the tie to business objectives and having a data-driven approach to modern learning."

LCD Model Drives Change. Some professionals are making dramatic changes based on their understanding of the LCD model. One says, "The model is excellent and certainly challenges L&D professionals to think differently and act differently. Rather than focusing on creating programs that simply check a box, focus on changing the job." Another relates, "Adopting the LCD model has encouraged me to think differently about how we deliver learning opportunities and the type of team members I will be focused on hiring." Still another comments, "The model has changed the way we develop content for our stakeholders."

Insights on What's Challenging for LCD Users. Twenty of the 22 respondents completed the question "What are the top 2–3 struggles you are navigating this year?" Their responses tell us what is top of mind. For some, it's the daily work, others the pandemic, and for some, it's related to implementing the LCD model. The following threads focus on the LCD-related comments.

Management Challenges. These showed up most commonly along the theme of buy-in: "Getting support for people to take time to learn"; "Leadership buy-in"; "The struggle to get buy in from senior leaders always looking to get results without committing to everything that is required to get those results. Can we do it quicker, spend less money, etc."

Learning Management Systems (LMS). The relationship between LMS and the LCD model will require additional investigation, given the ambiguity of the comments. For example "LMS evaluation" as a challenge may mean the respondent is implementing a new LMS, or that integrating the Track Action measures into the existing LMS is the challenge. Likewise, "changing LMS mindset" could reference the traditional thought process that LMS are used to house and track formal programs. If the learning asset isn't fully linked to and measured on the LMS, it's not worth L&D's time to work on.

L&D Growth and Development Challenges. These were articulated like this: "L&D team growth"; "team's organizational structure and the need to build new capabilities to support a holistic learner journey"; "Growing team and offerings". Growing L&D capability to meet modern learning design needs is a challenge. Though there are many avenues to upskilling in the LCD model, many might find investment as a barrier, though further investigation is needed.

Change Management. Some describe it like this: "Shifting organizational mindset to a new way of designing"; "constant change and transition."

Learners themselves are viewed as potential barriers to change. "Learner buy-in"; "Packaging the training in ways that 'sell' and meet these (learner) needs"; "Learner engagement"; "Acceptance of social learning"; "Learner/manager compliance."

Time is a barrier for most of us and was expressed by respondents like this: "Time to put into L&D work"; "Ability to produce fast"; "Competing priorities preventing timely development of complementary learning assets."

In summary, challenges to LCD Model are typical of most change management projects and span needs of leadership stakeholders, learners, L&D professionals themselves, and technology, specifically LMS.

3.2 Quantitative Data

Given that the survey contained just seven quantitative items, each was written to answer a specific question, as shown below.

Net Promoter Score. The first item is a Net Promoter Score (NPS) question, "How likely are you to recommend the LCD model to a friend or colleague?" Results (Table 1) indicate a 77% NPS, a high score. For reference, any NPS score above 0 is "good." It means that your audience is more loyal than not. Anything above 20 is considered "favourable." Bain & Co, the source of the NPS system, suggests that above 50 is excellent, and above 80 is world class.

Table 1. "How likely are you to recommend the LCD model to a friend or a colleague?"

Rating (1–10)	# (Total sample size: 22)
10	14
9	3

(continued)

Table 1. *(continued)*

Rating (1–10)	# (Total sample size: 22)
8	4
7	1
1–6	0

NPS = ((14 + 3)/22) – 0 = .77

Exposure to Model. The second question was used to determine the level of exposure respondents had to the LCD Model, with the possibility of eliminating anyone without significant exposure. All but one of the respondents had either attended a live or virtual multi-day workshop or read the book. One respondent was taught the model by another user and, based on open-ended responses, is a valid user of the model. Seventeen, or 77%, of respondents used multiple learning assets, including reading related blogs and getting personal consultations. There are five (23%) respondents who not only took the workshop, but also read the book.

This is a testament to the LCD model itself in that it demonstrated that by providing multiple learning assets, people will use the ones that meet their needs to help them learn how and when they want to learn (Table 2).

Table 2. "What learning assets have you used to learn about the LCD model?"

Learning asset	# (Total sample size: 22)	%
Book	17	77.3%
Blog(s)	14	63.6%
Webinars	12	54.5%
Workshop	8	36.4%
Personal consultation	1	4.5%
None used	1	4.5%

Experience with the Model. The third question seeks to understand the depth of experience with the model in the workplace. Nineteen (86%) reported applying the LCD model in more than one of the ways listed in Table 3.

Table 3. "How have you applied the LCD model?"

How LCD was applied	# (Total sample size: 22)	%
Apply the principles	18	81.8%
Shared the model with others	16	72.7%
Establishing LCD as the norm	12	54.5%
Created a cluster at work	8	36.4%
Not applied yet	2	9.1%

Benefit of Model to L&D. The fourth question asks what the LCD model helps the respondent do in an effort to understand why the LCD model is a benefit to these L&D professionals (Table 4). Of the 22 respondents, more than half (54%) selected four or more of the five items from the list provided. Three selected just one item, with two of them saying the benefit was to help them better meet learner needs and one saying it provided an approach for continuous learning for the company.

Table 4. "What does the LCD model help you do?"

Benefit	# (Total sample size: 22)	%
Better meet the needs of learners or clients	18	81.8%
Provides approach for continuous learning	18	81.8%
Streamlines work/thought-process for designing training	15	68.2%
Better meet the needs of the business	13	59.1%
Language improves communication and/or increases support for L&D	13	59.1%

What Users Value. The fifth question seeks to understand what aspects of the model are most valued by the users, with a request to select no more than three items (Table 5). While most respondents selected multiple items, one respondent was unsure what they valued, and two valued only the model itself. In addition, two respondents valued only one of the Actions – the Change Action for one respondent and the Upgrade Action for the other.

LCD Language Usage. The language for the LCD model was strategically selected with the intent to help L&D professionals signal a true change in what they were doing, and to elicit support during the change effort. Question 5 seeks to understand which terms are being used (Regularly, sometimes, not used).

The term "learning asset" is the most used term. A learning asset is defined as a general term describing a wide range of things that help people learn. It might be

Table 5. "What aspect of the LCD model is most valuable to you?"

Most valued parts of LCD (up to 3)	# (Total sample size: 22)	%
The model as a whole	15	68.2%
The LCD principles	15	68.2%
The surround action/learning cluster tool	12	54.5%
The upgrade action/9 elements tool	10	45.4%
How LCD integrates with other ID models	7	31.8%
The change action/strategic performance objective tool	6	27.3%
The learn action/learner personas tool	6	27.3%
The track action/measures tool	6	27.3%
Not sure	1	4.5%

content to read, an online search, a class (face-to-face or online), a discussion, a video, or even a motivational poster. It can be as small as a 30-s audio recording or as large as a three-month class. In traditional training, a learning asset most often takes the form of a class, an e-learning course, or a blended learning program.

Other terms that get high levels of usage include learning cluster, modern learner, and learning touchpoints.

Leadership Support. The last quantitative question was focused on understanding the level of leadership involvement for the new model (Table 6). Encouragingly, in all but one instance, respondents report that their leaders were, at a minimum, aware of the LCD model or supported the L&D professions in their efforts to try it. In 12 cases, leaders were more involved, either supporting the cost of a workshop, actively learning about the LCD model themselves, or having been the person to identify the model and bring it to the L&D professional.

Table 6. "At what level have your organizational leaders or L&D stakeholders/sponsors engaged with the LCD model?"

Level of support	# (Total sample size: 22)	%
Supports trying out the model	17	77%
Aware of the LCD model	9	41%
Invested in L&D attending workshop	6	27%
Engaged personally in learning about the model	5	23%
Encouraged L&D to find modern ways of learning	3	14%
Brought the LCD model to L&D	2	9%
None of the above	1	5%

Demographics. The demographic questions showed that the respondents represented the full range of L&D professions (manager, digital designer, trainer, consultant, designer/developer). Years of experience in L&D also covered the gamut with five having three or fewer years in L&D, six having four to eight years, and eleven having nine or more years' experience. Industries in which respondents work include healthcare (3), technology (9), energy (1), media (1), foods (1), and consulting (4), with three respondents not specifying their industry. With this base size, there is no correlation between responses and demographic differences.

4 Conclusions

The LCD model and LCD principles have been strongly accepted (77% NPS score) by L&D professionals who represent a broad spectrum of industries, experience levels, and types of L&D jobs. People are using the model in their workplace and are planning on continue to implement this program for the foreseeable future. The LCD approach is helping them meet learner and business needs and helping them create a continuous digital learning culture. The LCD approach has provided the guidance needed to help them implement the LCD model, streamline L&D work, and showcase a new language to better communicate with other and garner support for their work. Surprisingly and encouragingly, leadership level sponsors of L&D seem aware and engaged with the LCD model as a new way of working.

The issues and barriers to LCD implementation at this point appear to be those things common to all change management projects. Ways we can help adopters more through these is to continue communicating the impact and benefits of the LCD model more broadly and, perhaps, starting to push the message to non-L&D organizational leaders and managers so that requesting a Learning Cluster instead of a training program is seen as the modern way to do business. Continuing to encourage practice, pilots, and enrolling internal champions for the LCD model are all important ways to drive successful change.

The authors intend to continue tracking implementation across the industry with the intent of growing the pool of users and searching out barriers in a larger group with the intent of bringing users together to help resolve any emerging issues as a social community. "Modern" changes every day, and it is our intent to evolve.

References

1. Kadakia, C., Owens, L.M.D.: Designing for Modern Learning: Beyond ADDIE and SAM, 1st edn. Association for Talent Development, Alexandria, USA (2020)
2. Degreed "How the Workforce Learns Report". https://get.degreed.com/en/how-the-workforce-learns-report-2019. Accessed 11 Mar 2021

Online Student-to-Student Interaction: Is It Feasible?

Jenny Pange(✉)

Laboratory of New Technologies and Distance Learning, School of Education, University of Ioannina, Dourouti, Ioannina, Greece
lncsjpagge@uoi.gr

Abstract. The COVID-19 pandemic has led to some important changes to teaching and learning. Many universities worldwide offered online learning courses. In an online course, the most challenging aspect is not only the teaching process, but also the student-to-student interaction. It is well documented that, during COVID-19 pandemic, online learning has changed the way students interact, communicate, and exchange ideas. Students experienced new ways of online communication and missed student-to-student interaction during the pandemic. University students, up to now, had the opportunity for socialization during their university studies. Nowadays, they face the challenge to interact only online without any contact with their closest friends. In a face-to-face environment, students build metacognitive skills and improve their communication skills, working in well-established partnerships. When students faced the transition to online learning, they missed the interaction with their peers, and they were engaged more in specific class activities rather than forming concrete project-oriented groups. To test the student-to-student interaction during the pandemic, undergraduate students were asked to form their own virtual communities, interact with their classmates, and make use of social media for accurate and continuous communication. According to the findings, third-year students, who had previous experience of face-to-face communication, were more comfortable to trust online communities, interact with their peers, compared with second-year students. A blended-learning environment with face-to-face communication and extra online activities using social media will keep and support the student-to-student interaction for newer students.

Keywords: Student-to-student interaction · Online learning · Communication

1 Introduction

Technology plays a vital role in education promoting communication and supporting the learning process [18, 35]. In the learning process, the main categories of interactions are the student-to-teacher, the student-to-student, and between student and content interaction [22, 26, 35]. There is a long literature in educational psychology concerning the teacher-to-student interaction, but the student-to-student interaction is less discussed [13, 14, 18]. Interaction between students and teachers encourages active collaboration, debating, and learning activities [24, 34].

D. Guralnick et al. (Eds.): TLIC 2021, LNNS 349, pp. 250–256, 2022.
https://doi.org/10.1007/978-3-030-90677-1_24

Nowadays, learning and teaching principles have moved mainly to online experiences, increasing internet use and changing the role of the teacher [3, 6]. In the learning process, students use technology for synchronous or asynchronous communication with their teachers and their peers. Information computer technologies (ICT) are used for asynchronous communications, like messages, blogs, discussion forums, or other types of communication. Synchronous communication is used mainly on online learning platforms and social media [8, 9, 20, 27]. Therefore, uninterrupted internet connection at school or at home has become a crucial factor for the communication of students with their teachers and classmates [30, 35].

In the last decade, the internet played a significant role in the academic community and had a strong impact on the way students learn [21, 25, 28]. Many universities worldwide deliver distance education courses and blended learning, but during the pandemic, several universities transformed into fully online. According to researchers [23, 25], the enrolment rate in distance education courses had risen during the last ten years. This tendency is expected to continue during and after the COVID-19 period, as the internet supports the learning and teaching process [4].

Education, as a system, includes all types of interpersonal relationships. In schools, students form relationships and structure their knowledge, developing at the same time their cognitive, emotional, and social skills. The interaction of students with their peers affects their socialization, promotes the feeling of belonging, reduces their isolation, upholds their learning attitudes, and shapes their learning behavior [22, 27]. Online cooperation among students provides self-esteem and diminishes competitiveness [2, 11]. Additionally, teamwork upholds individual accountability, the equal participation of team members, and simultaneous interaction.

Although student-to-student interaction has been identified as one of the key components for successful face-to-face learning, few studies refer to student-to-student interaction in online learning courses [34, 35]. Researchers [31, 32] found that students can interact with other students when there is not any isolation of learners. But other researchers [16] reported in their study that their participants found the interaction with their teachers was more useful than the student-to-student interaction. In a distance learning environment, student-to-student interaction is influenced by gender, personality type, year of studies, and motivation of students for online courses [15, 19].

Until the COVID-19 pandemic, many research studies focused on blended learning and the online interaction in distance education [1, 33]. The ongoing coronavirus lockdowns have led globally to the unexpected closure of higher institutions and the shift to fully online learning [7]. Almost all universities, worldwide, to overcome the consequences of the health crisis and continue their learning and teaching process, shifted their curricula to fully online courses [5, 9, 10]. Particularly, the Greek Ministry of Education and Religious Affairs, soon after the first COVID-19 lockdown in March 2020, supported the shift of the Greek Universities towards online synchronous learning and provided for free the LMS platforms to accommodate online educational material for interaction and learning.

Thus, in COVID-19 lockdown, online learning has been in the center of interest for educators. Students nowadays spend a considerable amount of time on the internet apart from working for their online learning activities [7, 25, 34]. During pandemic, they had to demonstrate adaptability to this new normalcy. Several studies reveal uplifted stress levels for undergraduates due to fear of a life-threatening virus, psychological impact of social isolation, disruptions of everyday normalcy, and quality ambiguity on online educational procedures [6, 14, 29]. Konstantopoulou and Raikou [16], in their study, reported the increased symptoms of depression among Greek university students caused by COVID-19 threat. Furthermore, other studies revealed that tension and isolation, related to problematic internet use, reduced academic performance and communication with their educators [12, 31, 32]. Therefore, due to social isolation measures, students were fully dependent on the internet and social media to maintain their interpersonal relationships [21, 30].

The aim of this study is to present students' views regarding student-to-student interaction during the COVID-19 pandemic—especially focusing on how undergraduate students perceive the interaction among students in small online groups.

2 Materials and Methods

The teaching in Greek Universities during COVID-19 was online, and all students had to participate in online lectures. The University of Ioannina utilized the online platform of MS Teams and more than 200 undergraduate students from the Department of Early Childhood Education, regularly attended the online lectures on ICTs and statistics. Both are compulsory undergraduate courses. In the teaching process of these online courses, all students were asked to form small subgroups under MS Teams platform, and they had to complete various assignments. Additionally, the teacher encouraged all students to use social media for extra communication with their peers. Students in these subgroups had to interact with all group members, share ideas, and prepare a final common presentation. In both courses, clear time tables, procedures, and deadlines were given to all students by the teacher. For the needs of this study, two self-selected groups of undergraduates presented their views on student-to-student interaction during an assignment.

The main limitations of this study were the sample selection, the time of the survey, and the gender of the participants, because students were self-selected, this study took place during the COVID-19 pandemic and most of the participants were female.

2.1 The Process

The first group of students consisted of 105 undergraduates, participating in the ICT course. This course is offered in the sixth semester of the undergraduate studies. Students in this group formed 15 groups of 7 students each. They were created randomly by MSTeams "breakrooms" command. Students had to prepare a project for children in kindergarten using a selection of ICT tools. All students had to complete the project in 15 working days. Guidelines and technical support were offered by the teacher continuously during the assignment. All students had to work online as a team and at the end of the assignment to deliver a common PowerPoint presentation. This presentation included the material provided by all students in the group.

The second group consisted of 150 undergraduate students participating in the statistics course. This course is offered in the fourth semester of the undergraduate studies.

Students in the second group split randomly into 15 groups of 10 students by MSTeams 'breakrooms' command. These sub-groups of students had to collect and analyze data using the PSPP free online statistical package and complete the full project in 15 working days.

During the assignments, students were free to interact informally with their classmates and exchange ideas for the project using social media or other online tools for cooperation (e.g., Viber, Google hangouts).

After the process, students in both groups replied to a questionnaire, with closed and open-ended questions, concerning their views on student-to-student interaction during these online meetings as a team member.

Students firstly had to rate the online cooperation on a Likert scale from one to three (very good, good, fair/difficult) and then to choose from a list of advantages and disadvantages, up to three main advantages and up to three main disadvantages related to this online cooperation. Specifically, students had to declare as advantages or disadvantages the positive interdependence (e.g., working together for a common goal, helpful participation of team members), the individual accountability (e.g., responsibility of every member in the group, rely on group members), the equal participation as a team members (e.g., exchange of ideas equally), the simultaneous interaction (e.g., active engagement, connection in chats), and the connection/cooperation (e.g., technical issues, internet connection, use of mobiles, other issues). The replies of the students collected and analyzed using the SPSS statistical package for quantitative data and qualitative methods applied where appropriate.

3 Results and Discussion

In the first group of undergraduate students, 100 students were female, and 5 students were male. As most of the students were female, gender differences in student-to-student interaction were not examined. All sub-groups of students had their own folders and online chat to post their ideas. During the assignment, the teacher used the chat to give feedback and explanations to all sub-groups of students. The advantages and disadvantages in student-to-student interaction in this study that appeared in all groups more frequently (greater than 70%) are listed below.

Firstly, all students were asked if the face-to-face cooperation with their classmates before COVID-19 had a positive impact on their current online cooperation. Seventy-three (69.5%) of them declared that face-to-face cooperation was very important in their online teamwork because they knew each other, and they were able to discuss the project, split tasks in the group and cooperate.

At the second stage, students reported about the online interaction. Sixty-five students (61.9%) declared that the student-to-student interaction amongst team members was *very good* during this course. They reported as main advantages equal participation (they were able to equally exchange ideas and different views), positive interdependence (split the workload and encouraged group members for the common task), simultaneous interaction (they interacted in chats), and individual accountability (they were able to rely on one another). The students said that they also used Facebook for extra communication. They considered technical difficulties as the main disadvantage.

Thirty-one students (29.5%) reported that student-to-student interaction was *good*. The main advantage was the equal participation and simultaneous interaction (active communication in chat of team members). They reported, as disadvantages, technical problems (internet connection and use of mobiles instead of PCs) and individual accountability.

Nine students (8.6%) said that the communication was *fair*. They considered, as main disadvantages, technical difficulties, simultaneous interaction, and non-equal participation.

In the second group of students, 140 participants were female and 10 were male. Sex differences were not considered in this group because most of the participants were females. Students in this group could not meet all their classmates before, on the campus, due to COVID-19 pandemic.

Sixty students of the second group (40%) said that student-to-student interaction was *very good*, and the main advantages were: positive interdependence, individual accountability (the activities of team members promoted the success of the whole team), and simultaneous interaction (frequent online chat). The main disadvantages were the lack of equal participation and technical difficulties. They declared also that they used social media (Facebook) very often for extra communication and active participation during the assignment.

Seventy-five (50%) students said that student-to-student interaction was *good*, and the main advantages were the positive interdependence (exchange of ideas among team members to work together) and individual accountability (the ability to rely on one another). The main disadvantages were the inability to have equal participation, technical problems, and sometimes poor online interaction.

Fifteen students (10%) said that the interaction was *fair*, because they were unable to join other members of their team due to technical problems (problematic internet connection, not available PCs or mobiles).

Students in this group finished their assignment in 20 working days, which means 5 more days than the scheduled date of the assignment.

According to the views of students in the two groups, it comes straightforwardly that working in teams makes each member a valuable one; it makes increasing positive interdependence, individual accountability, and student-to-student interaction feasible, provided that all individuals are responsible and accept to cooperate equally online.

Focusing on the first group of students, equal participation was a main advantage, but on the contrary, in the second group, equal participation was reported as disadvantage. An explanation to this outcome is that students in the second group did not know each other very well from face-to-face meetings, to split the tasks accordingly. So some experienced students in the team had to carry out more workload than others to achieve the final goal.

This study also reveals the importance of social media communication. Online communication supports the interaction with students' classmates. Moreover, students said that regular face-to-face meetings or online meetings with teachers for additional advice is important for successful collaboration in the online teams. Thus, teachers must not underestimate the importance of online communication with their students and the role of social media for continuous, uninterrupted, and active online learning.

4 Conclusions

Student-to-student interaction is an important factor for educational success. When students work in teams online, they develop equal responsibility and eliminate controversies. Small online groups of students support the interaction, eliminate the stress, provide a safe place for communication, and promote active learning. Competition or individualistic learning is eliminated in student-to-student online cooperation because they work for a common task.

Additionally, the role of the teacher in online learning should not underestimated, as behavior management and quality feedback from teachers in online learning, stimulate and support the learning process. The teacher sets the learning goals and controls the online teamwork systematically, and the students support the student-to-student interaction when they are ready to accept their commitment to positive interdependence, individual accountability, equal participation, and simultaneous interaction.

According to our findings, older and more experienced students support better student-to-student interaction in online courses. Elder students, who have the experience of face-to-face teamwork support active learning and implement online teamwork more equally than their younger peers. So teachers must encourage student-to-student online interaction offering at the same time the required emotional and instructional support.

References

1. Agustino, S.: Toward a course conversion model for distance learning: a review of best practices. J. Int. Educ. Bus. **5**(2), 145–162 (2012)
2. Ali, W.: Online and remote learning in higher education institutes: a necessity in light of COVID-19 pandemic. HES **10**, 16 (2020)
3. Angelo, J., Moreno, M.A.: Screening for problematic internet use. Pediatrics **145**(2), 181–185 (2020)
4. Bakkar, M.: The role of COVID-19 spread in increasing the use of electronic applications. SSRN Electron. J. (2020). https://doi.org/10.2139/ssrn.3636868;10.2139/ssrn.3636868
5. Brooks, S.K., et al.: The psychological impact of quarantine and how to reduce it: rapid review of the evidence. Lancet. **395**(10227), 912–920 (2020). https://doi.org/10.1016/S0140-6736(20)30460-8
6. Costa, R.M., Patrão, I., Machado, M.: Problematic internet use and feelings of loneliness. Int. J. Psychiat. Clin. Pract. **23**(2), 160–162 (2019)
7. Crawford, J., et al.: COVID-19: 20 countries' higher education intra-period digital pedagogy responses. JALT **3** (2020). https://doi.org/10.37074/jalt.2020.3.1.7
8. Ellis, R.: Designing a task-based syllabus. RELC J. **34**(1), 64–81 (2003)
9. Garrison, D.R., Archer, W.: A Transactional Perspective on Teaching and Learning: A Framework for Adult and Higher Education. Elsevier Science Inc., Amsterdam (2000)
10. Gokhale, A.A.: Collaborative learning and critical thinking. In: Seel, N.M. (ed.) Encyclopedia of the Sciences of Learning, pp. 634–636. Springer, Boston (2012). https://doi.org/10.1007/978-1-4419-1428-6_910
11. Hsu, A., Malkin, F.: Shifting the focus from teaching to learning: rethinking the role of the teacher educator. J. Contemp. Issue. Educ. Res. **4**(12), 43–50 (2011)
12. Johnson, D.W.: Human Relations and Your Career, 3rd edn. Prentice Hall Publishing, Englewood Cliffs (1990)

13. Johnson, D.W., Johnson, F.P.: Joining Together: Group Theory and Group Skills, 10th edn. Allyn & Bacon Publishing, Boston (2008)
14. Johnson, D.W., Johnson, R.T.: An educational psychology success story: social interdependence theory and cooperative learning. J. Educ. Res. **38**(5), 365–379 (2009)
15. Kelsey, K.D., D'souza, A.: Student motivation for learning at a distance: does interaction matter? J. Dist. Learn. Admin. **7**(2) (2004). https://www.westga.edu/~distance/ojdla/summer72/kelsey72.html
16. Konstantopoulou, G., Raikou, N.: Clinical evaluation of depression in university students during quarantine due to covid-19 pandemic. Eur. J. Publ. Health Stud. **3**(1) (2020). https://doi.org/10.46827/ejphs.v3i1.65
17. Lee, J.: Patterns of interaction and participation in a large online course: strategies for fostering sustainable discussion. J. Educ. Technol. Soc. **15**(1), 260 (2012)
18. Leonard, P.E., Leonard, L.J.: The collaborative prescription: remedy or reverie? Int. J. Leadersh. Educ. **4**(4), 383–399 (2001)
19. Louis-Jean, J., Cenat, K.: Beyond the face-to-face learning: a contextual analysis. Pedagogic. Res. **5**(4), em0077 (2020). https://doi.org/10.29333/pr/8466
20. Mallya, J., Lakshminarayanan, S.: Factors influencing usage of internet for academic purposes using technology acceptance model. DESIDOC J. Lib. Inf. Technol. **37**(2), 119–124 (2017)
21. Moor, M.: Three types of interaction. Am. J. Dist. Educ. **3**(2), 1–7 (1989)
22. Moore, G.E., Warner, W.J., Jones, D.W.W.: Student-to-student interaction in distance education classes: what do graduate students want? J. Agric. Educ. **57**(2), 1–13 (2016). https://doi.org/10.5032/jae.2016.02001
23. Pange, J., Toki, E.I., Lekka, A.: Distance learning: a myth or a necessity for educators? In: Proceedings of the IADIS International Conference on International Higher Education, IHE 2011. IADIS (2011)
24. Rahimpour, M.: Current trends on syllabus design in foreign language instruction. Procedia Soc. Behav. Sci. **2**(2), 1660–1664 (2010)
25. Rivera, C.I., Baik, C., Lodge, J.M.: Teacher and student interactions in the first year of university. J. Furth. High. Educ. **44**(8), 1130–1142 (2020). https://doi.org/10.1080/0309877x.2019.1664731
26. Shea, P., Li, C.S., Swan, K., Pickett, A.: Developing learning community in online asynchronous college courses: the role of teaching presence. J. Asynchron. Learn. Netw. **9**(4), 59–82 (2005)
27. Tomczyk, Ł, Szyszka, M., Stošić, L.: Problematic internet use among youths. Educ. Sci. **10**(6), 161 (2020). https://doi.org/10.3390/educsci10060161
28. Welch, M.: Collaboration: staying on the bandwagon. J. Teach. Educ. **49**(1), 26–38 (1998)
29. Xu, X., Wang, J., Peng, H., Wu, R.: Prediction of academic performance associated with internet usage behaviors using machine learning algorithms. Comput. Hum. Behav. **98**, 166–173 (2019)
30. Yu, E., Sunny, C.: Student-inspired optimal design of online learning for generation. J. Educ. Online **17**(1), EJ1241579 (2020)
31. Zirkle, C.: Identification of distance education barriers for trade and industrial teacher education. J. Ind. Teach. Educ. **40**(1), 20–44 (2002)
32. Zirkle, C.: Distance education programming barriers in career and technical teacher education in Ohio. J. Vocat. Educ. Res. **29**(3), 157–179 (2004)
33. LNCS Covid-19. https://campustechnology.com/articles/list/covid-19.aspx. Accessed 10 Mar 2021
34. LNCS StaffWriter. https://www.reference.com/world-view/internet-education-567763d7d50b5636. Accessed 10 Mar 2021
35. LNCS UNESCO Internet in Education. https://iite.unesco.org/pics/publications/en/files/3214612.pdf. Accessed 10 Mar 2021

Supporting the Development of Critical Thinking Skills Through Work-Based Learning Activities: A Pilot Experience in the Educational Science Context

Antonella Poce[1(✉)], Maria Rosaria Re[2], Carlo De Medio[2], Mara Valente[2], and Alessandra Norgini[2]

[1] Department of Education and Humanities, University of Modena and Reggio Emilia, Reggio Emilia, Italy
antonella.poce@unimore.it
[2] Department of Education, University Roma Tre, Rome, Italy

Abstract. The present paper aims to illustrate the learning results of a pilot experience focused on the development of critical thinking (CT) skills through work-based activities, as an example of continuous instructional improvement in higher education institution (HEI) contexts. Specifically, more than 100 university students of two master's degree programs at University Roma Tre participated in two different online learning paths, carried out by the Centre for Museum Studies research group during the academic year 2020–2021, containing activities of analysis, interpretation, argumentation, and critical evaluation of work-based experiences in the educational fields. Specific online meetings with stakeholders were carried out within the paths, together with experiences based on the following teaching methodologies: problem solving, oral dissertation, digital storytelling, analysis, and critical reflection. Formative and summative evaluation activities were carried out in the pilot experience in order to collect and analyze data in relation to the promotion of professional and CT skills. Results from the pilot experience show a statistically significant improvement in some CT indicators within students participating in the activity and an overall good evaluation of the learning courses, stakeholder meetings, and assigned work-based activities. In addition to CT, collaboration and creativity skills were also self-assessed by the students, as stimulated by the online activities.

Keywords: Critical thinking · University · Work-based education

1 Introduction

Nowadays, a debate regarding the role that higher education is supposed to play in the broader society is present at an international level. The debate refers to a dialectical conflict between two different stances: should university prepare students to fulfil the job market needs? Or is the university supposed to transfer knowledge without considering

D. Guralnick et al. (Eds.): TLIC 2021, LNNS 349, pp. 257–269, 2022.
https://doi.org/10.1007/978-3-030-90677-1_25

professional skill training? An education system that focuses on developing higher-order skills, especially critical thinking (CT), could overcome such a conflict.

As stated by Facione, CT is "purposeful, self-regulatory judgement which results in interpretation, analysis, evaluation, and inference, as well as explanation of the evidential, conceptual, methodological, criteriological, or contextual considerations upon which that judgement is based. CT is essential as a tool to inquiry" (p. 2) [1]. In order to define CT, it is necessary to focus on the dimensions that characterized it, and that are basically divided into cognitive skills (such as interpretation, analysis, assessment, inference, explanation, and self-regulation) and dispositions (such as curiosity, appreciation of individual differences, skepticism, and flexibility) — positive approaches that favor its development [1–4]. CT skills are more and more considered pivotal for human and social progress in terms of innovation, economic and knowledge growth by educational policy [5, 6]. Also, CT provides students with tools to be autonomous thinkers, active citizens [7], and critical users of digital technologies [8]. As pointed out by several researchers, even during the COVID-19 pandemic, CT is defined as a necessary competence in order to support people in logical reasoning, thoughtful analysis, adaptability to new contexts, problem-solving, logical solutions, and social cohesion [9–11].

The complexity of the CT construct requires careful and rigorous pedagogical design and interpretation of data in the educational field in order to identify methodologies and tools that are truly effective in promoting CT in learners. Moreover, the promotion of CT learning and teaching methods should be considered as an urgent need in all the formal educational context, taking into consideration the different dispositions and cognitive skills to be promoted at a university level. University teaching activities aimed at promoting CT in students can be implemented through the use of real-world situations and/or workplace-based scenarios, as stated by Dominguez et al. [12]. In particular, active learning methodologies, such as authentic situations and problem-based activities, engage students in problem-solving and decision-making, thus promoting some of the six core CT skills set by the experts [13]: interpretation, analysis, evaluation, inference, and explanation. In particular, educational practices based on problem-based learning methodology can improve social science students' abilities to analyze, compare, and share experiences, as well as improve engagement with content and self-assessment [12]. In general, although research on the effectiveness of instructional interventions is inconsistent [12], if CT skills are solicited in disciplinary settings, outcomes appear improved [14].

In addition, great attention should be paid to the methods and tools for evaluating CT, which are considered an open challenge in the educational and professional field. As already defined in previous publications [15], CT assessment tests can be classified in different ways. Multiple-choice (MC) tests or questionnaires are those most widely used because — in general terms — they best meet the reliability criteria of an assessment test [16]. However, some authors point out that the MC measures are not suitable for higher-order skills assessment, such as CT, but they may be answered merely by low-level processing, such as factual recognition and selection [17]. Moreover, MC tests can never assess students' skills to synthesise ideas or written text [18]. Lastly, all the tests based on MC are not free, which limits their accessibility and their use in educational contexts. To address the limitations of MC tests, researchers and authors of the present paper have developed alternative assessment methods, which involve the adoption of open-ended tasks.

Starting from these assumptions, the Centre for Museum Studies (CDM) research group, coordinated by Antonella Poce, developed in the academic year 2020–2021 a series of higher education institution (HEI) learning paths aimed at promoting CT through specific activities carried out within workplace context. A specific evaluation rubric, already validated in previous publications [16], was used to assess CT skills solicited by the students during the activities.

2 Methodology

2.1 Goals of the Research

According to the above-mentioned premises, the main goal of the herewithin described pilot experience is the promotion of CT skills in HEIs students through work-based learning activities. The research group also tried to answer the following research questions:

- How do student CT indicator levels change in a university course designed to support students' CT levels?
- What is the level of CT perceived by the students at the end of the activities?
- What other transversal competences are solicited by a university course foreseeing work-based activities?
- Did the online activities, including assignments and meetings with educational professionals, improve the levels of CT in the participating students?

In order to answer the above Research Questions, the researchers designed different phases of monitoring and evaluation of the learning activities carried out by the students and two different evaluation tools were used.

2.2 Learning Activities

During the academic year 2020–2021, the research group developed a series of HEI learning paths aimed at promoting CT through specific activities carried out within workplace context.

Specifically, 125 students from two master's degree programs at University Roma Tre participated in two different learning paths containing activities of analysis, interpretation, argumentation, and critical evaluation of work-based experiences in the educational field. The pilot experience can be considered an example of continuous instructional improvement, in which students are continuously solicitated in analyzing, interpreting, evaluating, and finding solutions through specific work-based activities.

The students participating in the activities are divided as follows:

- Seven students from the master's degree course in pedagogical sciences attending the second-year course in "Experimentalism, Museum and Reading" from October to December 2020 (12 ECTS).

- A total of 118 students from the master's degree course in primary school education attending the first-year course in "Research Methodology in Education" from October to November 2020 (4 ECTS).

Within the above modules, specific online meetings with stakeholders were held, and experiences based on the following teaching methodologies were discussed: problem solving, oral dissertation, digital storytelling, analysis, and critical reflection.

In total, students were asked to participate in all meetings organized with stakeholders from the education sector and to accomplish at least three out of five activities assigned by the professionals themselves. The activities were organized throughout the courses and concerned contents in line with the courses' objectives.

For organization reasons and due to the different numbers of participating students, the activities of the course in "Research Methodology in Education" were carried out collaboratively by the students, who worked in 17 groups of 6–10 members each. For the "Experimentalism, Museum and Reading" course, students had the choice of working individually or in groups of up to three participants.

Table 1 gives a summary of all the online activities' topics and meetings organized for each course.

Table 1. Table of online meetings and activities organised for the two HEI courses participating in the pilot experiment.

Course	Meeting	Related activity
Research methodology in education	The Inclusive Memory (IM) project questionnaire: design, implementation, administration and results Participating stakeholders: museum education researchers, educators, trainees	Adaptation of the IM questionnaire to a different category of respondents + oral dissertation
	Music of oral tradition as intangible Cultural Heritage of humanity: an opportunity for inclusive teaching Participating stakeholders: museum education researchers, university researchers	Reading grid for empirical research production + oral dissertation

(*continued*)

Table 1. *(continued)*

Course	Meeting	Related activity
	Narrating the museum: the E-trouria VR app and digital storytelling for heritage education Participating stakeholders: museum education researchers, university researchers, graphic designers, computer engineers, educators	Collaborative storytelling for heritage education realization + oral dissertation
	Empirical studies of museum education. Promoting social inclusion of different categories of museum users Participating stakeholders: museum professionals, university researchers	Recognizing strategies and stages of education research + oral dissertation
Experimentalism, Museum and Reading	Creative approaches to explore the museum collection Participating stakeholder: J. K., the National Gallery – UK	Designing a museum exhibition for a specific community
	Cultural heritage and wellbeing in Flanders, a grassroots approach Participating stakeholder: B. D. N., FARO – NLB	Community and public engagement programme through Cultural heritage + oral dissertation
	An inquiry approach to museum education Participating stakeholder: S. Bailin, Simon Fraser University	Inquiry into the restoration of Palmyra site and analysis of scientific articles + oral dissertation
	Building relationships through heritage Participating stakeholder: L. E., Communities and Diversity National Lottery Fund – UK	Case study analysis + oral dissertation

The stakeholders who took part in the online meetings are organizations, associations, and companies linked to University Roma Tre through partnership agreements and work in the field of education and research in education. The work-based activities related to the online meetings were designed by the stakeholders, with the support of the teachers and tutors of the university courses, in order to ensure a connection with the specific learning objectives of the courses themselves, as well as a continuous implementation of the learning path according to the monitoring results collected during the process.

In line with to the European Commission document *Work Based Learning in Europe: Practices and Policy Pointers* (2013) [19], work-based learning can be identified by three main models:

- Model 1: alternance schemes or apprenticeships, also known as "dual system," in which students spend a significant amount of time training in companies and acquire general and occupation-related knowledge in VET schools or other education/training institutes.
- Model 2: school-based VET, which includes on-the-job training periods in companies.
- Model 3: work-based activities are integrated in a school-based programme, through on-site labs, workshops, simulations or real business/industry project assignments.

The pilot experience here presented was designed taking into consideration the last model, in which the simulation of a professional and entrepreneurial environment was carried out with the cooperation of the participating stakeholders.

2.3 Evaluation Tools and Phases

During the two courses, formative and summative evaluation activities were carried out in the pilot experience in order to collect and analyze data in relation to the promotion of professional and CT skills. In particular, the evaluation phases were divided as follows:

1. Evaluation of the work-based activities carried out by the students or groups of students participating in the pilot experience, aimed at monitoring the solicitation of CT skills during the courses.
2. Final summative evaluation through a questionnaire, aimed at assessing the perception of the levels of transverse competence reached by the students at the end of the pilot experience and the quality of the activities and meetings organized.

Two different evaluation tools were used for the two phases of monitoring and evaluation of the pilot experience.

- *CT evaluation rubric*, already validated in previous publications [16], and composed by six macro indicators on a scale from 1 to 5: use of the language, argumentation, relevance, importance, critical evaluation, and novelty.
- *Questionnaire for self-assessment of transverse and professional skills and evaluation of the quality of the activities and online meetings* carried out in the pilot phase. The questionnaire was created ad hoc for the experience, starting from indicators already validated and used in Poce, Agrusti, and Re [20].

In this analysis, 87 work-based products, realized by the student participating in activities, were evaluated in order to analyze the CT levels during the courses. Sixty-one products are part of the 4 work-based activities proposed for the "Research Methodology in Education" module and 26 for the "Experimentalism, Museum, Reading" module. Three evaluators blindly assessed each work-based activity using the CT evaluation rubric. The mean scores obtained were analyzed through basic statistical analysis.

The questionnaire for self-assessment of transverse and professional skills and evaluation of the quality of the activities and online meetings was filled in by all the 125 students participating in the pilot experience. This evaluation tool is composed by different section and indicators, as summarized in the following Table 2.

Table 2. Structure and indicators of the final evaluation questionnaire of the pilot experience.

Section	Indicators	Type of questions
Evaluation of the course	Professional skills, educational gaps, studies and work expectation, motivation, inspiration, teachers and tutors support, teachers and tutors feedback, online communities	Likert scale
Evaluation of the online meeting with stakeholders	Quality of the contents of the meetings in terms of ease of understanding, content adherence, clarity of the structure and the content, activities difficulty, exhaustiveness, digital tools effectiveness	Likert scale
Professional skills self-assessment (educational area)	Planning educational path, creating educational tools, identifying students' needs, evaluating educational activities, strengths and weaknesses, defining future development	Likert scale
Transverse skills self-assessment	Creativity, Innovation, Communication, Critical thinking, Problem solving, Research abilities, Collaboration	Multiple-choice, open-ended and Likert scale

Data from the final evaluation questionnaire were analyzed through basic statistical analysis.

3 Results

3.1 Analysis of the Data from the Final Evaluation Questionnaire

Generally, the data from the overall evaluation of the modules are positive. On average, 92% of the students agreed that the course promoted their professional skills. Most of the students strongly agree that they found new inspiration interacting with colleagues during meetings and work-based activities (63.2%), they perceived the support of the university team of professors and tutors during the course (64%), and they received feedback and useful answers during the online sessions (64.8%). Furthermore, more than 90% of the students agreed that the course will allow them to perform better in their studies (90.4%) and in their work activities (91.2%). Again, more than 90% of the students quite agree (62.4%) or strongly agree (29.6%) that their point of view was enhanced by the tutors and professors during the online meetings.

With regard to the quality of the content of the meetings, the data also show extremely positive results. A total of 56% of the participating students assign the maximum score of 5 to the didactic tools used as support during the online meetings, such as videos, PDFs, and slides. The language used by the stakeholders is defined as extremely clear for 45.6% of the students; 22.4% of the participants assigned a score of 5, and 46.4% assigned a score of 4 to the quality of the course in terms of "Exhaustiveness of content." However, only 8% of the students stated that the assigned work-based activities were not difficult at all, assigning the minimum score of 1 in terms of "Difficulties of the activities carried out." Actually, the activities had a difficulty level of 3 out of 5 for 32% of the participants and of 4 out of 5 for 20.8% (see Fig. 1).

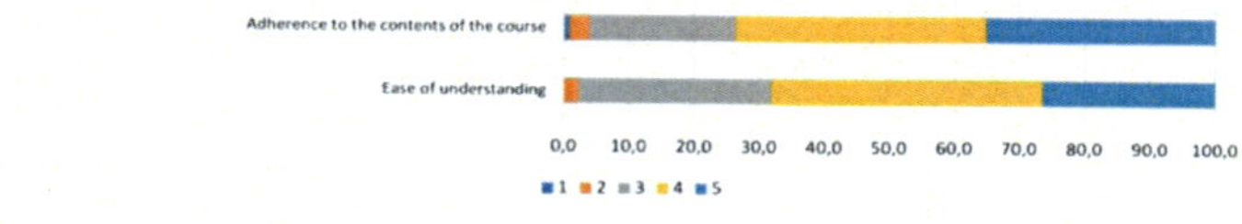

Fig. 1. Evaluation data on the quality of online meetings with stakeholders (%).

Concerning self-assessment of the professional competences promoted during the course, respondents give the highest mark to the competence of "Organising activities, workshops and educational paths," followed by "Independently carrying out research to create educational activities" and "Identifying students' needs, problems and barriers to learning" where thy consider themselves as very competent (33.6%, 32.8%, and 29.6%, respectively). The competence level of "Creating materials functional to educational interventions" is also self-assessed with high scores by the students: 23.2% of the participants perceive a maximum level (4 out of 4) of this professional competence at the end of the course and 58.4% perceive a level of 3 out of 4. The competence of "Team planning educational and research activities" is also self-assessed with high scores by the students (36% self-assessed the maximum level of competence): the reason is associated with the type of group activity set up for most of the participating students (118 out of 125), which proved to be particularly effective also in online education contexts. The competence related to "Planning educational interventions in partnership with other institutions" is the one which receives the lowest self-assessment scores from the students: 38.4% of the respondents score this competence 2 (out of 4).

Interesting results emerge from the analysis of the data concerning self-assessment of transverse competences. In the questionnaire, students had the possibility to select the 3 transversal competences that they felt had been most demanded during the course, either through online meetings with stakeholders or through the work-based activities. Collaboration, creativity, and CT were rated as the most supported skills, having been selected respectively by 91%, 67%, and 63% of students. These are 3 of the 4C skills, already defined by Trilling and Fadel [21] as the basic skills necessary for the promotion of more complex competences, skills, and attitudes. Moreover, the constructs of collaboration, creativity, and CT skills present similar characteristics and indicators: for example, the "Novelty" indicator, used in CT assessment, is closely related to the competence of creativity and divergent thinking. Furthermore, Collaboration is defined as an essential condition for the promotion of CT, considered as a competence that is realised living in society [22]. Research abilities and communication skills were also rated positively: 59% and 58% of students, respectively, selected them as the three most promoted skills during the course. The skills least selected by the students were problem solving (30%) and innovation (21%), even if some work-based activities were designed to foster such skills (see Fig. 2).

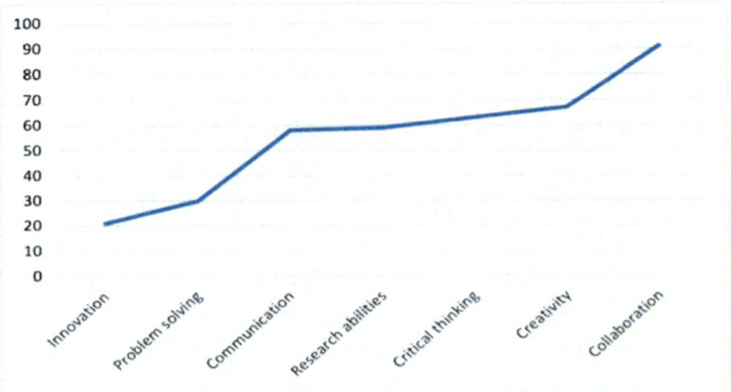

Fig. 2. Self-assessment of transversal competences most stimulated during the modules (%).

3.2 Evaluation of CT Skills

In total, 125 students participated to the learning experience. The first group of participants is composed of 7 students from the master's degree course in pedagogical sciences attending the second-year module in "Experimentalism, Museum and Reading."

The averages of the assessments carried out in the double-blind mode by three evaluators were analyzed through basic statistical analyses in order to verify the level of solicitation of the following CT indicators: use of language, argumentation, relevance, importance, critical evaluation, and novelty. Analyses show that the average scores assigned to 4 out of 6 indicators increase from the first activity to the last one: the mean of the scores assigned to the first group of students increases respectively in the indicators of *justification* (from 3.6 to 3.8 points), *relevance* (from 3.5 to 3.9 points), *importance* (from 2.3 to 3.7 points), and *critical evaluation* (from 3.4 to 3.6 points). The average score obtained by the participants changes from 21 points in the first activity to 22 points in the last activity (first activity ds = 2.98; last activity ds = 2.76). The average score in outgoing activity is always higher than the average score in incoming one, except for the indicators *use of language* and *novelty*.

Students participating in the four activities slightly increase their ability into justify their opinions (*argumentation*) and also slightly improve the ability to evaluate sources, data, and background knowledge using a personal and critical elaboration (*critical evaluation*). The Wilcoxon signed rank test shows that the improvement in participants' *importance* levels from the first to the last activity is statistically significant ($P = 0.048$).The second group of participants is composed by 118 students from the master's degree course in primary school education attending the first year course in "*Research Methodology in Education,*" who were divided in 17 groups. The average score obtained by the participants increased from 18.7 points (ds = 2.4) in the first activity to 22.5 points (ds = 2.7) in the last activity. The data analysis shows that the average score given by the participants increased in all the indicators: language use (pre-test = 3.14, ds = 0.6; post-test = 3.7, ds = 0.4), argumentation (pre-test = 2.9, ds = 0.5; post-test = 3.7, ds = 0.5), relevance (pre-test = 3.3, ds = 0.5; post-test = 3,8, ds = 0,4), importance (pre-test = 3, ds = 0,3; post-test = 3,8, ds = 0,4), critical evaluation (pre-test = 3,2, ds = 0,6; post-test = 3,9, ds = 0,5), novelty (pre-test = 3, ds = 0,7; post-test = 3,7, ds = 0,6). As shown in Fig. 3, students participating in the four activities increase their writing skills, demonstrating improvements in the *use of the language* indicator. They also show a higher capability of justifying their opinion (*argumentation*). They learned to be more relevant and consistent (*relevance*) in their expressions and to better link their previous knowledge with new situations (*importance*). Moreover, they show a higher ability into evaluating sources, data, and background knowledge using a personal and critical elaboration (*critical evaluation*), and using unusual terms to elaborate their ideas (*novelty*).

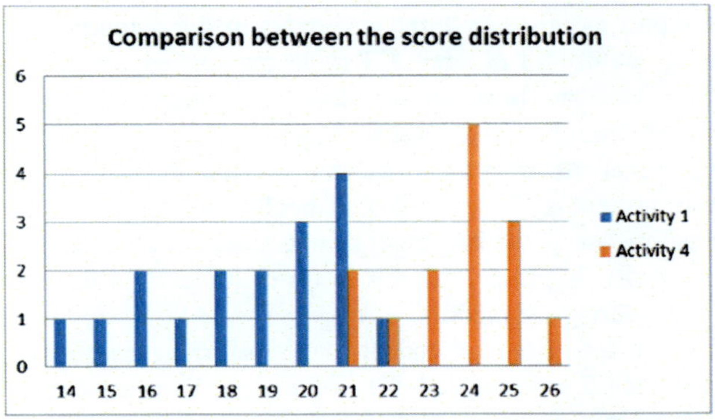

Fig. 3. CT score distribution in the first and in the last activity made by students attending the first-year module in "Research Methodology in Education."

In order to analyze whether the differences in the scores assigned to the indicators between the first and last activity were statistically significant, the Wilcoxon signed-rank test was used: the results show a significance $P < 0.01$ for all six indicators analyzed (see Table 3).

Table 3. Results of the Wilcoxon signed-rank test for the mean scores assigned to CT indicators in the first (T1) and last (T4) task.

	T1_UoL – T4_UoL	T1_Jus – T4_Jus	T1_Rel – T4_Rel	T1_Imp – T4_Imp	T1_CritEv – T4_CritEvt	T1_Nov – T4_Nov
Z	$-2{,}692^{b}$	$-3{,}293^{b}$	$-2{,}898^{b}$	$-3{,}470^{b}$	$-2{,}560^{b}$	$-2{,}510^{b}$
Sign. asint (two-tailed)	0,007	<0,001	0,004	<0,001	0,010	0,012

4 Discussion and Final Remarks

CT skills are increasingly considered as pivotal for human and social progress. For this reason, the creation of learning modules aimed at promoting CT should be enhanced at all levels of education. The present paper aims at illustrating some results of a pilot experience carried out by CDM research group based at University of Roma Tre for the purpose of increasing CT skills in HEI students through work-based activity and online meetings with stakeholders from the educational field. The pilot experience can be considered an example of continuous instructional improvement.

Data analyses show a generally positive evaluation of the pilot experience carried out in the academic year 2020–2021 by 125 university students attending two different modules, "Experimentalism, Museum and Reading" and "Research Methodology in Education." The analysis of CT levels carried out using a specific evaluation rubric shows that the scores assigned by three evaluators to the outgoing activity is higher than the incoming activity for different CT indicators: as regards the students participating in "Experimentalism, Museum and Reading" module (N. 7), the improvement of *importance* levels from first to the last activity is statistically significant ($P < 0.05$). Regarding the students attending the "Research Methodology in Education" (N. 118) module, the improvement in CT levels is statistically significant for all the CT indicators under evaluation: use of the language, justification, importance, relevance, critical evaluation, and novelty. In general, the CT levels of students participating in the pilot experience improve during the module.

At the end of the activities, the students participating in the pilot experience self-assessed their level of CT as very good: CT is one of the three most selected transverse competences defined by the students as more solicited through the online meetings with stakeholders and work-based activities, together with collaboration and creativity. In addition, some professional competences were also positively self-assessed by the students, such as "Organising activities, workshops and educational paths," "Independently carrying out research to create educational activities," and "Identifying students' needs, problems and barriers to learning." Stakeholder meetings and work-based activities are also evaluated positively by the students, especially in terms of didactic tools, language, and exhaustiveness of content.

The pilot experience has large room for improvement, but it presents some important indications for the implementation of online learning paths aimed at enhancing CT skills with the support of external stakeholders and through continuous instructional learning design.

Acknowledgments. A. Poce coordinated the research presented in this paper. Research group is composed by the authors of the contribution that was edited in the following order: A. Poce (1 and 4), Re M. R. (2.2 and 3.2), C. De Medio (2.3), M. Valente (2.1) and A. Norgini (3.1).

References

1. Facione, P.A.: Critical Thinking: What it is and Why it Counts. California Academic Press, San Jose (1992 updated to 2015)
2. Bailin, S., Case, R., Coombs, J.R., Daniel, L.B.: Conceptualizing critical thinking. J. Curric. Stud. **31**(3), 285–302 (1999)
3. Willingham, D.: Critical thinking. Why is it so hard to teach? Am. Educ. **31**(2), 8–19 (2007)
4. Grant, A.: Think Again: The Power of Knowing What You Don't Know. Penguin Publishing Group, London (2021)
5. World Economic Forum: The Future of Jobs. Employment, Skills and Workforce Strategy for the Fourth Industrial Revolution. World Economic Forum, Geneva (2016)
6. Scott, C.L.: The Futures of Learning 2: What Kind of Learning for the 21st Century? ERF Working Papers Series, No. 14. UNESCO, Paris (2015)
7. Davies, M., Barnett, R. (eds.): The Palgrave Handbook of Critical Thinking in Higher Education. Springer, New York (2015)
8. Poce, A. (ed.): Contributions for the Definition of a Critical Technology: An Assessment Study. FrancoAngeli, Milan (2012)

9. Chew, S.: What Pandemics Can Teach Us about Critical Thinking and Metacognition (2020). https://www.improvewithmetacognition.com/pandemics-and-metacognition/. Accessed 12 Mar 2020
10. Facione, N.: Examples of critical thinking strategies used by leadership. How are our best leaders are using critical thinking during this COVID-19 pandemic? (2020). https://www.insightassessment.com/blog/critical-thinking-strategies-for-the-covid-19-pandemic. Accessed 12 Mar 2020
11. Hage, G.: The haunting figure of the useless academic: critical thinking in coronavirus time. Eur. J. Cult. Stud. **23**(4), 662–666 (2020)
12. Dominguez, C. (coord.): A European Review on Critical Thinking Educational Practices in Higher Education Institutions. UTAD, Vila Real (2018)
13. Facione, P.A.: Critical Thinking: A Statement of Expert Consensus for Purposes of Educational Assessment and Instruction—The Delphi Report. California Academic Press, Millbrae (1990)
14. Halpern, D.: Manual Halpern Critical Thinking Assessment. Schuhfried GmbH, Mödling (2016)
15. Poce, A., Amenduni, F., Re, M.R., De Medio, C., Norgini, A.: Correlations Among Natural Language Processing Indicators and Critical Thinking Sub-Dimensions in HiEd Students. Form@re - Open Journal Per La Formazione in Rete **20**(3), 43–67 (2020)
16. Poce, A.: Verba Sequentur: Thinking and Writing for a Critical Development of Competences at Secondary School. FrancoAngeli, Milan (2017)
17. Nicol, D.: E-assessment by design: Using multiple-choice test to good effect. J. Further High. Educ. **31**, 53–64 (2007)
18. Popham, W.J.: Test Better, Teach Better. The Instructional Role of Assessment. ASCD, Alexandria (2003)
19. European Commission: Work-Based Learning in Europe: Practices and Policy Pointers. European Union
20. Poce, A., Agrusti, F., Re, M.R.: Sviluppo di uno strumento di valutazione delle risorse aperte (OERs). Analisi dei dati raccolti: abitudini nell'uso della tecnologia e di scrittura. CADMO Giornale Italiano di Pedagogia sperimentale. Int. J. Educ. Res. **23**(2), 86–92 (2015)
21. Trilling, B., Fadel, C.: 21st Century Skills Learning for Life in Our Times. Wiley, San Francisco (2009)
22. Ennis, R.H.: A taxonomy of critical thinking dispositions and abilities. In: Baron, J.B., Sternberg, R.J. (eds.). Series of Books in Psychology. Teaching Thinking Skills: Theory and practice, pp. 9–26. W H Freeman/Times Books/Henry Holt & Co (1987)

Teaching Programming Skills to Blind and Visually Impaired Learners

Sarantos Psycharis[1,5], Paraskevi Theodorou[2,5](✉), Ypatios Spanidis[3], and Pantelis Kydonakis[4]

[1] School of Pedagogical and Technological Education (ASPETE), Athens, Greece
[2] Department of Digital Systems, University of Piraeus, Piraeus, Greece
[3] Department of Surgery, School of Medicine, University of Patras, Patras, Greece
[4] MSc in Informatics, Department of Informatics, University of Piraeus, Piraeus, Greece
[5] Hellenic Education Society of S.T.E.M. (E3STEM), Nea Filadelfia, Greece

Abstract. Learning to program is a challenging process for children with visual impairment (VI). Despite the development of technology, there are still several social and technical barriers that each learner with VI has to overcome. Indeed, many programming environments use a screen reader, thus, making them inaccessible for students with VI. In this work, we aimed to examine and discuss practical issues, as well as suggest potential alternatives options for teaching programming to children with VI. For this purpose, we conducted a review of literature published up to March 2021. We have highlighted that the use of block-based, language-based, and auditory languages seems to be more appropriate. Moreover, special attention should be given to the inability of learners with VI to understand the overall structure of a code. Furthermore, we present the first from a series of applications under development by our team. Finally, we provided evidence that supports the use of tangible-based applications, robots, and microcontrollers that could help the programming learning process. The main outcome of our review is that further research is needed to examine and develop new and more effective strategies to introduce children with VI to the world of programming.

Keywords: Visually impaired · Programming skills · Computational thinking · Educational robots · Mobile game-based learning

1 Introduction

An important issue in contemporary society is to ensure inclusive and equitable education and learning opportunities for all citizens, including those with disabilities, such as visual impairment (VI), in the framework of the Universal Design Learning (UDL). This term includes both those who are blind and those with low vision [1]. According to the World Health Organization, approximately 285 million individuals worldwide have VI [2]. Despite the fact that many of them try to socialize and have equal education opportunities, they still face several difficulties, which prevent them from making use of these opportunities.

D. Guralnick et al. (Eds.): TLIC 2021, LNNS 349, pp. 270–279, 2022.
https://doi.org/10.1007/978-3-030-90677-1_26

Nowadays, an increasing number of countries have included programming and computational learning into education and national curricula [3]. To achieve this goal, many educational tools are needed, such as block-based languages [4]. The rapidly evolving technological development can be helpful in our effort to provide equal educational opportunities in individuals with VI. Nevertheless, the formalization of computational learning highlights the lack of appropriate tools for many children with disabilities. Indeed, although programming is challenging for children, for learners with VI, there is still a number of additional barriers, which need to be overcome. Interestingly, previous works have examined the accessibility [5], bias [6], navigation misinformation [7], and usability [8] of technology for such individuals. However, in several countries, the inclusion of individuals with VI in education procedures was found to be limited because of the limited financing, involvement of people with disabilities in decision-making, and knowledge regarding the needs of these individuals [9]. Therefore, it seems that there is an increasing need to generate accessible learning units oriented toward the development of computational thinking practices for VI learners, which is at the same time important to better equip them for living and working in an increasingly digital world, while it is also challenging, as there is a tendency to rely on visual representations to convey abstract concepts.

In this study, we conducted a review of literature on education, computational thinking, and educational technologies for students with VI. Our main aim was to provide an overview of the different strategies that have been used to make learning programming accessible to VI learners. To this end, we performed an extensive search in Google Scholar and the PubMed databases for works published up to March 2021. Additionally, we present the first of a series of educational applications that are being developed by our team, incorporating existing technologies for the Android operating system to be used by students with VI. Can the adoption of mobile-assisted learning tools help visually impaired students to acquire knowledge on equal terms as those without such an impairment when more emphasis is placed on the graphic presentation?

2 Physical Programming Languages

2.1 Accessibility

Recent works have demonstrated that several programming environments are not friendly for individuals with VI, as they do not provide auditory support or screen reading [10, 11]. In contrast, Visual Studio (2010) is technically accessible. However, there is no sound to indicate when the user switches between tabs. Thus, it is very challenging for children with VI to develop an overall understanding of code structure and the relationships among the different elements [12, 13]. The use of a standard text editor alongside a screen reader has been attempted, but it lacks debugging tools, which are standard in most programming environments [11].

Moreover, the choice of programming language is also of great importance. The most frequently used languages, such as C and Java, use non-alphanumeric characters (i.e., brackets and curly braces), making it challenging to work while using a screen reader [14]. In addition, many languages have a complex syntax, which can increase the possibility of errors [14].

Based on the above, it is clear that programming for children and, especially, those with VI is challenging and demands great effort to overcome several obstacles. Block-based languages create a useful and efficient tool in programming courses. Nevertheless, many of them that have a visual nature are not accessible to students with VI. Conversely, physical programming languages using physical blocks or pods to represent a command are more accessible, as each block could be distinguishable by touch [9, 16]. Therefore, a question could be raised regarding the use of text-based languages, as alternatives of block-based languages.

Indeed, languages that use text and limit the use of non-alphanumeric characters, such as "Python," are preferable [14]. It should be noted that the accessibility of text-based languages is largely dependent on the programming environment, with many being inaccessible. Interestingly, Kane et al., in their study, enrolled 12 patients with VI and found that they could successfully write programs in "Ruby" [14]. Moreover, other text-based languages (i.e., Audio Programming Language [APL]) have been developed especially for individuals with VI [15]. The latter language uses a reduced set of commands, which can be accessed and selected through a circular command list, with no requirement for memorization [16]. In addition, an American team performed a study and evaluated the accessible programming environment "Sodbeans," along with the developed "Hop" programming language. The first language focuses on middle to high-school students and uses audio cues for navigation along with an auditory debugger. The researchers showed that their participants presented increased self-efficacy after participating in a programming workshop that used these two languages [17]. Interestingly, the Hop language was further developed and led to the creation of "Quorum," which is accessible for all users, even those with VI [17].

2.2 Code Navigation

A common issue reported in many studies is the difficulty that the individuals with VI face in navigating their code and in understanding the structure when they use a screen reader [10, 14]. This issue could lead to mistakes when inserting code, as in some cases, the learners insert it at an incorrect position [18]. Therefore, it is very challenging for children to develop an overall understanding of code structure and the relationships among the different elements [18].

However, this issue was taken into consideration by the programmers when developing "StructJumper." The aforementioned is a plugin for the Eclipse programming environment that enables individuals with VI to navigate through a program written in Java [19]. Especially, this language uses nested structures, which enables the user to easily jump between each nested structure in the code [19].

"Code Jumper," a tangible, music-coding environment, is also another code-learning toy that is developed by Microsoft for children with VI, especially those aged between 7 and 11 years [20]. However, it can also be used for those without VI. Specifically, it uses physical pods to represent commands that can be joined together to construct programs that produce sound in the form of music, stories, and poetry; therefore, it introduces foundational programming knowledge to learners [20]. Among its features are the following: it supports screen readers and Braille display for students who are blind and large font and high contrast for those with low vision; provides an auditory feedback on all components (i.e., Code Jumper application); enables teachers and parents who do not have any prior programming experience to teach students the basics of coding; includes a curriculum with student activities and teacher guidance, such as educational videos and activities; includes assessments to gauge students' skills and understanding; and accompanies the required Code Jumper kit, which is composed of a set of physical "pods" that are connected to build computer programs [21] (Fig. 1).

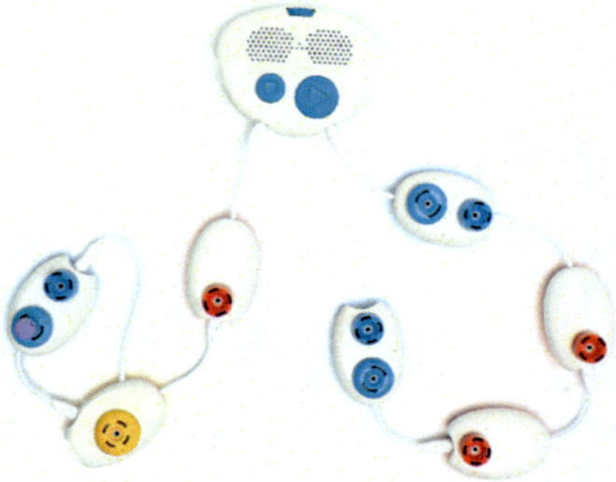

Fig. 1. Code jumper.

2.3 Physical Computing Devices

Physical computing devices (i.e., robots and microcontrollers) can be effectively used by amateur young programmers who have VI. Especially, the BBC micro:bit is extensively used in the United Kingdom (UK) since 2016 because it has a low cost and plug-and-play nature. Therefore, it is an accessible choice that has influenced the lives of several children with VI and introduced them into the world of programming [22]. Nevertheless, Python is more preferable compared to BBC micro:bit, as it is the most popular text-based programming language in UK and the micro:bit implementation of Python has a basic speech synthesizer [22].

Another programming environment, named Mu, has been recently introduced to help students program the micro:bit using Python. However, it does not work with screen readers; thus, it could not be used by children with VI. Nevertheless, a command-line tool (named uFlash) can be used to transfer a Python program onto a connected micro:bit. This tool is used in combination with NotePad++ (i.e., a text editor) in order to define keyboard shortcuts [23, 24]. The users write their Python program in Notepad++ and press a keyboard shortcut when they are ready to test it on micro:bit [23, 24].

3 Tangible-Based Applications

As mentioned previously, children with VI use the sense of touch and hearing in their education. Therefore, the use of an education tool should be primarily based on these two senses to be accessible [25]. A tangible user interface-based software has been suggested for teaching tactual shape perception and spatial awareness sub-concepts in small-scale space in children with VI [25]. It provides three-dimensional-printed objects that feature the place values in Braille along the top and a series of slots that "bits" can be placed into [25]. Children can feel these objects, place them on the surface, and receive audio feedback regarding shape and spatial relationships in the context of various learning activities. Each "bit" corresponds to a separate tile with the Braille representation for 0 on one side and 1 on the other. Interestingly, each student can represent numbers by flipping the "bits" to the correct side and placing them under the appropriate place value. Thus, such applications can be effectively used to design programs [25].

4 Game-Based Learning

Game-based learning not only creates games for students to play, but designs learning activities as an integral part of the games that can gradually introduce concepts and guide users to new skills [26]. Games offer self-affirmation (in other words, students know when they are correct by the way the game progresses) and prepare them for the next stage of their development. Through games, people expand their capabilities. They can help develop social interaction, self-expression, affirmation, problem-solving, experimentation, and strategy development. These abilities and skills, of course, can then be applied to the real world [27].

In recent years, there has been a push to introduce coding and computational thinking to children, and robotics is a great tool to achieve this [28]. Educational robots can help children develop computational thinking and develop new skills that can apply to a variety of situations to solve problems in almost all areas of their lives [29].

Programmable floor robots, such as Bee-Bot, are specially designed robots to be used even by preschoolers, but especially in the first grades of elementary school. They are programmed with keys (on-board) by entering commands (forward, backward, turn left and turn right), which are then executed when the start button is pressed. In this way, they can be programmed to move the robot precisely in a predetermined space [30].

Through this process, children learn by "playing" at important programming principles, such as coding — the (logical) series of individual commands, which are then processed one after the other (sequentially) when the program is called. This approach promotes problem-solving skills in children, as sequences are one of the three basic algorithm structures in coding, along with selections and loops.

However, there are practical problems when visually impaired children use games to develop skills — even more so when it comes to mobile video games, where there is a great deal of emphasis on visual dimensions, which makes them most inaccessible to the visually impaired.

5 Our Ongoing Implementation

Our goal is to make a sound-based game accessible to the visually impaired, which is based on the last category. We have developed a mobile game that can be played with keys as well as with voice commands, allowing children with VI to play on equal terms and share the same gaming experience.

The game concept is based on the logic of the educational robot "Bee-Bot," as mentioned above, and is planned to deploy as an Android application. The development environment selected is App Inventor because of the built-in features it provides for speech recognition and text-to-speech (TTL) conversion.

The student should guide a character (a turtle) to a target (a flower) by specifying a path on a grid. The turtle always starts from the upper left corner, while the position (row and column) of the target is always stated. Therefore, the students with VI should mentally plan each step (direction and number of steps in this direction) as they move in the grid.

As mentioned above, commands can be given either via buttons or by voice commands (forward, backward, turn left, turn right). These commands are given sequentially. Once the route has been determined, the player must press the "Go" button or give the voice command "Start" to execute the commands.

During the execution and as the turtle moves in the grid, the commands are reported aloud. If the turtle exceeds the limits of the grid, the game starts from the beginning. In each iteration, our target changes position randomly (Fig. 2).

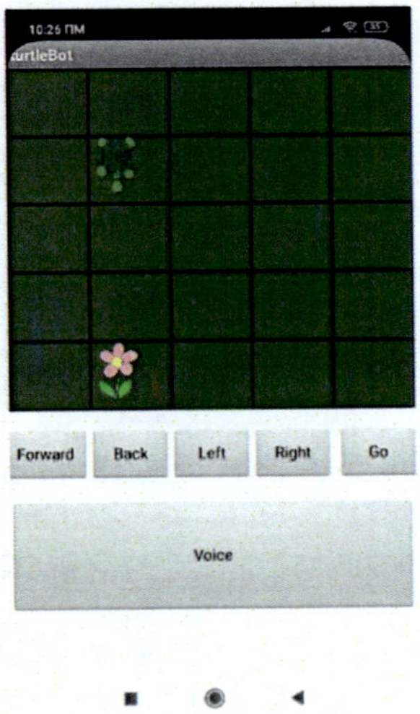

Fig. 2. First application screenshot.

A differentiated and improved version of the application that is being developed includes some obstacles, which students should avoid in trying to reach the target. As a result, another level of difficulty is added to the game. These obstacles are not announced at the beginning of the game, so the students have to discover them while trying to find the path to the goal. This is the first attempt of our team for the design and the implementation of the applications in this field. A series of applications will follow (Fig. 3).

Fig. 3. Second application screenshot.

Similar to individuals with VI who create concept maps to explore an environment, the environment, in these games, is revealed through exploration, actions, and decisions whose results are not known in advance. They are expected to employ algorithms similar to the ones they use in daily life navigation. The ultimate goal is to equip students with VI with skills to tackle the difficulties they face in real life.

6 Pilot Study

Putting ourselves in the user's position, we took the time to think about the people who will use our application and how they will interact with the app environment. Although we intended to test the app in the field, due to the constraints of the COVID-19 pandemic, we had difficulty in visiting special education schools. This is now, therefore, a future plan.

As a pilot study, the first step after the implementation of the application was to ask some experts to use and test this app and its differentiated versions in order to check its feasibility and its usability. Through this method, we tried to understand how they perceive the capabilities that they are offered from the app and how we will improve the interaction with the app interface. Additionally, we noted the steps that a user should take using the app and created a manual to assist him in better comprehension. Additionally, we tried similar apps in order to gain a clear understanding of how users will interact with the interface and how they will improve their programming skills. Finally, we gradually developed the user interface, taking into account the comments of these experts and the aforementioned similar apps.

7 Conclusions

Nowadays, the need to provide equal educational opportunities for young individuals with disabilities, such as VI, remains on the spotlight. Although the scientific research has focused on interventions and student engagement in several educational fields, the knowledge of teaching programming to children with VI remains limited. In this study, we aimed to summarize the currently available options for learning programming that seem to be suitable and accessible to children with VI.

According to the literature that our team has reviewed, several guidelines have been proposed for educators and developers who work with children with VI. The use of block-based languages is suggested as the more suitable alternative for teaching programming, while 3D artefacts can be used to teach simple programming concepts (i.e., writing an algorithm to move a bee in a tactile grid). More specifically, abstract concepts taught using visual representations can often be effectively taught to children with VI using 3D artefacts.

In addition, the choice of language is important when the learners are children with VI. Languages with limited use of non-alphanumeric characters seem to be more suitable. Moreover, the programming environment should be accessible using a screen reader. Furthermore, in cases of an inappropriate environment for learners with VI, the use of a plain text editor seems to be efficient.

Finally, tasks that involve programming of a physical device (i.e., a robot) are more engaging. Indeed, the choice of an appropriate theme for programming is crucial and supports the engagement of learners with VI.

In this research, we present an effort, taking advantage of innovative technologies such as TTL and voice recognition, and create a mobile game adapted and accessible on equal terms to blind and visual impaired students. Taking into account the different ways they learn, we propose this method with successful results after trials on the field with experts.

The biggest barrier to integrating children with VI into computer science-related activities is that existing methods rely heavily on the use of text-based or graphics-based educational environments, making them inaccessible to people with VI. In addition, these educational environments require teachers to spend a lot of time designing and adapting to the needs of children with VI [31].

However, in recent years, screen readers, as well as voice commands, have been integrated functions in all mobile devices that are widely used by children with VI. The design of mobile apps for computer science concepts has a significant supplemental learning effect as a mobile-assisted learning tool. The more user-friendly and practical these learning objects are, the easier it is to access computer science. According to [32], it seems that parameters such as ease of use, consistency, operability, perceived usefulness, minimal memory load, system usability scale were found to have a positive effect on user satisfaction and experience of the user.

In the future, the studies should focus on the differences observed concerning the effectiveness of the proposed methods between the two sexes. The research should be performed with an aim to help bridge the digital divide. We hope that our work would help and motivate the development of new tools in order to enhance the education of students with VI in the field of programming and, especially, in exploring, analyzing, and visualizing data.

References

1. Corn, A.L., Erin, J.N.: Foundations of Low Vision: Clinical And Functional Perspectives. American Foundation for the Blind (2010)
2. Shakespeare, T., Officer, A.: World report on disability. Disabil. Rehabil. **33**(17–18), 1491 (2011)
3. Grout, V., Houlden, N.: Taking computer science and programming into schools: the Glyndŵr/BCS turing project. Proc. Soc. Behav. Sci. **141**, 680–685 (2014)
4. Maloney, J., Resnick, M., Rusk, N., Silverman, B., Eastmond, E.: The scratch programming language and environment. ACM Trans. Comput. Educ. **10**, 16 (2010)
5. Giraud, S., Thérouanne, P., Steiner, D.: Web accessibility: filtering redundant and irrelevant information improves website usability for blind users. Int. J. Hum Comput. Stud. **111**, 23–35 (2018)
6. Du, J.T., Haines, J.: Working with Indigenous communities: reflections on ethical information research with Ngarrindjeri people in South Australia. Proc. Assoc. Inf. Sci. Technol. **55**(1), 794–796 (2018)
7. Guerreiro, J., Ahmetovic, D., Kitani, K., Asakawa, C.: virtual navigation for blind people: building sequential representations of the real-world (2017)
8. Tekli, J., Issa, Y.B., Chbeir, R.: Evaluating touch-screen vibration modality for blind users to access simple shapes and graphics. Int. J. Hum Comput. Stud. **110**, 115–133 (2018)
9. Alotaibi, H., Al-Khalifa, H.S., AlSaeed, D.: Teaching Programming to students with vision impairment: impact of tactile teaching strategies on student's achievements and perceptions. Sustainability **12**(13), 5320 (2020)
10. Ludi, S.: Position paper: Towards making block-based programming accessible for blind users. In: 2015 IEEE Blocks and Beyond Workshop (Blocks and Beyond), pp. 67–69 (2015)
11. Cheong, C.: Coding without sight: teaching object-oriented java programming to a blind student, pp. 1–12 (2010)
12. Stefik, A., Siebert, S., Stefik, M., Slattery, K.: An empirical comparison of the accuracy rates of novices using the Quorum, Perl, and Randomo programming languages. In: Proceedings of the 3rd ACM SIGPLAN Workshop on Evaluation and Usability of Programming Languages and Tools, pp. 3–8 (2010)
13. Morrison, C., Villar, N., Hadwen-Bennett, A., Regan, T., Cletheroe, D., Thieme, A., Sentence, S.: Physical programming for blind and low vision children at scale. Hum. Comput. Interact. **36,** 1–35 (2019). https://doi.org/10.1080/07370024.2019.1621175
14. Kane, S.K., Bigham, J.P.: Tracking @stemxcomet: teaching programming to blind students via 3D printing, crisis management, and twitter. In: Proceedings of the 45th ACM Technical Symposium on Computer Science Education, pp. 247–252 (2014)
15. Sánchez J., Aguayo F.: APL: audio programming language for blind learners BT - computers helping people with special needs, pp. 1334–1341 (2006). https://doi.org/10.1007/11788713_192
16. Sánchez, J., Aguayo F.: Blind learners programming through audio. In: CHI 2005 Extended Abstracts on Human Factors in Computing Systems, pp. 1769–1772 (2005)
17. Stefik, A.M., Hundhausen, C., Smith, D.: On the design of an educational infrastructure for the blind and visually impaired in computer science. In: Proceedings of the 42nd ACM Technical Symposium on Computer Science Education, pp. 571–576 (2011)

18. Hadwen-Bennett, A., Sentance, S., Morrison, C.: Making programming accessible to learners with visual impairments: a literature review. Int. J. Hum. Comput. Stud. **2**(2), 3–13 (2018)
19. Baker, C.M., Milne, L.R., Ladner, R.E.: StructJumper: a tool to help blind programmers navigate and understand the structure of code. In: Proceedings of the 33rd Annual ACM Conference on Human Factors in Computing Systems, pp. 3043–3052 (2015)
20. APH: Code Jumper (2020). https://codejumper.com/
21. Kesselman, M.A., Wilson, E.: Conference report, consumer electronic show, Las Vegas, part 2. Libr. High Tech News **37**(6), 15–18 (2020)
22. Physical computing and visually impaired learners – Computer Science Education @ King's. https://blogs.kcl.ac.uk/cser/2018/01/08/physical-programming-and-visually-impaired-learners/
23. Schmidt, A.: Increasing computer literacy with the BBC micro: bit. IEEE Pervasive Comput. **15**(2), 5–7 (2016)
24. Ball, T., Protzenko, J., Bishop, J., Moskal, M., Halleux, J.D., Braun, M., Hodges, S., Riley, C.: Microsoft touch develop and the BBC micro: bit. In: 2016 IEEE/ACM 38th International Conference on Software Engineering Companion (ICSE-C), pp. 637–640 (2016)
25. Jafri, R., Aljuhani, A.M., Ali, S.A.: A tangible interface-based application for teaching tactual shape perception and spatial awareness sub-concepts to visually impaired children. Proc. Manuf. **3**, 5562–5569 (2015)
26. Pho, A., Dinscore, A.: Game-based Learning. Tips and Trends. http://acrl.ala.org/IS/wp-content/uploads/2014/05/spring2015.pdf (2015)
27. Trybus, J.: Game-Based Learning: What it is, Why it Works, and Where it's Going. New Media Institute. http://www.newmedia.org/game-based-learning-what-it-is-why-it-works-and-where-its-going.html (2012)
28. Bers, M.U., González-González, C., Armas-Torres, M.B.: Coding as a playground: promoting positive learning experiences in childhood classrooms. Comput. Educ. **138**, 130–145 (2019)
29. Catlin, D., Woollard, J.: Educational robots and computational thinking (2014)
30. Papadakis, S.: Robots and robotics kits for early childhood and first school age. Int. J. Interact. Mob. Technol. **14**, 18 (2020)
31. Milne, L.R., Ladner, R.E.: Blocks4All. In: Proceedings of the 2018 CHI Conference on Human Factors in Computing Systems (2018)
32. Khan, A., Khusro, S., Alam, I.: BlindSense: an accessibility-inclusive universal user Interface for blind people engineering. Technol. Appl. Sci. Res. **8**, 2775–2784 (2018)

At Your Best! Artificial Intelligence, People and Business Analytics, Highly Realistic Avatars, Innovative Learning and Development

Fernando Salvetti[1,2,3](✉), Barbara Bertagni[1,2,3](✉), Filippo Negri[4], Matteo Bersanelli[4], Rosalinda Usai[5], Alessandro Milani[5], and Dalila De Rosa[5]

[1] Centro Studi Logos, 10143 Turin, TO, Italy
{salvetti,bertagni}@logosnet.org
[2] Logosnet, 6900 Lugano, CH, Switzerland
[3] Logosnet, Houston, TX 77008, USA
[4] Glickon, 20129 Milan, MI, Italy
{filippo,matteo.bersanelli}@glickon.com
[5] Cattolica Assicurazioni, 37100 Verona, VR, Italy
{rosalinda.usai,alessandro.milani,
dalila.derosa}@cattolicaassicurazioni.it

Abstract. This article is about a training project within a leading Italian insurance company, based on people analytics and a blended learning program enhanced by e-REAL® and artificial intelligence. The intervention started with an Organizational Network Analysis regarding informal and spontaneous communication from which skills and ideas arise: this is the starting point that allows the HR Team to take decisions and outline strategies to support growth. Then, a Sentiment Analysis was performed, in order to track and monitor the climate in different clusters within the company, by mapping sentiment through natural language processing algorithms (all data anonymized). This step also opened to a Tensions Analysis targeting expressions of anger or misalignment with the corporate culture. As a further step, an Employees Potential Chart was developed. As an output of this phase, the HR Team was provided with essential metrics such as performance peaks, relational networks, communication and collaboration among teams and departments, key competencies, and skills. Then, a learning and development plan based on people analytics was designed: the main elements were interactive tutorials and immersive online experiences made by a multimedia and visual storytelling approach, as well as an extensive interaction within an escape room animated also by realistic avatars. A number of other methodologies were introduced as a further step: first of all training on-the-job, coaching, and mentoring. All the interventions were based on a large amount of people analytics collected from gamified activities, allowing for micro-targeted and highly precise interventions that are very innovative in the field of learning and development.

Keywords: People analytics · e-REAL · Interactive learning

D. Guralnick et al. (Eds.): TLIC 2021, LNNS 349, pp. 280–289, 2022.
https://doi.org/10.1007/978-3-030-90677-1_27

1 At Your Best! People and Business Analytics, Data Science, Avatars, Immersive and Interactive Learning

1.1 Organizational Network Analysis

This article is about a training project for insurance claims liquidators, developed within a leading Italian insurance company, based during a first stage on people analytics and data science applied for human resource development, then on multimedia—immersive and interactive—learning tools.

The intervention started with a first phase designed and implemented by Glickon, an Italian consulting firm in people analytics and HR assessment and development. A first step was performing an Organizational Network Analysis (ONA) regarding informal and spontaneous communication from which skills and ideas arise: this is the starting point that allows the HR Team to take decisions and outline strategies to support growth. ONA is a structured way to visualize how communications, information, and decisions flow through an organization. Organizational networks consist of nodes and ties, the foundation for understanding how information in your organization is flowing, can flow, and should flow.

In every organization, people build informal "go-to" teams. They rely on that one person who always knows "how we do things here." They find someone in finance who can answer any budget question. These spontaneous, critically important connections are the lifeblood of organizations worldwide. Every organization has people (nodes) who serve as critical conduits for the exchange of ideas and information.

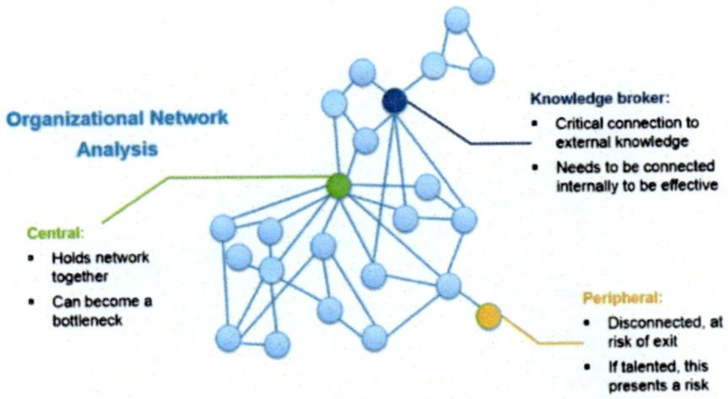

Fig. 1. Representative actors from the organizational network.

In the ONA framework, a connection delivers value when needed information is exchanged (Fig. 1).

- *Central node:* These are the people who seem to know everyone. Central nodes share lots of information and influence groups quickly. Central nodes can be anywhere in the hierarchy of an organization, are often well liked, and are highly engaged in company news and developments.

- *Knowledge broker:* These people create bridges between groups. Without knowledge brokers, information and idea sharing grinds to a halt.
- *Peripheral:* Easily overlooked and unconnected to the rest of the company, high-potential peripherals can be a risk to organizations. Exceptional Java coders who don't teach others best practices not only stagnate product development, but they are also easily convinced to take their talents elsewhere.
- *Ties:* Ties are the formal and informal relationships between nodes. Establishing optimal relational ties between central nodes and knowledge brokers helps ensure useful information moves easily between and within groups.

Nodes and ties are the core elements of ONA. Visualizing the relationships between nodes and ties makes it much easier to identify critical connections and potential barriers to information flow and collaboration. ONA reveals where central nodes could have the most impact—in groups that "don't talk much," for example—or where you should build stronger connections to streamline information sharing and eliminate redundancy. ONA is a method for studying communication and socio-technical networks within a formal organization. This technique creates statistical and graphical models of the people, tasks, groups, knowledge, and resources of organizational systems. It is based on social network theory and more specifically, dynamic network analysis [1].

1.2 Organizational Climate, Sentiment and Tensions Analysis

Second, a Sentiment Analysis (SA) was performed, to track and monitor the climate in different clusters within the company. Organization climate (OC) (Fig. 2) usually does not have a clear definition and is explained as a perception that every employee in

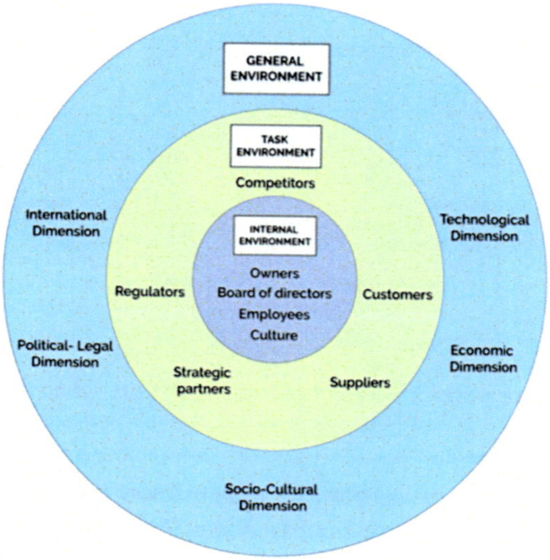

Fig. 2. Main factors related to organizational climate.

an organization feels, understands, and shares. OC is determined by factors from both internal and external environments. Internal environment refers to the factors that affect the perception of the employees that occur within the organization, and external environment factors are from outside the organization, which is difficult to predict and control. OC is the "weather" that a particular organization is having, reflecting how its' system deals with the members—which is something that is difficult to be measured because it is senses or feelings of employee perception toward the organization, and this cannot be evaluated using tools. The combination of the perceptions reflect the actual workplace of the organization. The SA was aimed at bypassing the climate measurement's issue and was performed by mapping sentiment through natural language processing algorithms [2].

SA, also known as opinion mining, is a process (automated and performed with different methodologies) aimed at recognizing positive, negative, and neutral emotions in social media comments, reviews, and survey responses—all anonymized. In SA, the accuracy can never be 100%: for example, a machine does not understand sarcasm. The machine's accuracy level is higher that standardized human perception, so manually analyzing is not an option [3] (Fig. 3).

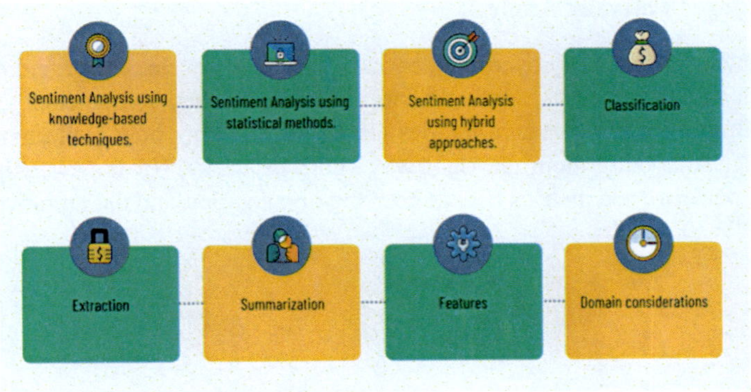

Fig. 3. Word breakdown structure of a sentiment analysis.

This step opened to a Tensions Analysis (TA), aimed at targeting expressions of anger or misalignment with the corporate culture (Fig. 4). Corporate people and executives may talk about problems in terms of dilemma, paradox, and other tensions generated by environmental forces such as culture. Such tensions usually pervade organizational and managerial cognitive and emotional frames [4].

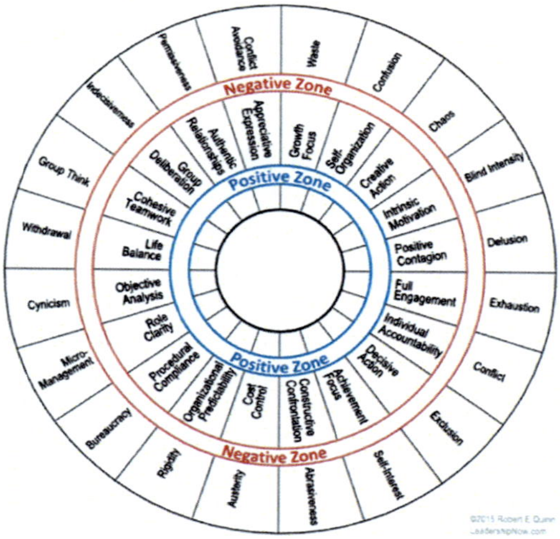

Fig. 4. A representative framework of organizational tensions.

1.3 Employee Potential Analysis

As a further step, an Employees' Potential Chart was developed, providing the HR Team with essential metrics, such as performance peaks, key competencies, and skills (Fig. 5). Employee performance metrics are key to tracking how well employees are performing. Analyzing them the right way could be tricky, but it's essential because employee performance metrics benefit both the organization and the employees [5].

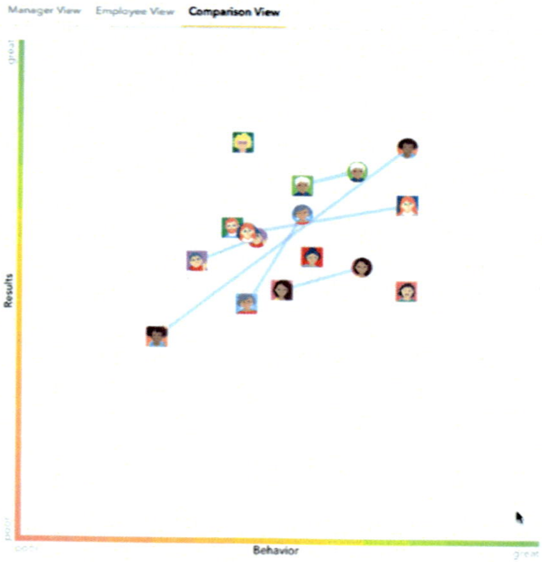

Fig. 5. A representative employees performance chart.

As an output of this project phase, the HR Team was provided with essential metrics such as relational networks, communication and collaboration among teams and departments, key competencies and skills, performance peaks, and correlations between performance and training or between performance and managerial styles (Fig. 6).

Glickon's algorithms organize and analyze data moving from an unusual starting point: quizzes, trivia games, and open questions to be addressed during gamified activities. Gamification creates engagement and produces essential information to get to know corporate people better, organized in a panel of data and insights with one main goal: to discover the potential of the human resources in order to grow the business.

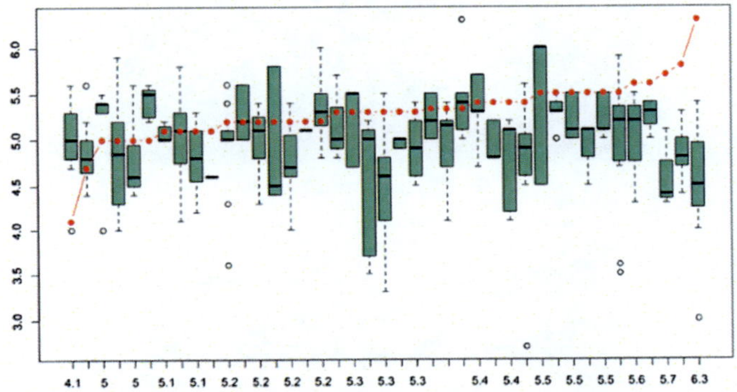

Fig. 6. Representative management and performance correlation chart.

2 The e-REAL® Blended Learning Program

2.1 Interactive Tutorials and Immersive Online Experiences

Then, as an ideal next step based on gamification, a blended learning program enhanced by the e-REAL® online platform and some artificial intelligence (mainly deep learning components for interactive digital humans) was designed.

The key components are interactive tutorials and immersive online experiences (first of all an online escape room) based on the people and business analytics from the project phase summarized above. Therefore, gamification is a key word that puts the two stages of the intervention into an ideal continuity.

Interactive tutorials consist of content along with interactive components for checking and reinforcing understanding (Fig. 7). They can include any or all of the following:

- Narratives, figures, illustrations, and equations
- Quiz questions
- Videos

These tutorials automatically preserve work done within them, so if a learner works on a few exercises or questions and returns to the tutorial later, they can pick up right where they left off. They are useful to cascade basic knowledge and know-how that all the learners have to share and master before approaching the next level.

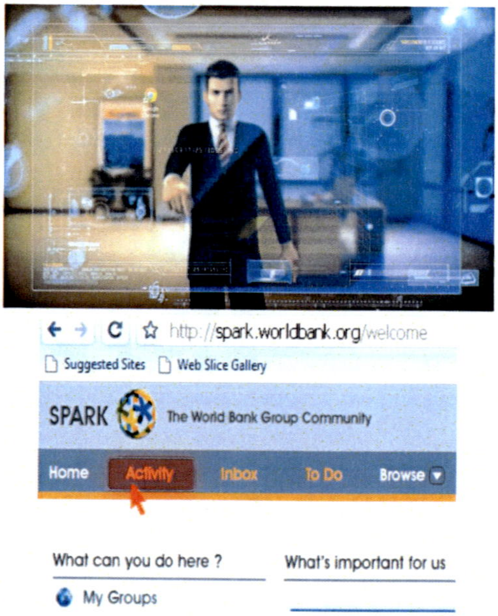

Fig. 7. Representative interactive tutorial with an avatar displaying virtual content (left) and opening a website (right) by the flick of its hands (© courtesy of Logosnet™ and World Bank Group™).

Online experiences are provided in the e-REAL® platform that is designed to engage learners in simulation scenarios and into an online escape room: for example, an office to meet—online—an actor and/or a digital human performing as colleagues to talk with, or as providers to negotiate with, or as professionals to meet in order to analyze complex damages to be avoided (i.e., to avoid a trial) (Figs. 8–10). An *actor* is a professional co-trainer performing live as a standardized character; a *digital human* (or *avatar*) is a human-like virtual being, sometimes powered by some artificial intelligence, that may interact with real human beings using verbal and non-verbal cues—like tone of voice, facial expressions, and/or body posture.

The e-REAL online platform submerges learners in an immersive reality where the challenge at hand is created by sophisticated, interactive computer animation—without the need to wear special glasses or other wearable augments. The system includes live and real-time interaction with peers, instructors, tutors, facilitators, and mentors—not only actors and digital humans. Thus, it adds a very important social component that enhances learning outputs, skills, and cognitive and metacognitive processes. Learning by doing within an immersive setting, based on multimedia and visual storytelling, leaves both

Fig. 8. e-REAL online platform: representative avatar introducing the learners within a multimedia scenario based on a Mr. Bean adventure and able to interact dialogically with the learners about specific topics (© courtesy of Logosnet™ and the Harvard Center for Medical Simulation™).

Fig. 9. e-REAL online platform: representative multimedia scenario with an avatar performing as an employee into an insurance company's office (© courtesy of Logosnet™ and Cattolica Assicurazioni™).

the learners and the trainers with memorable experiences, and produces robust learning outputs—all directly linked to the data coming from the first project phase.

The e-REAL teaching and learning approach is designed to have the learners working on tasks that simulate an aspect of expert reasoning and problem-solving, while receiving timely and specific feedback both from fellow students and the trainers. These elements of deliberate practice and feedback are general requirements for developing expertise at all levels and disciplines, focusing on specific topics and key-performance indicators collected during the first project phase [6].

Fig. 10. e-REAL online platform: representative avatar displaying and performing as the CEO of a company, in a direct talk with people invited online about specific topics (© courtesy of Logosnet™).

2.2 The e-REAL "Phygital" Classroom Experience and Further Methodologies to Use

At the end of the online immersive and interactive experience, the training project is designed to allow learners to enter a "phygital" e-REAL classroom (Figs. 11 and 12), that is a mixed-reality environment for hybrid simulation where physical and digital

Fig. 11. Representative e-REAL® "Phygital" Classroom: mixed-reality environment for hybrid simulation where physical and digital objects are available for tactile, visual, or vocal interaction in a real, brick-and-mortar, learning setting.

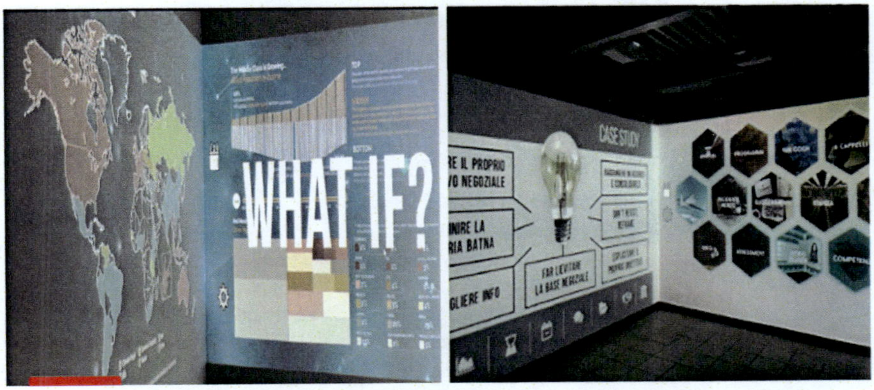

Fig. 12. Representative e-REAL® content designed for a "phygital" environment.

objects are available for tactile, visual, or vocal interaction in a real learning setting (without the use of virtual reality glasses).

Other learning and development methodologies are available, as well as a third stage of the intervention: training on-the-job, coaching, and mentoring—just to name a few. All the learning methodologies and the related content were designed starting from the large amount of data about people and business analytics collected and analyzed during the initial project's phase—based on gamified activities, as the further steps of the project. This model of intervention is allowing for micro-targeted and highly precise "answers" to the key-points, analyzed during the initial phase of the project, that are very innovative in the field of learning and development.

References

1. Merrill, J., Bakken, S., Rockoff, M., Gebbie, K., Carley, K.M.: Description of a method to support public health information management: organizational network analysis. J. Biomed. Inf. **40**(4), 422–428 (2007). https://doi.org/10.1016/j.jbi.2006.09.004
2. Hellriegel, D., Slocum, J.: Organizational climate: measures, research and contingencies. Acad. Manag. J. **17**(2), 255–280 (1974)
3. Feldman, R.: Techniques and applications for sentiment analysis. Commun. ACM **56**(4), 82–89 (2013)
4. English, T.: Tension analysis in international organizations: a tool for breaking down communication barriers. Int. J. Organ. Anal. **9**(1), 58–83 (2001)
5. Shields, J., et al.: Managing Employee Performance & Reward: Concepts, Practices, Strategies. Cambridge University Press (2015)
6. www.e-real.net

A Digital Mindset for the Society 5.0: Experience an Online Escape Room

Fernando Salvetti[1,2,3](✉), Cristiano Galli[4], Barbara Bertagni[1,2,3], Roxane Gardner[5,6], and Rebecca Minehart[5,7]

[1] Centro Studi Logos, 10143 Turin, TO, Italy
{salvetti,bertagni}@logosnet.org
[2] Logosnet, 6900 Lugano, CH, Switzerland
[3] Logosnet, Houston, TX 77008, USA
[4] Global Aerospace Logistics/Air Defense Institute of the Emirates Air Force, Abu Dhabi, UAE
cristiano.galli@nolec.training
[5] Center for Medical Simulation, Boston, MA 02129, USA
[6] Brigham and Women's Hospital/Children's Hospital/Massachusetts General Hospital, Harvard Medical School, Boston, MA 02114, USA
rgardner1@bwh.harvard.edu
[7] Massachusetts General Hospital, Harvard Medical School, Boston, MA 02114, USA
rminehart@mgh.harvard.edu

Abstract. Escape rooms are a great way for teams to work together to practice quickly solving problems under pressure and time constraints — all while simultaneously having fun. This article is about an online escape room designed to require cooperation and conflict, rational choice, decision making, truth-functional logic, causality, hypothesis testing, problem solving, cognitive and metacognitive challenges, teamworking and interpersonal communication under time pressure and stress, and organizing a strategic response that can later translate to the efficient and effective facilitation of real-world actions — all within a totally digitalized environment called e-REAL®. Cognitive biases, fixation errors, and other obstacles to overcome are assured, but evitable thanks to the effective learning experience are also assured. This escape room is designed to present lively, interactive activities aimed at understanding challenging scenarios like decision-making chains, as well as thrillers or cybercrime scenarios, Blockchain-based operations or digital and cryptocurrencies, but also data visualization and interpretation, preventative maintenance operations, and crisis management in healthcare or telemedicine.

Keywords: Escape room · Digital mindset · e-REAL

D. Guralnick et al. (Eds.): TLIC 2021, LNNS 349, pp. 290–303, 2022.
https://doi.org/10.1007/978-3-030-90677-1_28

1 An Online Escape Room to Enhance Effective Thinking, Unbiased Decision Making and Digital Mindset

1.1 An Agile Mindset for the Digital Society

This article is about an online escape room designed for its participants to experience cooperation and conflict, rational choice, decision making, truth-functional logic, causality, hypothesis testing, problem solving, cognitive and metacognitive challenges, teamworking, and interpersonal communication under time pressure and stress — all within a totally digitalized environment.

The main instructional target and learning output that is grounding our escape room is fostering an agile mindset for the digital society we're living in, because in our opinion, the ongoing digital transformation — that we can name by two key-words: Industry 4.0 and Society 5.0 — is first of all about mindset.

Society 5.0 is a very interesting trend expected to balance economic advancement with the resolution of social problems by a system that highly integrates cyberspace and physical space. For example, a huge amount of information (big data) from sensors in physical space is expected to be accumulated in cyberspace. In cyberspace, information is analyzed by artificial intelligence, and the analysis results are fed back to humans in physical space in various forms. The vision of Society 5.0 requires us to reframe two kinds of relationships: the relationship between technology and society and the technology-mediated relationship between individuals and society. According to the early research about Society 5.0, that so far are focusing mainly on Japan, it is a human-centered society, merging cyberspace with physical space: both a knowledge-intensive society and a data-driven society [1].

Society 5.0 is expected to go beyond Society 4.0 as an information society characterized mainly by production processes usually called Industry 4.0, which is mainly based on various typologies of artificial intelligence aimed at transferring the decision-making capability that belongs to humans to electronic systems and machines. Industry 4.0 is the technology revolution that is changing the way we live, work, and relate to one another. Billions of people are connected by mobile devices, with unprecedented processing power, storage capabilities, and access to knowledge. Ubiquitous super-computing, intelligent robots and "cobots" are boosting human-machine cooperation, self-driving cars, neuro-technological brain enhancements, genetic editing, and the confluence of artificial intelligence, robotics, and the Internet of (every-) things. The evidence of dramatic change is all around us, and it's happening at exponential speed [2–4] (Fig. 1).

The digital society, both in its current 4.0 configuration, as well in the next 5.0 configuration, needs teamwork that is more and more diverse, dispersed, digital, and dynamic — with frequent changes in membership and contributors with multiple affiliations, from "within" the same organization or, moreover, from the "outside" world. The process of digital transformation is inherently uncertain: changes need to be made provisionally and then adjusted; decisions need to be made quickly; and usually different groups need to get involved. As a result, traditional hierarchies get in the way, and it's best to adopt a flat organizational structure. Silicon Valley start-ups are still an interesting benchmark, as they are known for their agile decision-making, rapid prototyping, and flat structures. Networks, communities, teams, and groups, and how they work

Fig. 1. Ongoing trends: digital society that is expected to evolve from 4.0 to 5.0.

together, are the most important factors and the best way for people to work together with agility, to respond quickly to change and innovation challenges.

"Digital" is on everyone's lips. But while there are arising questions like "what's our strategy in relation to mobility, social media, the cloud or big data or telemedicine," the answers are often limited to only addressing the impact of each trend on a certain aspect of running current business. What is really needed is a better grasp of the deeper, ongoing mega-shift, which is utterly transforming the very nature of our societies and businesses, and of which mobility, social media, the cloud, and big data or telemedicine are mere manifestations [5].

1.2 An Online Escape Room to Foster the Key Factors Enabling a Digital Mindset Within the Workforce

Industry 4.0 and Society 5.0 are disrupting the "analogic" and "siloed" mindsets that are so easy to find within our work environments. The online escape room that we developed within the e-REAL® platform is aimed at fostering a digital mindset within people and their companies, international organizations, governments, schools, museums, hospitals, and more [6].

Escape rooms are a great way for teams to work together to practice quickly solving problems under pressure and time constraints — all while simultaneously having fun [7]. The virtualized escape room within the e-REAL setting, which is highly immersive and interactive both in this online configuration and also into the original "phygital" configuration (that is available as well, but more difficult to use today due to COVID-19 restrictions), has been designed to foster collaborative analysis, knowledge sharing, effective teamwork, and dynamic decision making. Cognitive biases, fixation errors, and other obstacles to overcome are assured, but evitable thanks to the effective learning experience is assured within the e-REAL setting.

According to the research performed by Centro Studi Logos, the key factors enabling a digital mindset are as follows [8]:

- Agility management
- Cognitive flexibility and adaptation to constantly changing environments
- Collaborative and inclusive work-style
- Data-driven decision making
- Digital literacy about the main 4.0 and 5.0 drivers
- Internal locus of control
- Metacognition
- Networking
- Virtual teamwork.

Agility management refers to tools, procedures, processes, and a mindset allowing participants to react quickly to rapidly changing circumstances and accelerate activities on critical paths. It enables organizational flexibility and the capability to act other.

Cognitive flexibility refers to the mental ability to switch between thinking about two or multiple different concepts, and to the ability to shift thoughts and adapt behavior to an ever-changing environment.

Collaborative and inclusive work style refers to the ability to work together with one or more individuals to complete a project or task, or to develop ideas or processes. Inclusivity further refers to the inclination to include individuals regardless of age, ethnicity, gender, sexual orientation, social classes or clusters, or disability.

Data-driven driven decision making involves making decisions that are backed up by hard data rather than making decisions that are intuitive or based on observation alone. It has become a fundamental part of all sorts of industries, including important fields like medicine, transportation, and equipment manufacturing.

Digital literacy refers to an individual's ability to find, evaluate, and compose clear information through writing and other media on various digital platforms.

Internal locus of control refers to the degree to which individuals believe that they, as opposed to external forces beyond their influence, have control over the outcome of events in their lives.

Metacognition refers to the awareness and understanding of one's own thought processes through self-reflection; it allows conscious changes in mental strategy toward solving problems and understanding new concepts.

Networking refers to the exchange of information or services among individuals, groups, or institutions.

Virtual teamwork refers to groups of geographically, organizationally, or time dispersed workers brought together — both synchronously and asynchronously — by information and telecommunication technologies to accomplish one or more organizational tasks [9] (Fig. 2).

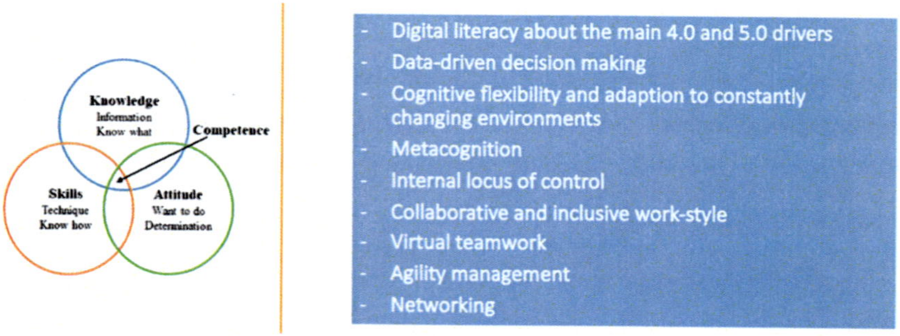

Fig. 2. Digital mindset and its key factors as a competence.

Moreover, there are at least five key psychological traits that make up a digital mindset and that are strongly needed in order to foster digitally minded leadership. They may seem paradoxical or contradictory, but they are crucial to managing the digital transformation upon us [10] (Fig. 3).

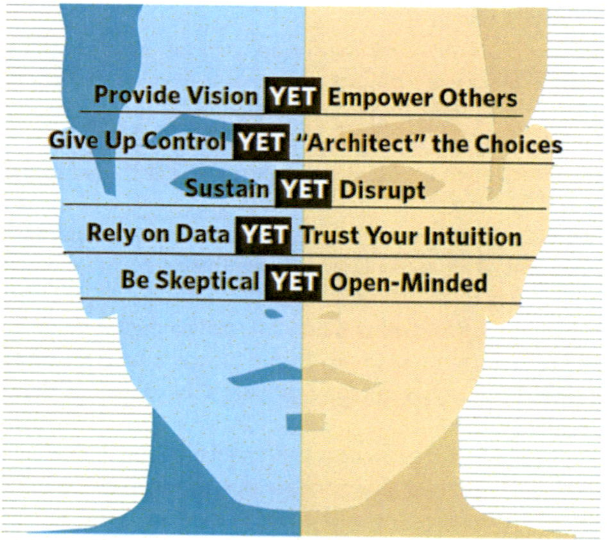

Fig. 3. The five key psychological traits to make up digital mindset and to foster a digitally minded leadership.

Digitally minded leaders need to cast the vision for this new age of digital transformation, while at the same time supporting the grass roots initiatives of teammates and colleagues to translate this high-level vision into on-the-ground action. This means fostering an organizational environment where people are safe to experiment with their ideas and where learnings from their experimentations are systematically captured, analyzed, and acted upon. Digitally minded leaders are expected to empower teammates and colleagues by giving up control; that doesn't mean leaders stop being in the driver's seat. Instead of sticking to rigid rules, leaders should seek to influence outcomes more through the way they design and present the choices to those best suited to make the decisions and carry out the tasks. Leaders become "choice architects." An easy way of understanding this is by making the option you want become the default choice, so people have to opt out of doing something instead of affirmatively choosing to opt in.

For most organizations, digital transformation will be a multi-step journey, with some steps enhancing and extending the firm's existing capabilities, while others will be disruptive, stirring up internal rifts and undermining skills and competencies that everyone worked so hard to master. Digitally minded leaders need to mitigate conflicts and serve as a bridge between the old and the new.

As a digitally minded leader, the goal is to move managerial discourse away from contentious turf wars of opinions, toward a reasoned conversation based on facts and experimental measurements. Data are key when it comes to decision-making. But the digital mindset doesn't rule out judgment and intuition. Data are often historical in nature and may not always serve as an accurate predictor of the future, especially in rapidly changing environments. Intuition plays a role in envisioning the future, helping leaders to formulate hypotheses and define assumptions to decide which data types and resources to focus on, and how to combine, analyze, and interpret them, in today's endless sea of information.

As well as encouraging experimentation throughout the organization, digitally minded leaders will embrace the try-it-and-see approach in their personal lives. They will jump on the next digital bandwagon, if only to realize it is not for you may not be such a bad idea, so long as there is a bigger objective — which is to engage in an exercise of active sense-making, forming new meanings of what technology affords us to do across contexts and experiences. So be skeptical, but see it and prove it for yourself. There is no substitute for sharing in the same experiences that your customers and employees are going through [11].

2 Complex Decision Making Within a Virtual Escape Room: Intuitive Decisions Are Better

2.1 The e-REAL Virtual Escape Room as an Immersive and Interactive Simulation Setting

Escape rooms can be used as a learning tool to organize a strategic response that can later translate to the efficient and effective facilitation of real-world actions. An escape room is essentially a game through which players can cooperatively discover clues, solve puzzles, and accomplish tasks in one or more rooms. By accomplishing these tasks, the players can progress to further achieve a specific goal in a limited amount of time. To "win" the game, the goal is often for the players involved to escape from the room.

The e-REAL escape room designed to enhance digital mindset is a highly immersive and interactive one: participants are attempting to simulate the decision-making and knowledge-to-application process, so are provided with conditions, including being under operational and psychological pressure — feeling a sense of urgency that more closely mirrors (even if at times vastly accelerated) what team members experience in real-world events in their professions [12–15]. It's an escape room designed to perform lively interactive activities aimed at understanding challenging scenarios like decision making chains as well as thrillers or cybercrime scenarios, Blockchain-based operations or digital and cryptocurrencies, but also data visualization and interpretation, preventative maintenance operations, and crisis management in healthcare or telemedicine (Figs. 4 and 6).

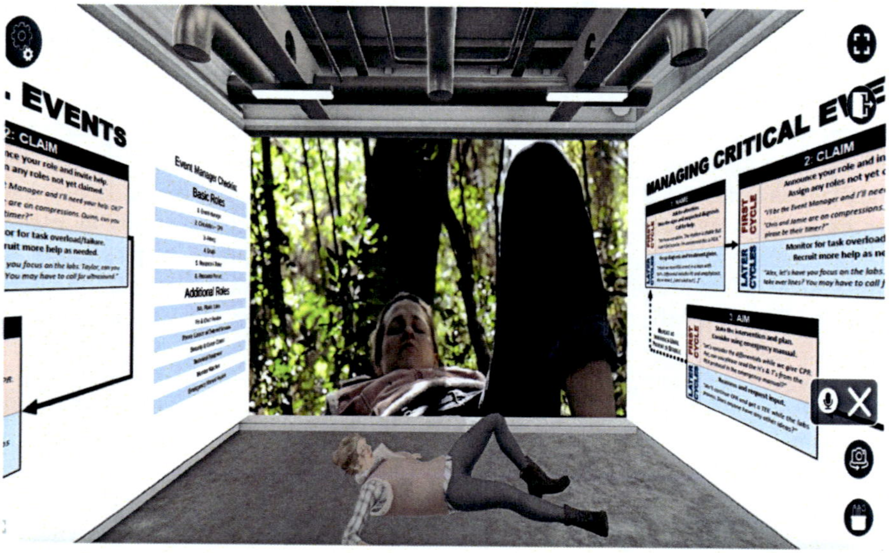

Fig. 4. e-REAL Escape Room: Managing critical events with the *Name-Claim-Aim©* mnemonic and check-list from the Harvard Center for Medical Simulation.

Instructional design and learning outputs are related to key issues such as: What are the best ways to decide, argue and solve problems? Why is analogy at the core of human cognition? How can a psychologist label an idea as a creative insight? What about judging right from wrong, subjective belief from intersubjective evidence? What do economists mean when they refer to a rational agent?

Answers to these questions are rooted in the critical — and sometimes counterintuitive — concepts experts use to make decisions [16, 17]. Their methods determine a number of relevant decisions: whether we are guilty or innocent, where we should invest our money, or whether a drug effectively treats a particular illness.

In the e-REAL online platform, several multimedia resources are available for incorporating escape rooms as an advanced simulation. One of the most valued experiences available among the e-REAL escape rooms focuses on the cases when you are not the only one choosing. Game theory is a rich and interesting paradigm, mainly if we look at the cooperative approach by looking at the Prisoner's Dilemma and the trust games related to the principle of reciprocity. Usually decision makers are generally less selfish and less strategic than game theory predicts, and they value social factors such as reciprocity, fairness, and relative social status more than the theory predicts.

Fig. 5. e-REAL Escape Room: Cognitive and metacognitive challenge within the *Simplification game*™ (further described below).

Fig. 6. e-REAL Escape Room: Blockchain, digital currencies, and cybercrime game designed within a training program for young leaders into the energy sector, in collaboration with Eni and Aramco under the umbrella of the World Economic Forum.

Other experiences within the e-REAL escape room, valued by learners as highly rewarding, are about how a subjective degree of belief should rationally change to account for evidence. Even if we are used to expressing information in terms of probabilities and percentages, our cognitive systems evolved over millions of years to process information in frequency — without even counting. We do it automatically.

What our perceptual and cognitive system registers is quantities or frequencies, not percentages: what do we do when we are required to make decisions based on probability information? Mostly, it seems that we tend to ignore prior probabilities. Fallacy is the rule, even with people who take advanced courses in probability and statistics. Usually, we are very good at classifying items according to similarity to a prototype, but not at all at making probability judgments. More broadly, we can say that our decisions are influenced more by how a problem is described (framed) than by the objective data contained in the problem: the strongest framing effects are usually found when probability information is pitted against a deep-seated bias in our cognitive architecture.

The best example of this is loss aversion bias, that is our tendency to strongly prefer avoiding losses to acquiring comparable gains. By the way, we are not the only ones who act this way because it seems that Capuchin monkeys do the same things. Moreover, we respond to changes in relative, rather than absolute, gains and losses (losing or gaining $10 means more if you have only $20 than if you have $200). This means that we make decisions by comparing them to a flexible reference point. Escaping from an e-REAL room understanding this is a highly valued outcome.

2.2 Intuitive Decisions are Better?

Research in the last 25 years indicates that human cognition involves two qualitatively different processing modes. The first one is fast, affective, parallel, associative, and holistic and has been attributed to the operation of an experiential/affective or intuitive system, while the second is slower, sequential, rule-based, and analytic and has been attributed to the operation of a rational or a deliberation system. As each of these processing modes has its strengths and limitations, for each cognitive activity, optimal performance requires an adequate mix of the two modes [18].

Decision-making is a cognitive activity, which has traditionally been thought of as the normative play-field of rational deliberation. Accordingly, one should consider the various aspects of the alternatives, consciously evaluate the risks and values, and decide through careful analysis the expected (or weighted) utility of each option. Initial research in decision-making has provided support for this view, in the form of a variety of decision biases that have been attributed to the shortcomings of the intuitive mode. Prominent examples are reference and decoy effects, loss-aversion, and violations of invariance and transitivity, which have been explained by assuming that intuitive decisions involve the deployment of a set of heuristics; similarly, affect has also been associated with decision biases thought to reflect a type of heuristic [19–21].

Recent work, however, indicates a need to re-evaluate this assessment. First, it has been proposed that the association between heuristics (such as lexicographic rules, responsible for violations of intransitivity) and intuition in decision-making is unwarranted. Researchers are pointing out that one can distinguish experimentally, between the deployment of such heuristics, which are sequential and rule based, and the deployment of fast/parallel but compensatory modes of decision-making, which characterize the intuitive mode. It seems to be possible that the traditional view has underestimated the power of the intuitive/affective mode [22]. Simon, who was the first to appreciate the capacity limitations of conscious processing in decision-making, has warned us that "there is a complete lack of evidence that, in actual choice situations of any complexity, these [rational] computations can be, or are in fact, performed... but we cannot, of course, rule out the possibility that the unconscious is a better decision maker than the conscious" [23].

Simon's warning is now supported by two lines of research. First, neuropsychological studies of decision-making and of social cognition have indicated the existence of an affective or somatic process, which under certain conditions plays a substantial (rather than a biasing) role in value integration and in intuitive judgments [24]. For example, affect is thought to provide a common currency, which is essential for value integration, while intuition can be characterized as the subjective experience of a mostly non-conscious process that is fast, a-logical, and inaccessible to consciousness that, dependent on exposure to the domain or problem space, is capable of accurately extracting probabilistic contingencies [25]. This process has a different (holistic) nature and can be distinguished experimentally, from the rule-based, sequential decisions that are subject to capacity limitations in working memory [26]. Experimental studies have also indicated that conscious deliberation (explicitly analyzing reasons when making decisions) can have a disruptive effect on attitudes and that distraction — a process likely to reduce the contribution of the deliberation process in decision-making — is not always detrimental for product evaluation [27].

More recently, the hypothesis that the processes associated with intuition can facilitate decision quality has been supported by a series of studies according to which decision quality is enhanced by setting the problem aside (from conscious scrutiny) after the initial (conscious) encoding, or in other words, by "sleeping on it" [28].

The scientific debate about distraction and intuition, the unconscious thought effect, and the so-called affective versus analytical mindsets is still vigorous. What we are trying to do is keep the e-REAL escape room within a framework that is based on two interacting subsystems of decision-making: an affective/intuition based system and an analytic/deliberation system.

2.3 Intuition is Decision Making

Intuition has been approached from different perspectives by many researchers: it is described in terms of expertise [29], heuristics [30], implicit learning and memory [31, 32], and individual differences in processing styles or decision-making modes [33], as well as lower level perceptual processing [34].

Kahneman and Tversky, since 1973, have investigated the role of heuristics in decision making and established the bases for common human errors in decision making, later renamed cognitive biases. In 1979, they also discovered and published The Prospect Theory, an interesting explanation on human brain risk assessment when facing gains and losses. Their model is based on two separate systems for decision making: the first, called System 1, is the brain decision making network responsible for automatic, unconscious, habitual decisions. This system is assumed to be extremely quick in response and efficient from an energy consumption perspective. Evolutionarily older than its rational counterpart, System 1 bases it's functioning on associative memory mechanisms; it is efficient, but very prone to mistakes — especially when facing decisions related to complex problem solving.

System 2 is, instead, the brain dedicated decision making network functioning at a conscious level. It is much slower and much less efficient in terms of energy consumption, compared to System 1. Is the system that we use for complex decision making and, differently from System 1, solidly error free. According to this model, almost 95% of our daily decisions are carried out by System 1 — whose decision making network is strongly related to intuitive decision-making processes.

In 1994, Bechara and Damasio [24], based on their Iowa Gambling Task experiment, argued the role of emotional processing in decision making. Participants were presented with four decks of cards. They were told that each deck holds cards that will either reward or penalize them, using game money. The goal of the game was to win as much money as possible. The decks differed from each other in the balance of reward versus penalty cards. Thus, some decks were "bad decks," and other decks were "good decks," because some decks were intended to reward the player more often than other decks.

Most healthy participants sampled cards from each deck, and after about 40 to 50 selections were fairly good at sticking to the good decks. Patients with orbitofrontal cortex (OFC) dysfunction, however, continued to persevere with the bad decks, sometimes even though they knew that they were losing money overall. Concurrent measurement of galvanic skin response showed that healthy participants manifested a "stress" reaction to hovering over the bad decks after only 10 trials, long before conscious sensation that the decks were bad. By contrast, patients with amygdala lesions never developed this physiological reaction to impending punishment. So Damasio and Bechara theorized a phase of intuitive decision making that they called "pre-hunch," in which emotional arousal is used by the brain to stimulate a rise in the level of anxiety in order to signal, at an unconscious level, effective choices, long before our brain becomes able to process the rational reasons behind our choices.

In 2008, Betsch defined intuition on the basis of three core components: inputs, processes, and outcome: "Intuition is a process of thinking. The input to this process is mostly provided by knowledge stored in long-term memory that has been primarily acquired via associative learning. The input is processed automatically and without conscious awareness. The output of the process is a feeling that can serve as a basis for judgments and decisions" [35]. Thanks to his work, we can figure out that intuition is, by all means, an effective and efficient decision-making process that is characterized by the following elements:

- It is *knowledge based,* in the sense that intuitive decision making cannot be diminished as a mere gambling activity producing results out of luck and causing excessive risk for the organizations. Properly addressed and nurtured, intuitive decision-making efficacy can be tailored to specific organizational *expertise.*
- It is *processed automatically and without conscious awareness.* This second element highlights the importance of intuitive decision making in modern organizations dealing with a volatile, uncertain, complex, and ambiguous environment (the so-called VUCA world).
- It is *correlated with associative learning* processes stored in long-term memory. The big insight about intuitive decision making is that it is something that can be *learned* and not merely a personal trait or belonging to specially gifted individuals.

- The output of the intuitive process is a *feeling*. Looking at the research outputs from Kahneman, Tversky, Damasio, and Bechara, intuition or "hunch" can be considered an emotionally influenced decision making process.

From this perspective, we can say that intuitive decision making can be trained in order to minimize bias driven errors. By e-REAL escape room training, people can retain all the benefits of intuitive decisions, like speed and efficiency, while mitigating the increased risk correlated with cognitive biases.

3 The Simplification Escape Room

We are frequently required to make decisions based on calculations that are difficult to do in our heads. Standard economic models based on rational choice presuppose the operation of a single rational information processor, while results of neuroscience research indicate that decisions are the outcome of two separate neural processors — and the two processors also have been found to compete with each other in decision-making.

Many researchers are pointing out a competition between two neural systems: an emotion-based system (amygdala) and a predominantly analytic system (orbital and prefrontal cortex). So, when people decide to make risky decisions, the reward areas of the brain become highly active just prior to making the decision. In other words, this neural signature shows that they are anticipating large payoffs and are not thinking about the probability of payoffs [36]. This is one reason gambling can be so addictive: the act of placing the bet can feel as rewarding as winning. Throw into the mix the fact that we are frequently deciding under conditions of uncertainty, with insufficient information either because that's all there is or because information is asymmetrical. It the end, we try to avoid risk, yet we make extremely risky decisions when facing potential losses.

And as we wrote above, we can't forget overconfidence bias: both novices and experts are more confident about their decisions than is justified given the environment; as a result, they frequently stop their search for answers before all available evidence can be collected. When System 1 is in charge, we can end up being overly confident in our decisions, and that's referred to as overconfidence bias.

Simplification Escape Room training (Fig. 5) has been designed to enhance the human brain's ability to conduct efficient intuitive decision making to problem solving, mitigating the negative effects of cognitive biases and time induced stressors on emotional arousal.

Simplification Escape Room is a multiplayer gaming activity presenting the players with multiple cognitive riddles to solve in a limited amount of time in order to escape a room. Each cognitive riddle has been embedded with intuitive shortcuts that allow the players to drastically reduce the time required to solve the riddles. The intuitive decision-making process in the game is complicated by the use of stressors related to limited time for riddle solutions and the requirement to manage teamwork dynamics. The escape room can only be solved using collective intelligence and knowledge shared inside the players' team, but this process also increases the individual emotional strain and complicates the effective use of intuitive decision processes.

Furthermore, the gaming activity is designed to collect and to provide, during the debriefing phase, individual and team related key-performance indicators.

As human beings, we want to maintain a knowledge base filled with true beliefs, particularly true beliefs about what causes what. This means that we have to test the truth of our beliefs via a process called hypothesis testing, based on assuming the belief is true, then on deducing what it predicts, and, finally, on testing those predictions. This process is very challenging because people are genuinely bad at this [37]. Even if we are aware that a number of pitfalls occur in our intuitions about hypothesis testing, mainly because our beliefs and expectations influence how we interpret the so called *facts*, we are mainly seeking evidence that confirms beliefs, and having ourselves engaged with trying to falsify beliefs and hypothesis is not so easy — despite that at the heart of scientific inquiry we find hypothesis testing by "disproof" and "falsification" [38]. Seeking evidence that confirms beliefs may also affect our problem solving procedures, creating the so called "framing effects" (the way a problem is described influences how we think about it) and making it harder and harder to think out-of-the-box and bridge different domains of expertise strengthening our analogical mind that is the core of cognition. It the end, this is what makes escaping the e-REAL room highly valuable and worthy to do.

References

1. Deguchi, A., et al.: Society 50. Springer, Singapore (2020). https://doi.org/10.1007/978-981-15-2989-4
2. Salvetti, F., Bertagni, B.: Leadership 5.0: an agile mindset for a digital future. Int. J. Adv. Corp. Learn. **13**(22), 57–66 (2020)
3. Polat, L., Erkollar, A.: Industry 4.0 vs. Society 5.0. In: Durakbasa, N.M., Gençyılmaz, M.G. (eds.) ISPR 2020. LNME, pp. 333–345. Springer, Cham (2021). https://doi.org/10.1007/978-3-030-62784-3_28
4. Onday, O.: Japan's Society 5.0: going beyond Industry 4.0. Bus. Econ. J. **10**(2), 1–6 (2019)
5. Kaganer E., Sieber S., Zamora J.: The 5 Keys To A Digital Mindset. IESE Business School. Forbes, March 2014
6. www.e-real.net
7. Baily L.: Escape rooms: active learning resources for passive gaming. In: Healthy Simulation Newsletter, August 2021
8. Salvetti, F.: Digital mindset and virtual leadership for the Italian Air Force. Selected readings by Fernando Salvetti, Nicola Bassetto, Luca Campisi, Lara Como, Chiara De Luca, Carlotta Denegri, Lorenzo Fessia, Carola Francione, Camilla Giaccone, Giusy Letizia, Sofia Lombardi, Thuy Linh Nguyen, Giulia Persich, Sara Ritondo, Olaomopo Shobowale, Junior Uwoghiren. Centro Studi Logos, Turin (2021)
9. Salvetti F.: Digital Mindset. Franco Angeli, Milan (expected 2021)
10. Thaler, R., Sunstein, C.: Nudge. Improving Decisions about Health. Wealth and Happiness. Yale University Press, London (2008)
11. Kaganer, E., Sieber, S., Zamora, J.: The 5 keys to a digital mindset. IESE Business School. Forbes, March 2014
12. Anderson, M., Lioce, L., Robertson, J., Lopreiato, J., Díaz, D.: Toward defining healthcare simulation escape rooms. Simul. Gaming **52**(1), 7–17 (2021)
13. Baily K. Escape rooms in medical simulation: Downloadable sepsis scenario & more. In: Healthy Simulation, January 2020
14. Díaz, D., Clapper, T.: Escape rooms: a novel strategy whose time has come. Simul. Gaming **52**(1), 3–6 (2021)

15. Kutzin, J., Strother, C., Sanders, E.: Virtual escape rooms to build teamwork and enhance communication. In: Conference Workshop at the 2020 International Telesimulation in Healthcare Conference: NYC (2020)
16. Salvetti, F.: Good thinking: book review, in philosophical practice. J. Am. Philos. Pract. Assoc. **10**(1) (2015)
17. Cummings, D.: Good Thinking: seven Powerful Ideas that Influence the Way we Think. Cambridge University Press, Cambridge (2012)
18. Usher, M., Russo, Z., Weyers, M., Brauner, R., Zakay, D.: The impact of the mode of thought in complex decisions: intuitive decisions are better. Front. Psychol. **2**, 37 (2011)
19. Tversky, A.: Intransitivity of preferences. Psychol. Rev. **76**, 31–48 (1969)
20. Tversky, A., Kahneman, D.: Judgment under uncertainty: heuristics and biases. Science **185**(4157), 1124–1131 (1974)
21. Gigerenzer, G., Goldstein, D.: Reasoning the fast and frugal way: models of bounded rationality. Psychol. Rev. **103**(4), 650–669 (1996)
22. Glöckner, A.: Does intuition beat fast and frugal heuristics? A systematic empirical analysis. In: Intuition in Judgment and Decision Making, pp. 309–325, xix, 335. Lawrence Erlbaum Associates Publishers, Mahwah (2008)
23. Simon, H.A.: A behavioral model of rational choice. Q. J. Econ. **69**, 99–118 (1955)
24. Bechara, A., Damasio, A.: The somatic marker hypothesis: a neural theory of economic decision. Games Econ. Behav. **52**, 336–372 (2005)
25. Lieberman, M.: Intuition: a social cognitive neuroscience approach. Psychol Bull. **126**(1), 109–137 (2000)
26. Glöckner, A., Betsch, T.: Multiple-reason decision making based on automatic processing. J. Exp. Psychol. Learn Mem. Cogn. **34**(5), 1055–1075 (2008)
27. Betsch, T., Plessner, H., Schwieren, C., Gütig, R.: I like it but I don't know why: a value-account approach to implicit attitude formation. Pers. Soc. Psychol. Bull. **27**, 242–325 (2001)
28. Dijksterhuis, A.: Think different: the merits of unconscious thought in preference development and decision making. J. Pers. Soc. Psychol. **87**(5), 586–598 (2004)
29. Burke, L., Miller, M.: Taking the mystery out of intuitive decision making. Acad. Manage. Executive **13**(4), 91–99 (1999)
30. Gigerenzer, G.: Gut Feelings: The Intelligence of the Unconscious. Viking, New York (2007)
31. Tversky, A., Kahneman, D.: Judgment under uncertainty: heuristics and biases. Science **185**, 1124–1131 (1981)
32. Loewenstein, G.: Out of control: visceral influences on behavior. Organ. Behav. Hum. Decis. Process. **65**, 272–292 (1996)
33. Epstein, S., Pacini, R., Denes-Raj, V., Heier, H.: Individual differences in intuitive-experiential and analytical-rational thinking styles. J. Pers. Soc. Psychol. **71**, 390–405 (1996)
34. Volz, K.G., von Cramon, D.Y.: What neuroscience can tell about intuitive processes in the context of perceptual discovery. J. Cogn. Neurosci. **18**, 2077–2087 (2006)
35. Betsch, T.: The nature of intuition and its neglect in research on judgment and decision making. In: Plessner, H., Betsch, C., Betsch, T. (eds.) Intuition in judgment and decision making, pp. 3–22. Lawrence Erlbaum, Mahwah (2008)
36. Knuston, B., Bossaerts, P.: Neural antecedents of financial decisions. J. Neurosci. **27**, 8174–8177 (2007)
37. Cummins, D.D.: Good Thinking. Seven Powerful Ideas that Influence the Way We Think. Cambridge University Press, New York (2012)
38. Popper, K.R.: Logik der Forschung (The Logic of Scientific Discovery). Mohr, Tübinghen (1934)

Digital Learning: Healthcare Training by Tele-Simulation and Online Cooperation

Fernando Salvetti[1,2,3](✉), Roxane Gardner[4,5], Rebecca Minehart[6,4], and Barbara Bertagni[1,2,3,4,5]

[1] Centro Studi Logos, 10143 Turin, TO, Italy
{salvetti,bertagni}@logosnet.org
[2] Logosnet, 6900 Lugano, CH, Switzerland
[3] Logosnet, Houston, TX 77008, USA
[4] Center for Medical Simulation, Boston, MA 02129, USA
rgardner1@bwh.harvard.edu
[5] Brigham and Women's Hospital/Children's Hospital/Massachusetts General Hospital, Harvard Medical School, Boston, MA 02114, USA
[6] Massachusetts General Hospital, Harvard Medical School, Boston, MA 02114, USA
rminehart@mgh.harvard.edu

Abstract. How can we design engaging and effective online learning in healthcare? How can we perform tele-simulation? How can we make digital learning authentic and not over-produced? How can we apply gamification strategies to enhance online cooperation? Tele-simulation and online cooperation are essential today in healthcare education and training, first of all for medical simulation: The e-REAL® online platform for healthcare simulation, aimed at fostering effective teamwork, knowledge sharing, and cooperation, is a solution that we designed to work within a highly immersive setting where learners and simulation instructors may interact all together in real time among themselves, as well as with virtual objects and avatars. This solution allows simulations in a virtual environment that display challenging situations; unlike other VR and AR solutions, e-REAL allows users to experience full immersion without the need for glasses or goggles, providing a memorable experience with a robust learning outcome. e-REAL is user-centric and easy to use: an Internet connection and a browser are the only requirements.

Keywords: Tele-simulation · Online cooperation · e-REAL

1 Healthcare Training by Tele-Simulation and Online Cooperation

1.1 Key Questions About Online Simulation and First Answers

How can we design engaging and effective online learning in healthcare? How can we perform tele-simulation? How can we make digital learning authentic and not over-produced? How can we apply gamification strategies to enhance online cooperation?

D. Guralnick et al. (Eds.): TLIC 2021, LNNS 349, pp. 304–311, 2022.
https://doi.org/10.1007/978-3-030-90677-1_29

Tele-simulation and online cooperation are essential today for medical simulation. The main challenge, in our experience, is really enhancing learners' online cooperation and proactive participation. To do so, we created a number of scenarios for healthcare simulation aimed at fostering effective teamwork, knowledge sharing and cooperation into a platform called e-REAL® Online.

The simulation setting is highly immersive, and learners may interact together in real time among themselves, as well as with virtual objects and avatars. Users can also take notes, overlay key concepts, rotate virtual objects, or move them point to point and complete questionnaires directly in the online environment that works as a sketch-board. Usability also extends to content, as it is easy to import and show a user's existing content, as well as create and edit new content. It is truly a totally customizable solution, partnered with a powerful multimedia editor that is provided on demand as a premium feature.

This solution allows simulations in a virtual environment that display challenging situations; unlike other VR and AR solutions, e-REAL [1] allows users to experience full immersion without the need for glasses or goggles. What this solution mainly provides is a memorable experience with a robust learning outcome. It is really easy to use: an Internet connection and a browser are the only requirements (Fig. 1).

Fig. 1. The e-REAL® Online Platform for the Center of Medical Simulation, Boston, MA. A case regarding an injured ankle within an alpine environment, made available online into the platform that is configured displaying a room, to recreate a setting similar to the one where the learners are in a brick-and-mortar e-REAL classroom that acts as an immersive and interactive "phygital" environment (both physical and enhanced by digital objects).

The feedback we receive most often — both by the learners and by the simulation instructors invited to coordinate and facilitate the online sessions — is that they didn't expect such easy access to an intensely engaging and meaningful experience. This is a very important feedback because our efforts are aimed at designing solutions that empower learning with a strong focus on the UX, both for learners and trainers.

We believe that flexible and user-friendly solutions are needed in learning, so we designed the e-REAL online platform to include ready-to-use courses, courses co-designed with our team, or creating and uploading existing content with a dedicated editor in a very simple way—even last minute.

Our immediate goal is to build significant learning solutions that help educators and "simulationistas" integrate into their practice new technologies to boost learning in easy and effective ways. Feedback and outputs from the field are pointing out that tele-simulation based on online cooperation within the e-REAL immersive learning environment makes for engaging and effective healthcare simulations — in particular, immersive and interactive experiences that can be performed without special glasses, joysticks, or other "augments" to allow authentic and hands-on experiences. e-REAL is specifically designed to be properly produced and easy to use; those are the two main concerns usually regarding online learning.

1.2 Further Answers to the Key Questions

Designing and delivering online learning as a cooperative tele-simulation, within an easy-to-use environment that works like a virtual reality setting but not requiring head-mounted displays or other augments to interact with, is essential for an effective experience. A good user experience (UX) is expected to be among the main targets for the instructional design team, and this means a focus on having a deep understanding of users, what they need, what they value, their abilities, and also their limitations.

Another essential success factor is mastering simulation techniques: the use of simulation is growing exponentially in academic and service settings because it can enable students, new graduates, and experienced clinicians to develop clinical competence and confidence in a number of clinical situations, far more than a learner can be exposed to in a live environment [2].

To be effective, the instructional design supporting the tele-simulation experience has to be state-of-the-art, which is unfortunately not the standard in the online learning industry.

1.3 Can Online Learning Be Effective?

What did you do during your recent online course?

"I fell asleep."

"I was playing an online game."

"I was watching TV."

"I was chatting with friends on social media."

"I haven't seen the last three lectures."

These could be the answers from student feedback [3]. As the use of online learning worldwide has continued to grow, the perception of most online learning has often been less than positive. As David Guralnick pointed out, while of course there are exceptions, the negative comments about online learning programs have been generally well founded.

There is no single way to create learning experiences that are effective and that the target audience enjoys and connects with, but there are common success factors and active online learning methods that have been successful [4].

How can we design engaging and effective online learning? First of all, we can do this by designing experiences to connect with, and not "to-do-lists" or "medical prescriptions" working one way, from the teacher to the student. It's all too easy for a subject matter expert to distill "core content" down into a list of abstract concepts, but even if there is someone who memorizes everything, this doesn't mean that he or she knows how to apply the concepts in real life [5, 6].

We suggest designing the online learning experiences by involving learners in authentic stories that address key issues and realistic concerns, because human beings respond well to things that are relevant to them, concrete, and context related. We suggest adopting the situation-impact-resolution (SIR) format to establish story context for each simulation. This derives from Aristotle's *Poetics* and focuses on the sequence drama, suspense, and resolution — to use with epistemological acumen, within a systemic paradigm.

How can we make digital learning authentic and effective? By encouraging learners to learn by doing and allowing them to cross conceptual and theoretical boundaries within a simulation's setting. In an online learn-by-doing simulation, the learners are asked to play particular roles and must make decisions in realistic situations, sometimes with the assistance of a coaching or mentoring component. It's critical that learners then receive feedback as to the results of their decisions.

In e-REAL, learners practice handling realistic situations, rather than learning facts or techniques out of context. Context refers to the circumstances that form the setting for an event, statement, or idea. Context related factors can be influential and even disruptive: for example, a loud background noise within a virtually recreated operating room in e-REAL negatively impacts the surgical team's ability to communicate and may consequently contribute to surgeons committing an error. The most effective learning occurs through being immersed in context, requiring the ability to understand the limits

Fig. 2. Representative situation into the e-REAL online platform.

of our knowledge and action, and to adapt that knowledge to an environment different from the one in which it was developed (Fig. 2).

Ultimately, as research in the area of video games has noted, there is another key factor that works as a motivator for participants: by providing a dramatic, realistic storyline with difficult decisions for the learners to make, and providing feedback so the learners can see how things are going, a simulation can create a true challenge for learners. Game-like elements can play a substantial role in the success of an online learning experience [4].

A context related experience within an e-REAL setting is similar to being immersed within a video game. Characteristics of games that facilitate engagement can be grouped into two general categories: those that create a rich mental model of the game environment and those that create consistency between the things in that environment. The richness of the mental model depends on having a cognitively demanding environment and a strong and interesting narrative. A bird flying overhead is good. Hearing it screech is better. Cognitively demanding environments in which players must focus on what's going on in the game will occupy mental resources. The richness of the mental model promotes engagement and immersion into the scenario, because if the brain power is allocated to understanding or navigating the world, players are too occupied to notice all of the game's shortcomings that would otherwise remind them that they're playing a game. Finally, good stories with interesting narratives (which are credible because they are as intrinsically congruent as possible) attract attention to the game and make the world seem more believable. Good stories tie up mental resources.

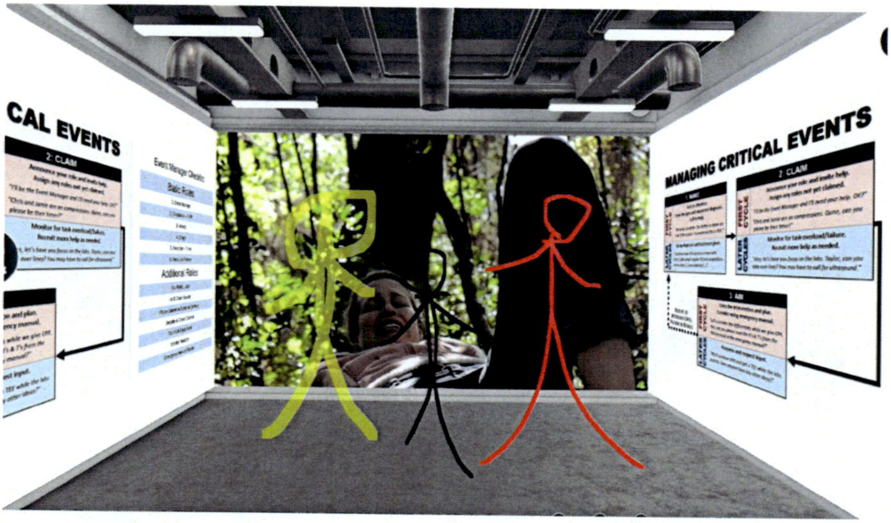

Fig. 3. Multimedia storytelling enhanced with features allowing mnemonics and checklists, as well as drawings: in this case the overlaying is about the positions taken by the first responders within the simulation scene in an e-REAL online tele-simulation platform.

Turning to game traits related to consistency, believable scenarios, and behaviors in the game world ensures that virtual characters, objects, and other creatures in the game world behave in the way learners expect. Game developers strive for congruence among all these elements. In this way, learners are challenged both cognitively and behaviorally in a fully immersive and multitasking learning environment, within interactive scenarios that usually present a wealth of information: the many levels of the situation are made available simultaneously, by overlaying multisource — e.g., words, numbers, images — within an environment designed by AR techniques [7, 8] (Fig. 3).

e-REAL submerges learners in an immersive reality where the challenge at hand is created by sophisticated, interactive computer animation. Importantly, the system includes live and real-time interaction with peers, instructors, tutors, facilitators, and mentors — as well as with avatars or digital humans. Thus, it adds a very important social component that enhances learning outputs, skills, cognitive processes, and metacognitive processes. The process of learning by doing within an immersive setting, based on multimedia and visual storytelling, leaves both the learners and the simulation instructors with memorable experiences and produces robust learning outputs.

Our contribution to this growing approach enables a multilayer vision, based on cooperative communication and learning; the many levels of the situation are made available simultaneously by overlaying multisource information — e.g., words, numbers, images and by visualizing relations between topics, contextual factors, cognitive maps, and dynamic cognitive aids — but without AR and VR head-mounted displays.

The e-REAL teaching and learning approach is designed to have the learners working on tasks that simulate an aspect of expert reasoning and problem-solving, while receiving timely and specific feedback both from fellow students and the trainers. These elements of deliberate practice and feedback are general requirements for developing expertise at all levels and disciplines [9–11].

2 Tele-Simulation and Online Cooperation as Enhanced Learning

We are working with a wide array of simulation instructors from all around the world. We understand that helping an instructor formulate an effective or innovative experience or assignment is only part of our job. We're also here to help the instructor bring that idea to life, whether that means offering specialized training to teaching fellows, a custom-built web resource, or workshops where learners receive direct training in the skills where they need to demonstrate their understanding of a key course concept.

If anything, this has become critical in these times of remote teaching. e-REAL is at the forefront of a trend that COVID-19 is accelerating: reality in the digital age is becoming more and more virtual. We are mixing up the latest trends from digitalization, virtualization, and artificial intelligence with the goal of building powerful solutions for medical simulation, including a focus on organizational behavior and leadership. As Pierre Lévy used to say, reality in the digital age is becoming more and more virtual [12]. In healthcare simulation, the dematerialization of the learning environment is allowed by new technologies that offer options to improve the usability of traditional e-learning methods. Sharing and mixing up the latest trends from digitalization and virtualization, neurosciences, artificial intelligence, and advanced simulation allows us to establish a new paradigm for education and training.

Nothing is revolutionary within a simple VR or AR headset, but if virtual and augmented content and scenarios are actualized — or enhanced — within an online simulation setting, this adds value to the digital learning process. In such a way, e-REAL can contribute to enhancing our learning experience (Fig. 4).

So far, the ongoing exploratory projects within the e-REAL set up at the Center for Medical Simulation in Boston are:

1. The further use of the e-REAL visualizations in Labor & Delivery programs and in Anesthesia programs.
2. The design of distance-based simulations to take care of COVID-19 related situations: logistics, team safety, relationship with patients and families.
3. The introduction of online learning modules where different types of virtual objects are co-existing: artificial but realistic avatars and real actors performing as standardized patients or family members or colleagues, photorealistic 3D tools, indoor or outdoor scenarios.
4. The development of self-learning solutions to improve results related to critical conversations, debriefing sessions, video-interviews and videoconferences.
5. The use of AR solutions to provide guidance during remote on-site simulations.
6. The visualization of interactive checklists and mnemonics to foster team performance.

Fig. 4. Representative multimedia visualization within the e-REAL platform.

References

1. Salvetti, F., Gardner, R., Minehart, R.D., Bertagni, B.: Enhanced reality for healthcare simulation. In: Brooks, A.L., Brahman, S., Kapralos, B., Nakajima, A., Tyerman, J., Jain, L.C. (eds.) Recent Advances in Technologies for Inclusive Well-Being. ISRL, vol. 196, pp. 103–140. Springer, Cham (2021). https://doi.org/10.1007/978-3-030-59608-8_7
2. Palaganas, J., Ulrich, B., Mancini, B. (eds.): Mastering Simulation. A Handbook for Success. 2nd edn. Sigma Theta Tau, Indianapolis (2020)
3. University World News. https://www.universityworldnews.com/post.php?story=20200515072512826. Accessed 30 Apr 2021
4. Guralnick, D.: Re-Envisioning online learning. In: Salvetti, F., Bertagni, B. (eds.) Learning 4.0. Advanced Simulation, Immersive Experiences and Artificial Intelligence, Flipped Classrooms, Mentoring and Coaching. Franco Angeli, Milan (2018)
5. Salvetti, F., Bertagni, B. (eds.). Learning 4.0. In: Advanced Simulation, Immersive Experiences and Artificial Intelligence, Flipped Classrooms, Mentoring and Coaching. Franco Angeli, Milan (2018)
6. Auer, M.E., Guralnick, D., Uhomoibhi, J. (eds.): ICL 2016. AISC, vol. 545. Springer, Cham (2017). https://doi.org/10.1007/978-3-319-50340-0
7. Aukstakalnis, S.: Practical Augmented Reality. A Guide to the Technologies, Applications, and Human Factors for AR and VR. Addison-Wesley, Boston (2017)
8. Troussas, C., Krouska, A., Sgouropoulou, C.: Collaboration and fuzzy-modeled personalization for mobile game-based learning in higher education. In: Computers & Education, vol. 144, p. 103698 (2020)
9. Rudolph, J.W., Simon, R., Raemer, D.B., Eppich, W.J.: Debriefing as formative assessment: closing performance gaps in medical education. Acad. Emerg. Med. **15**(11), 1010–1016 (2008). https://doi.org/10.1111/j.1553-2712.2008.00248.x
10. Schapiro, M., et al.: Defining team performance for simulation-based training: methodology, metrics, and opportunities for emergency medicine. In: Academic Emergency Medicine, vol. 15, pp. 1088–1097 (2008)
11. Lyons, R., et al.: Enhancing the effectiveness of team debriefings in medical simulation: more best practices. In: The Joint Commission journal on Quality and Patient Safety, vol. 41 no. 3 pp. 115–123 (2015)
12. Lévy, P.: *Qu'est-ce que le virtuel ?* (What is the Virtual?) La Découverte, Paris (1998)

An Approach for Continuous Supervision of Bachelor's and Master's Theses in Engineering Studies

Sigrid Schefer-Wenzl(✉) and Igor Miladinovic

Computer Science and Digital Communications, University of Applied Sciences Campus Vienna, Vienna, Austria
{sigrid.schefer-wenzl,igor.miladinovic}@fh-campuswien.ac.at

Abstract. Bachelor's and Master's theses are excellent ways for students to dive deep into a research question and gain in-depth knowledge about a research topic. At the end of a curriculum, theses act as a kind of bridge between the educational stage and work or further study. Thesis supervision is a resource-intensive task and often a critical factor for the quality of a thesis. After decades of experience with different methods, we have developed a concept for continuous thesis supervision and applied it to two specific degree programs at our university, the Bachelor's program in Computer Science and Digital Communication and the Master's program in Software Design and Engineering. Compared to traditional supervision methods, this concept has led to a higher adherence to deadlines and to a higher quality of the final theses. In this paper, we present our findings on different types of supervisors and our concept for continuous thesis supervision, which can be adapted to each identified supervisor type. This concept comprises several methods, intended to inspire other supervisors to choose the set of methods that best suits their needs.

Keywords: Supervision · Higher education · Bachelor thesis · Master thesis

1 Introduction

At the end of a study program, students are usually required to write a final scientific thesis to demonstrate their maturity and readiness to investigate a complex topic in depth. This requires two major changes in student behavior. First, they must move from guided learning in larger groups to self-organized learning with an individual topic. Second, the topic and research question of a Bachelor's or Master's thesis is usually new and the outcome is unpredictable, which is different from other courses in a study program. Thus, students work on a complex problem, while most other courses deal with complicated topics [1]. The main difference between complex and complicated problems is that solving complicated problems requires expert analysis, followed by the application of one of several good practices. For solving complex problems, we need to rethink existing solutions, explore and sense their implications, and find new solutions iteratively and evolutionarily.

D. Guralnick et al. (Eds.): TLIC 2021, LNNS 349, pp. 312–320, 2022.
https://doi.org/10.1007/978-3-030-90677-1_30

The supervision process has repeatedly been shown to be an important aspect for the success of a thesis project [8]. Different types of supervisors have been identified [12], and there exist several recommendations and guidelines for supervising scientific theses [6]. In this paper, we analyze goals and challenges of scientific theses, present a categorization of supervisors along the two dimensions "responsibility" and "autonomy," and derive a recommendation for a supervising process addressing all types of supervisors. We already have applied this process at our university for several semesters and experienced an increase in the number of students finishing their thesis on time, as well as an increase in the quality of theses.

The remainder of this paper is structured as follows. Section 2 presents related work in the field of supervising Bachelor's or Master's theses. In Sect. 3, we analyze the main goal of theses, as well as corresponding critical success factors and necessary conditions. Main challenges for writing theses and their potential root causes are described in Sect. 4. Section 5 presents our concept for supervising scientific theses, including a classification of four different supervisor types, and Sect. 6 concludes this paper.

2 Related Work

Several authors discuss that the supervision process is highly influenced by the personal relationship between supervisors and students. In Ref. [2], the different responsibilities of students and supervisors are analyzed, and their match in expectations according to who is responsible for which task in the supervision process is evaluated. Their study shows that there exists a significant mismatch between students' and supervisors' expectations regarding who is primarily responsible for which elements in this process (e.g., topic selection, initiation of meetings, checking on progress, methodology discussion or requesting feedback). An alignment between expectations is highly recommended to improve results. Similarly, Grant [3] discusses the complex relationship between students and supervisors and derives three different supervision styles: traditional-academic, techno-scientific, and psychological. The traditional-academic style is formal and asymmetric, where students have a passive role in the supervision process. The techno-scientific style focusses on the development of students' research skills. This also is an asymmetric style resulting in a high degree of dependency of students on their supervisors. Psychological supervision styles are characterized by an equal power distribution between students and supervisors.

In Ref. [4], an alternative model to the traditional one-to-one supervision is presented. This model aims at utilizing the benefits of group supervision and peer feedback to overcome the negative effects of students being supervised by only one person. They emphasized the benefits of student colloquia where students discuss their ideas, positive and negative experiences, as well as other thesis-related problems. In addition, early individual and group feedback is recommended. Holmberg [5] also discusses three different supervisor roles. According to this research, supervisors either act as coaches, who see a thesis as a joint project, and the supervisor primarily acts as a trainer to support the student. Other supervisors take the role of a consultant, where supervisors act as a resource that students have the responsibility to use according to their individual needs. The third type of supervisor acts like a parent who carries the whole responsibility of the supervision process.

The authors of a study presented in Ref. [6] conclude that effective supervision of theses should focus on the following two parts: support at the beginning to clarify the aim of the thesis and support of students in applying the methodology.

The main challenges in supervision processes are analyzed in Ref. [7]. They found that one of the main difficulties is to find the right balance between student autonomy and supervisor direction, as well as between authorship and responsibility. In addition, the diversity of students implies that there is no perfect supervision process that fits the needs of all students. Supervisors also need to keep in mind to allocate time between students equally. Furthermore, giving feedback is a crucial point.

The importance of metacommunication in the supervision process is discussed in Ref. [8]. Metacommunication hereby refers to the communication between supervisors and students about certain aspects of the supervision process. Again, the authors emphasize the importance of early-phase metacommunication, where students clarify the roles of supervisors and students, expectations, and how the process is structured. In addition, regular metacommunication contributes to a better communication between supervisors and students.

Several authors evaluate how blended learning approaches can facilitate and improve the supervising process. The integration of agile methods from software development, such as using a Trello board, is proposed in Ref. [9]. In Ref. [10], several parts of the supervision process are supported by an online tool. Particularly, tasks concerned with thesis administration, exchange of information, and collaboration have a high potential for blended learning tool support.

3 Goal of Bachelor's and Master's Theses

Successful completion of a thesis demonstrates several skills. First, students show a deep understanding of a specialized topic, which in our study programs is in the field of engineering. Second, they demonstrate the ability to work in a self-organized manner, because there is no strict process to follow in order to be successful, as it is the case in most other courses in a study program. Finally, social skills are necessary to cooperate with supervisors and stakeholders in the company (for final theses that are done in collaboration with our industrial partners). All these skills are also highly demanded by the industry, so we can define as an ultimate goal of a thesis to prepare students for their careers in industry, as shown in the goal tree in Fig. 1.

In order to develop these skills required to achieve the goal (also called critical success factors), several necessary conditions must be met. For example, students must build advanced knowledge of a particular topic based on what they have learned in their studies. They must conduct an extensive literature review and formulate a scientific research question. These three conditions are necessary for students to deepen their knowledge of an advanced engineering topic (see Fig. 1). Additionally, in order to work in a self-organized manner, they must define and follow a project plan, which also serves as a commitment for the completion of each task. Finally, students also need a solid level of communication skills to meet stakeholder expectations.

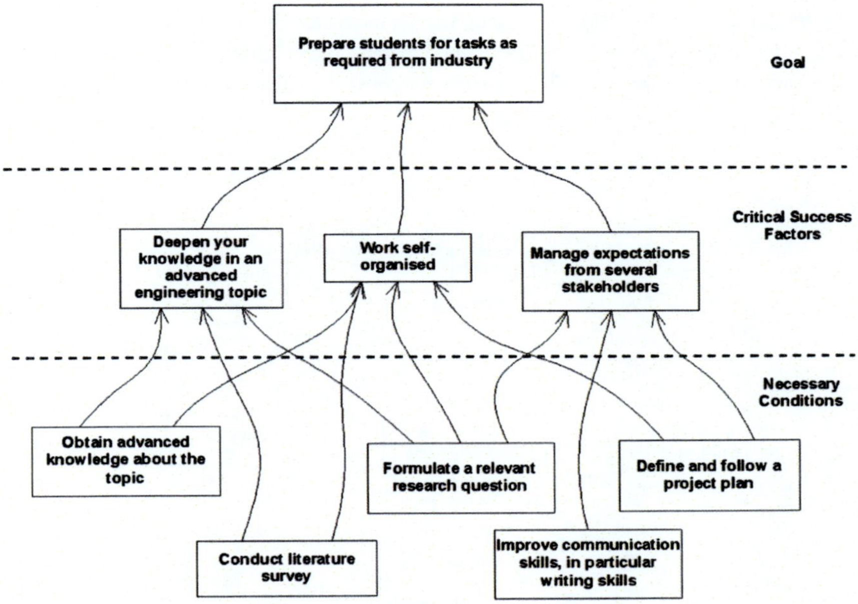

Fig. 1. Goal tree of a bachelor's or master's thesis.

4 Challenges of a Thesis

As we have seen, a Bachelor's or Master's thesis is expected to be the final preparation for the students' further career. However, every year at our university, we experience a number of students who do not successfully complete the thesis, even though they have successfully completed all other parts of their studies. In this section, we will discuss the reasons for this, based on the feedback we have received from students.

At our university, there are firm deadlines for students to submit their final theses. If students miss this deadline, they cannot finish their studies. Therefore, the most undesirable effect (UDE) in the context of supervising Bachelor's and Master's theses is that students cannot meet their deadline. In Fig. 2, we analyze the main reasons leading to this effect in a current reality tree focusing on this UDE. We identified three causes of UDE, inadequate writing, inadequate self-study skills, and lack of time. As shown in Fig. 2, there are other causes for these three deficits, all of which lead to two main causes: a strong focus on technical skills in an engineering curriculum and insufficient thesis supervision.

After identifying the root causes, we considered ways to address them. First, we redesigned our curriculum to include more professional skills courses. For example, we introduced a "Scientific and Technical Writing" course in the third semester of our Bachelor's program and some courses in almost every semester that require students to write a term paper. To show the practical relevance of the courses, we have combined them with the engineering courses into modules, with a common module grade for both courses. This mitigates the first root cause. More about the structure of our Bachelor program can be found in Ref. [11]. Second, we have developed and implemented a concept for the supervision of Bachelor's and Master's theses that supports the elimination of the second cause, insufficient supervision. In this paper, we introduce this concept and discuss its benefits and challenges.

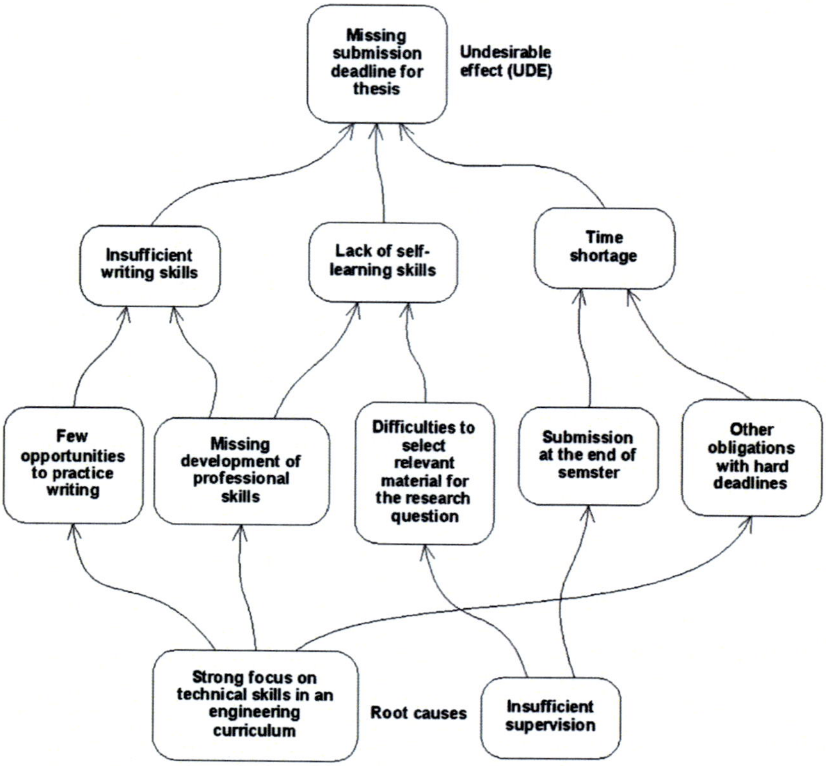

Fig. 2. Focused current reality tree for an unsuccessful Bachelor's or Master's thesis.

5 Concept for a Supervision Process

As mentioned in Sect. 2, there are different types of supervisors. We have experienced the greatest differences in the perception of responsibility for the thesis, as well as in the perception of students' ability to work in a self-organized manner. On this basis, supervisors can be divided into four categories, as shown in Fig. 3:

1. **Controller:** The supervisor assumes the main responsibility for the thesis and also for the entire process. The supervisor defines the topic, sets deadlines, and usually has a clear expectation of the outcome of the work.

2. **Gatekeeper:** The supervisor works with students who take responsibility of the thesis writing process and are self-organized. The supervisor has a clear expectation of the results and feels responsible for the results.
3. **Supporter:** The supervisor defines the topic of the thesis and a clear process for the supervision. During the process, students have defined submissions to track the progress of the work and receive feedback. Students have primary responsibility for the outcomes of the thesis.
4. **Coach:** The supervisor allows students to work in a self-organized manner. Coaching sessions are provided if students request them. Again, students have the primary responsibility for results.

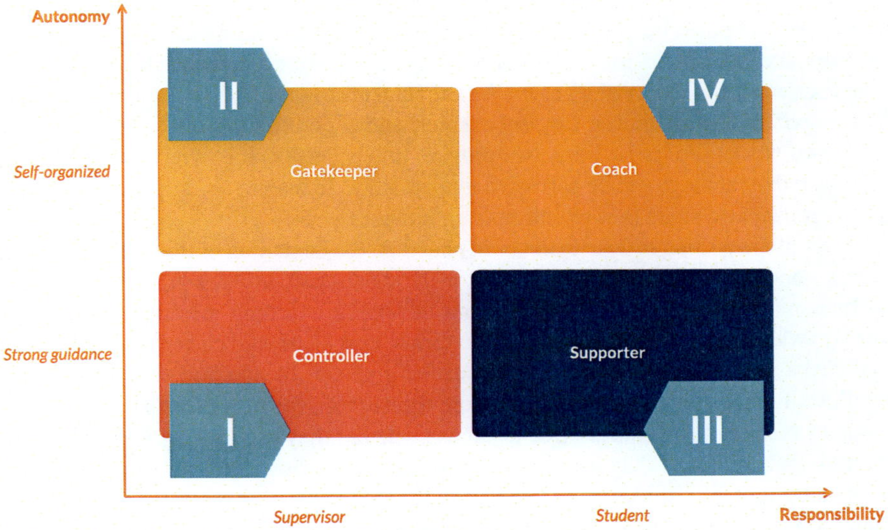

Fig. 3. The four supervisor types.

We designed and implemented a supervision process that can be adjusted to each supervisor type. The process defines a clear and continuous guidance to support students according to their needs. The supervisor ideally applies the process individually for each student to support different student types.

Figure 4 summarizes our supervision process of theses that addresses the goals and challenges of theses as described in Sects. 3 and 4. We have divided this process into four successive phases. The first phase is about topic selection. For this purpose, we organize a kick-off meeting where we present potential topic pools of the respective supervisors. During this event, students can clarify questions and also propose and discuss their own ideas directly with the supervisors. Students can then vote for three different topics and are assigned to the priority that best suits them. Each faculty member supervises a group of students who are all working on related topics. In this way, students can also inspire each other while being supervised.

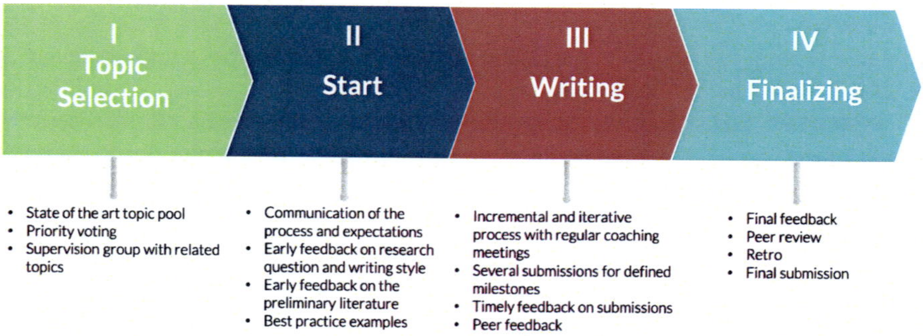

Fig. 4. The four phases of continuous thesis supervision.

The second phase begins with a meeting with the corresponding supervisor in which we introduce the supervision process and communicate our expectations for a successful thesis project to the supervised group. We ask students to submit their research question, a first draft of the content, a first subchapter, and a preliminary bibliography within about two weeks. For these parts, we provide timely feedback so that potential errors and misunderstandings can be corrected immediately. To inspire and support students at this stage, we also share best practices from previous years.

Most of the writing is done in the third phase. At the beginning of this phase, students have a good, but almost blank, outline for the thesis and, at the end, a draft version of the complete thesis. This phase consists of regular coaching meetings (usually bi-weekly) where we discuss the status of the thesis, next steps, and current and upcoming challenges with the respective student coaching group. In addition, we set optional deadlines for the submission of parts of the thesis (e.g., the next 10 pages or the next chapter). Note that students are free to choose whether they submit their progress at the given deadlines and if they participate at the meetings. For their submissions, which are usually monthly, students receive timely feedback to be incorporated into the next submission. All submissions are set and done in our Moodle course. In the coaching sessions, where we discuss feedback on these papers, students also receive peer feedback from other students. Therefore, it is important that all students work on related topics in a coaching group.

In the fourth phase, students submit their preliminary final theses in our Moodle course. The deadline for this submission is usually 2–3 weeks before the official deadline. This allows us to provide final feedback on the complete thesis and also to organize a peer review, where each student not only reviews and provides feedback on the theses of two other students, but also receives feedback from two peers. At this stage, there is also a final coaching meeting where we discuss the feedback on the preliminary final version and conclude the supervision with a retrospective reflecting on what worked well, what could be improved and why. After that, students are ready to officially submit their final theses.

In particular, by providing feedback early in Phase I, we can mitigate difficulties in selecting relevant material for the research questions. By introducing regular deadlines and gradual submission of theses, we ensure that students work continuously on their theses and do not postpone the work until the final submission at the end of the semester. This concept thus eliminates "insufficient supervision" as the cause of the current reality tree (Fig. 2).

6 Conclusion

We developed and implemented the presented concept for continuous theses supervision for the first time in the summer semester of 2017. We started with four groups of students and were the only two lecturers with this concept. In total, 40 out of 50 students were able to complete their thesis on time (80%). In our groups, we supervised 22 students according to the given concept, of which 19 finished on time (86.4%). In the other groups, there were a total of 28 students, of whom 21 finished on time (75%). After this year, we started to transfer our concept to other groups and share it with other supervisors. Today, more than 75% of our students are supervised following this concept.

Our preliminary findings are that there are a few key success factors for a supervision of a Bachelor's or Master's thesis. First, the process and expectations need to be clearly communicated at the beginning. Second, support and feedback at the early stage of a thesis helps students to find an appropriate topic and formulate a research question, which is a precondition for a successful thesis. Third, continuous work on thesis and intermediate feedback from the supervisor increases on-time delivery rate of theses.

In our future work, we plan to identify an optimal supervision process for different supervision styles, to further evaluate this concept, as well as to investigate reasons for a failure to complete theses. According to these reasons, we will introduce further changes in the curriculum or adaptions in the proposed supervision process.

References

1. Snowden, D.: What Cynefin is in brief. In: Greenberg, R., Bertsch, B. (eds.) Cynefin – Weaving Sense-Making into the Fabric of Our World, Cognitive Edge - The Cynefin Co (2020)
2. Stappenbelt, B., Basu, A.: Student-supervisor-university expectation alignment in the undergraduate engineering thesis. J. Technol. Sci. Educ. **9**(2), 199 (2019)
3. Grant, B.M.: Fighting for space in supervision: fantasies, fairytales, fictions and fallacies. Int. J. Qual. Stud. Educ. **18**(3), 337–354 (2005)
4. Dysthe, O., Samara, A., Westrheim, K.: Multivoiced supervision of master's students: a case study of alternative supervision practices in higher education. Stud. High. Educ. **31**(3), 299–318 (2006)
5. Holmberg, L.: Coach, consultant or mother: supervisors' views on quality in the supervision of bachelor theses. Qual. High. Educ. **12**(2), 207–216 (2006)
6. Strebel, F., Gürtler, S., Hulliger, B., Lindeque, J.: Laissez-faire or guidance? effective supervision of bachelor theses. Stud. High. Educ. **46**(4), 866–884 (2021)
7. Vehviläinen, S., Löfström, E.: 'I wish I had a crystal ball': discourses and potentials for developing academic supervising. Stud. High. Educ. **41**(3), 505–524 (2016)
8. Baltzersen, R.K.: The importance of metacommunication in supervision processes in higher education. Int. J. High. Educ. **2**(2), 128–140 (2013)
9. Brodtkorb, A.R.: Agile supervision of Bachelor, Master, and PhD. Theses. In: Proceedings of the MNT-konferansen, Tromso, Norway (2019)

10. Karunaratne, T.: Blended supervision for thesis projects in higher education: a case study. Electr. J. e-Learn. **16**(2), 79–90 (2018)
11. Miladinovic, I., Schefer-Wenzl, S., Hirner, H.: Curriculum of a telecommunications study program—a matter of trends? In: Proceedings of the 15th International Conference on Telecommunications (ConTEL), Graz, Austria (2019)
12. Schmolitzky, A., Schümmer, T.: Patterns for supervising thesis projects. In: Proceedings of the 13th Annual European Conference on Pattern Languages of Programming, Irsee, Germany (2008)

Hybrid^x Higher Education – A Multidimensional Overlay of Hybrid Forms of Learning and Teaching

Christian-Andreas Schumann[1] (✉), Anna-Maria Nitsche[1,2], Claudia Tittmann[1], and Kevin Reuther[1,2]

[1] University of Applied Sciences Zwickau, Kornmarkt 1, 08056 Zwickau, Germany
christian.schumann@fh-zwickau.de
[2] University of Leipzig, Augustusplatz 10, 04109 Leipzig, Germany

Abstract. The diversification of study programs serves to impart knowledge and skills in a way that is appropriate to the target group and to individualize learning paths in a student-centered manner. Traditional hybrid forms of learning such as the combination of face-to-face and distance learning (blended learning) are extended by new hybrid approaches. However, these often appear rather by chance than on a sound basis due to the diversity of opportunities and digital tools available today. Hybrid forms of university teaching are increasingly combined in different contexts and degrees of digitization. The resulting increase of higher education teaching complexity can only be managed through structured, transparent, and evidence-based approaches when designing learning and teaching models. The suggested Hybrid^x university teaching reference model describes the combination of several teaching and learning dimensions. It builds a foundation for systematically restructuring university teaching using innovative, digital approaches. Next to the conceptualization of the novel Hybrid^x reference model, the paper provides case study examples and recommendations for using hybrid forms of learning and teaching.

Keywords: Hybrid higher education · Hybrid learning · Learning strategy · Learning in practice

1 Introduction – Knowledge Transfer in the Network Society

The transfer of information and knowledge has gained in importance with the development of the industrial society via the knowledge society to the network society [1]. The growing complexity of knowledge resources forces new forms of knowledge transfer while learning. Networking is mainly driven by new organizational structures, whereby individual knowledge nodes, containing either individuals or complex organizational units, form connections with each other [2]. A knowledge node is a small knowledge unit, the smallest possible knowledge carrier. It may be a single person, a group of persons, a part of an organization, a whole organization, or a company (see Fig. 1).

D. Guralnick et al. (Eds.): TLIC 2021, LNNS 349, pp. 321–332, 2022.
https://doi.org/10.1007/978-3-030-90677-1_31

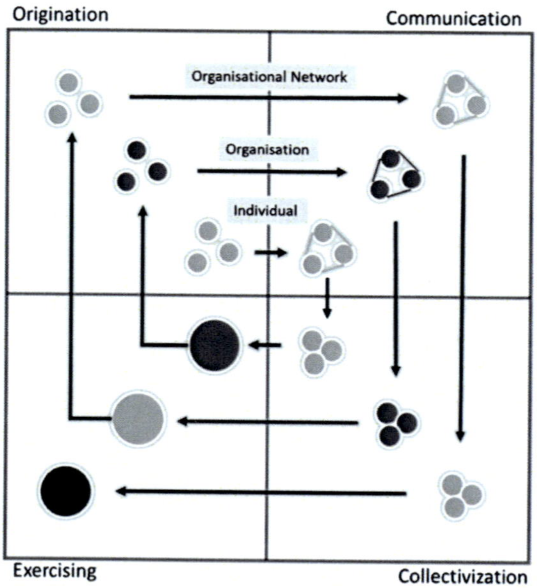

Fig. 1. Cycle of knowledge node growth – from the individuals to organizational networks [3].

Organizational linkages are significantly facilitated and pushed by the possibilities of digitization in the context of new media and platform strategies. The growing complexity of knowledge resources requires a more selective approach to knowledge resources and increases the compulsion to specialize, whereas the growing networking of knowledge nodes technically and organizationally promotes access to both specialized and general knowledge. Implicit knowledge, thus, generally grows in the course of possible individualization on the basis of constantly improved access to explicit knowledge in networked systems. Individuals have constantly expanded opportunities to access extensive, networked knowledge stocks. However, this also increases the requirements for having competencies that enable access to global knowledge. Knowledge is based on medial and informal literacy, which at the same time is associated with new forms of teaching and learning.

These dynamic processes of new forms of teaching and learning for knowledge transfer in the network society will be intensified and accelerated in the future by closer human-machine cooperation. The increasing intertwining of natural and machine intelligence creates new opportunities and risks for knowledge transfer, which will be more strongly characterized by the application of Artificial Intelligence (AI) in the future. The existing use of webbots already demonstrates how the ways of life and learning of people in the network society will be different. Due to the multitude of influencing factors (dimensions) and their continuous or discrete expression in different contexts, hybrid approaches are becoming more and more significant [4], because the diversification for the target-group-oriented transfer of knowledge and skills and the accompanying individualization of learning paths for learners is progressing. Learning forms and units, as well as knowledge transfer and learning processes, are cross-linked [3].

2 Knowledge Transfer via Hybrid Forms of Teaching and Learning

Teaching and learning forms can be distinguished according to whether learning processes take place through teaching in the sense of externally controlled impulses or whether teaching intends self-directed learning. This distinction between self-directed and externally directed learning is not absolute, but a fluid continuum. Therefore, different social forms as well as methods of teaching and learning emerge. Teaching is pedagogical action for the purpose of a successful learning process and, thus, not directly knowledge transfer. Learning, on the other hand, is directly aimed at the individual acquisition of knowledge, skills, and competencies. Teaching is classically divided into several basic forms, for which there are different scientific approaches. A division into twelve basic forms [5] differentiates in telling and referring, showing, and demonstrating, looking, and observing, dealing with texts, writing texts, working out the course of action, forming concepts, building up problem solutions, working through, practicing and repeating, applying, and independently discovering learning. Another structuring according includes four basic forms: Class work, group work, partner work, and individual work [6]. Jank and Meyer [7] distinguish individualized, instructional, and cooperative instruction.

Already in this context regardless of the type of classification, each basic form can occupy different degrees between two poles in a continuous or discrete spectrum. For example, either pure group work or pure individual work as extremes (poles) or any mixed forms in a defined but variable ratio of the two basic forms. Each basic form can also be considered and hybridized individually, because, for example, between the extremes, no or absolute individualization of teaching and learning between the poles a range of 0%–100% can be chosen. Classical hybrid learning forms such as the combination of face-to-face and distance learning (blended learning) are overlaid by new hybrid approaches [8], such as the mixture of face-to-face and online teaching [9] in the face-to-face part or digitized learning and classical literature study in distance learning. Theoretical approaches and practical solutions of hybrid forms of teaching and learning [10] have existed for many years for changes of a feature or an object between two extreme states or for the change of the ratio when combining two features or two objects. As far as only between two states or combinations is to be differentiated, in the following, the term dimension is used for it. While solving complex teaching and learning scenarios for successful knowledge transfer, several dimensions have also been combined. Due to the digitalization of all areas of life [11], including education, further dimensions are added, so that the degree of multidimensionality is constantly increasing [12]. This multidimensionality is shown by examples in Table 1.

Table 1. Pragmatic attempt to connect learning dimensions with digital applications [13].

Selected main dimensions	Examples of digital sub-dimensions
Self-learning	e-Books, CBT/WBT, Chatbots, Learning Nuggets, e-Paper, Video Instructions, Badges

(continued)

Table 1. (*continued*)

Selected main dimensions	Examples of digital sub-dimensions
Teacher centered learning	Online-Tutoring & Coaching, e-Assessments, Online Seminars, interactive Whiteboards
Group centered learning	Interactive Gaming, online Barcamps, Internet Collaboration, Online group work
Group centered communication	Social Networks, Chats, Newsgroups, Instant Messaging, Micro Blogging
Universal tools	Virtual Reality, Augmented Reality, Moocs, Learning Videos, Podcasts, Simulation Tools

The multitude of possible teaching and learning forms from different classification views in connection with diverse and increasingly also digital learning scenarios, all of which can be mapped via hybrid dimensions, shows the dilemma of growing complexity. This challenge is already evident in the small excerpt of possibilities [13]. Therefore, insights from systems and software engineering should also be used for modeling hybrid teaching and learning worlds. If the principles, concepts, and models of well-structuredness can also be applied to educational systems with hybrid dimensions of teaching and learning, there is a chance to solve the dilemma of constantly increasing complexity through systematic modeling.

3 Knowledge Transfer in the Network Society

The basis of hybridism is the mixing and varying of system or object states, which are characterized by the variety of possible attributes and variable attribute values of entities and the associated complexity and combinatorics [14]. In the context of describing forms of teaching and learning in their various combinations, their manifestations can be described by attributes with mutable attribute values and can be conceived as dimensions. The possibility that dimensions can thus be hybrid opens the chance of dynamic changes of attribute values in a continuous or discrete spectrum, which can continuously increase adaptability and flexibility of teaching and learning systems.

Minimally, hybrid forms lead to simplifications and reductions in complexity, but also to more static and inflexible forms of learning and thus knowledge acquisition. Maximum hybrid forms extremely enable learner-centered scenarios with a high degree of individualization of knowledge transfer, but also increase complexity and, thus, reduce controllability, as well as transparency of learning processes.

Hybrid is the property of being able to adopt mixed states or generate mixed forms. Selected states are shown in Table 2.

Table 2. Typical forms of one or more hybrid states.

Hybrid forms	Solution forms	Graphic representation
A possible state between two opposite states of one thing	One-dimensional line	
A state by combining different states of two different things	Two-dimensional area	
A state by combining different states of three different things	Three-dimensional cube	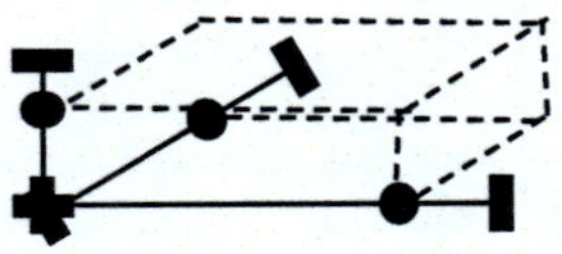
A mixed state composed of different states of n different things (n>3)	Multidimensional space	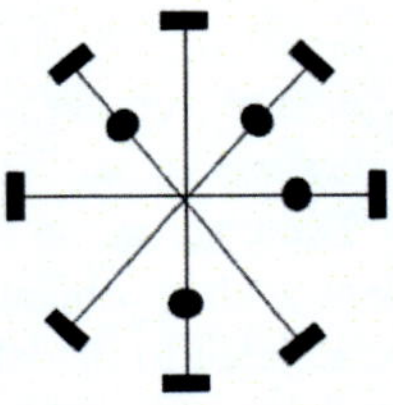

Thus, in a holistic view in the context of systemic modeling, 1 to n dimensions are to be considered for hybrid systems, which in turn can be hybrid themselves. Starting from a single dimension, by means of which hybrid states of a characteristic between two opposite extreme states of the characteristic are described, an infinite number of dimensions with an infinite number of hybrid states can be included in the modeling. Furthermore, it must be possible in the modeling to combine 2 to n dimensions characterized by hybrid states. The successful combination of hybrid dimensions is an application field of poly-optimization for the purpose of multicriteria goal achievement.

4 Reference Model Approach for Multidimensional Hybrid Forms of Teaching and Learning

To better analyze and model complex systems, the method of object orientation was developed. Objects are elements of the real world that exist physically or conceptually. They are uniquely identifiable and distinguishable from other objects, have attributes with attribute values, and can have methods for describing their behavior in the context of state changes [15]. If the dimensions are conceived as objects, then their interaction can be described using this method. Similar objects or dimensions can be grouped into classes, which abstractly serve as a means of ordering. A class is derived by recognition of similarities of objects. Similar objects are summarized in classes. Objects as well as classes can have relations among themselves. To be able to describe dimensions, objects, and order by classes, the object-oriented analysis is accomplished, whose result is a conceptual model. Afterwards, the object-oriented design takes place, as described in Table 3.

Table 3. Application of object orientation in the dimensional approach of hybrid systems.

Object-oriented analysis	Object oriented design
(1) Identify the dimensions (2) Ordering the dimensions (3) Assigning attributes of the dimensions (4) Assigning behavior of the dimensions (5) Describing interactions of the dimensions	(1) Restructuring the classes (2) Implementing attributes of dimensions (3) Implementing behavior of dimensions (4) Implementing the control (5) Implementing the associations

Typical objects are, for example, the lecturer Miller, the student Brown or the PC at Mr. Jones' workplace. They are objects of classes such as "lecturer," "student," or "computer." Dimensions, on the other hand, are a different, conceptual type of classes. They represent abstract objects where one or more attributes can take on different attribute values. Related to modeling hybrid teaching and learning, the primary interest is how classes behave, for which the concept of dimension was chosen. Class formation is exemplified by the example of networking (see Fig. 2).

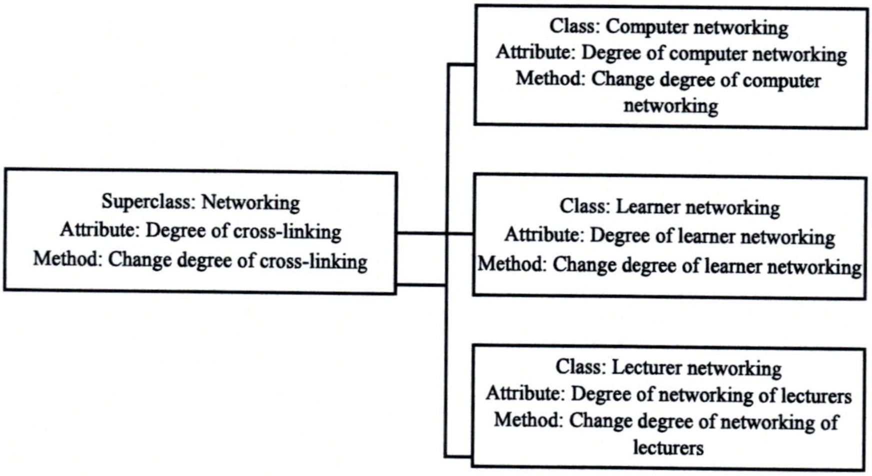

Fig. 2. Reference model class approach for dimensions of hybrid teaching and learning forms.

The attributes of the classes can take attribute values between zero (minimum) and one (maximum) either continuously or discretely, depending on how it is defined. Thus, each dimension forms a class with attributes having hybrid values. In addition, classes can also have relationships among themselves.

5 Hybrid^{3} Higher Education Multidimensional Cases

Hybrid learning is defined by Singh and Reed [16] as learning that uses more than one learning model, with the aim to optimize learning outcomes. Furthermore, Singh & Reed state that hybrid learning is mixed learning that focuses on optimizing the achievement of learning goals by applying learning technology [16].

Multidimensional hybrid modeling has not yet been explored in a complex theoretical way, but it has been practiced in research and teaching for several years. Due to the increasing digitization and rising complexity of global educational activities, clearly pushed by the contact restrictions in the wave of the global pandemic in 2020 and beyond, the focus on hybridization in teaching and learning is becoming apparent, which also reveals the increasing need for research. This can be done at a level based on successful concepts and models of multidimensional, hybrid educational projects. Two three-dimensional, hybrid approaches are presented as examples.

Case 1. Hybridity in the Generalization and Specialization of Teaching for Digitization

In the classical T-shape model [17], the vertical bar of the letter T means specialization—i.e., representation of the depth of the acquired competences in a subject area—and the horizontal bar means generalization—i.e., representation of the breadth of the acquired knowledge for the purpose of interdisciplinary cooperation with other specialists based on generic competences. Both dimensions are modifiable and can thus be hybrid. During the growing digitization of education, it was examined what influence it has on the individual subjects. The consequence is that each subject has to be digitized in a specific way depending on the respective field of knowledge. This makes it necessary to introduce a third dimension, digitization, which in turn is also hybrid in nature (see Fig. 3).

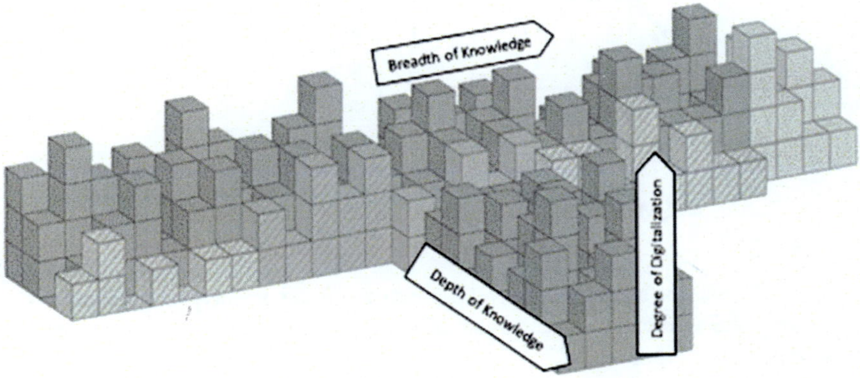

Fig. 3. Hybrid3 – Impact of digitalization of subjects on T-shape model [18].

This three-dimensional hybridization of the extended T-shape model improves the planning of digitized study content in the curriculum by conceiving the depth and breadth of the competencies to be taught as well as the degree of digitization as hybrid and thus freely variable within certain limits to be able to act in a target group-specific manner. The model was, for example, the basis for complex digitization steps for study content in the Chinese-German University of Applied Sciences [18].

Case 2. Individualized Content Delivery from a Complex Knowledge Base
To be able to individualize learning more strongly, complex knowledge pools on one or several subject areas must be granulated—i.e., broken down into elementary knowledge modules, which are then compiled as needed for the concrete demand, be it a short learning sequence or an entire teaching unit for individual learners or learning groups. For the mainly automated realization of such learning systems based on a knowledge base, three dimensions were considered in the hybrid model: Difficulty, Methodology, and Presentation. The states of each dimension were stored as discrete values. Each state of a dimension can be combined arbitrarily with the respective states of the other two dimensions. This results in hybrid forms with respect to the characteristics of the individual dimensions, as well as in the combination of the states of several dimensions (see Fig. 4).

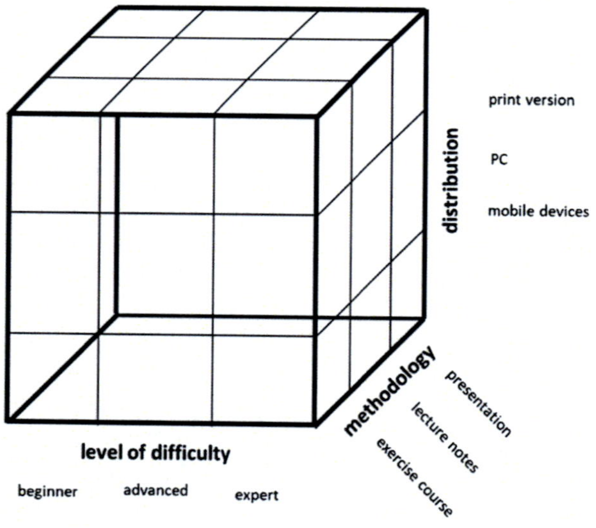

Fig. 4. Hybrid3 – Three-dimensional selection of content modules.

The model was the basis for the development of a digital prototype for the multivalent target group and application-oriented use of knowledge modules from a complex body of knowledge (see Fig. 5).

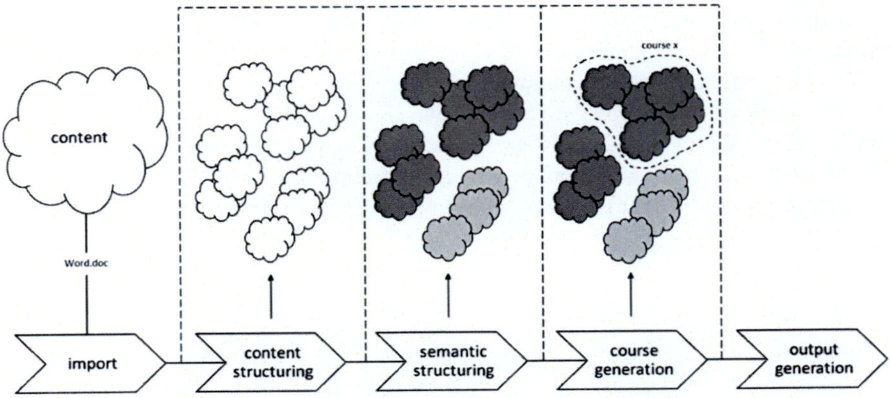

Fig. 5. Hybrid3 – Course generation of structured content using a semantic model.

Monolithic knowledge bases are imported and then decomposed into elementary knowledge components that are stored as objects in a semantic hybrid network. Semantic networks are causally related to applied AI. The discrete state of each of the three hybrid dimensions is chosen so that a generation of the learning unit appropriate for the target audience can be generated, which is then used with the output format of the chosen type of output device [19].

6 Hybridism and Learning Theories

Learning processes are scientifically described and explained by means of learning theories. An all-encompassing learning theory has not yet been developed, but special theories exist that relate to specific forms of learning and learning situations. Learning theory research is based on two main branches: behaviorism and cognitivism. Humanistic learning theory should also be mentioned. From the pedagogical point of view, instructionism and constructionism should be emphasized as learning theories. Around this basic set of scientific explanations of learning, numerous other learning theories are grouped [20–22].

One of the learning theories that builds on constructivism is connectivism, whose basic principle is that knowledge is distributed in networks and can be socially generated by individuals. Learners must be able to access distributed knowledge and make it useful to themselves. In a self-organized process, learning takes place in real and virtual networks, and the acquired knowledge is constructively incorporated into one's own thought model. The advantage of this approach is that increasing medialization and technologization in the context of digital transformations are an essential part of this learning theory. Learning, thus, refers not only to the pure acquisition of knowledge from networked sources, but also to how selection, access, and participation in relevant knowledge take place in the network. A limiting factor is that humans primarily have an influence on the type, form, and characteristics of the relations in the network, without being able to specifically plan and externally control them. Due to the extensive decentralization of the learning processes as an interaction between the participants, essential options of a central control are omitted.

The structural foundations of connectivism show parallels to knowledge networks, which consist of a set of knowledge nodes and knowledge edges. The knowledge of the individual nodes, which can be single individuals, groups of individuals, or organizations, can be transferred via edges if the nodes are networked by such edges (see Fig. 6).

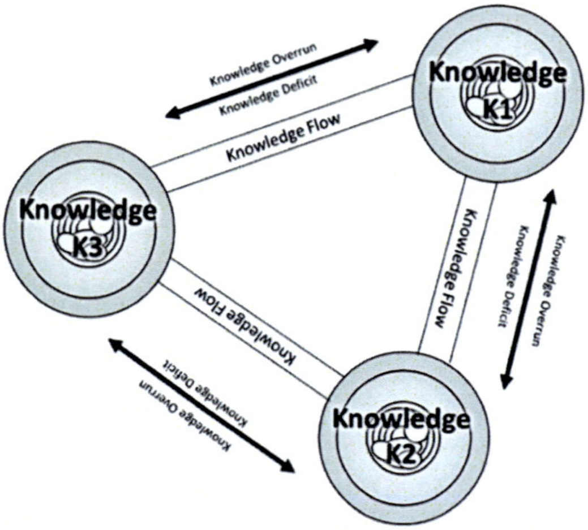

Fig. 6. Triggering of knowledge flows through different knowledge potentials [3].

In this model approach, as in the potential equalization method, flows are generated between nodes when the difference in the potentials of two connected nodes exceeds a certain level. In analogy to connectivism, this equalization occurs as an internal process without the possibility of external control options. Learning is controlled as an internal process within the node itself.

Furthermore, it is possible to influence from each node which edges (relations) to other nodes are established and in which form. Further control options arise when, as in real systems of knowledge transfer, socially, organizationally, and/or technically justified limitations or extensions of the knowledge flows are made by the representative of the knowledge node. In nodes themselves, as well as within the edges connected to them, throttling and increases of knowledge flows can be achieved by filters.

In learning processes based on these knowledge flows, filter parameters can be continuously or discretely influenced and thus changed. They can assume hybrid states. This principle underlying hybridization enables both self- and external control of knowledge transfer and learning processes in the network. Control can also be transformed into regulation when one or more control parameters are introduced. There are three basic possibilities of hybrid control (possibly regulation) of learning processes by means of hybrid states in knowledge networks: self-control, socially determined external control, and machine-determined external control.

To summarize, hybridization offers the option of a new learning theory approach for the explanation of learning in social and digital networks. Building on other basic and special learning theories, hybridization opens a possibility to explain learning in networked systems in the digital age as a partially controllable system with self- or external control via hybrid states. Moreover, using hybridism as a learning theory can also provide explanations for the collaboration of human and machine learning.

References

1. Castells, M.: The Rise of The Network Society. Blackwell Publishing, Oxford (2000)
2. von Krogh, G., Back, A., Enkel, E.: Knowledge Networks for Business Growth. Springer, Heidelberg (2007)
3. Schumann, C.-A., Tittmann, C.: Chapter V: Potentials for externalizing and measuring of tacit knowledge within knowledge nodes in the context of knowledge networks. In: Harorimana, D. (ed.) Cultural Implications of Knowledge Sharing, Management and Transfer: Identifying Competitive Advantage. IGI Global (2010)
4. Kraidy, M.: Hybridity in Cultural Globalization. Commun. Theory **12**(3), 316–339 (2002)
5. Aebli, H.: Zwölf Grundformen des Lehrens: Eine Allgemeine Didaktik auf psychologischer Grundlage. Medien und Inhalte didaktischer Kommunikation, der Lernzyklus. Stuttgart, Klett-Cotta Verlag (2019)
6. Grunder, H.U., Ruthemann, U., Scherer, St., Singer, P., Vettinger H.: Unterricht: Verstehen - planen - gestalten - auswerten. Baltmannsweiler: Schneider Verlag (2012)
7. Jank, W., Meyer, H.: Didaktische Modelle. Cornelsen Schulverlage GmbH, Berlin (2014)
8. Horn, M.S., Crouser, R.J., Bers, M.U.: Tangible interaction and learning: the case for a hybrid approach. Pers Ubiquit. Comput. **16**, 379–389 (2012)
9. Doering, A.: Adventure learning: transformative hybrid online education. Distance Educ. **27**(2), 197–215 (2006)
10. Olapiriyaku, K., Scher, J.M.: A guide to establishing hybrid learning courses: employing information technology to create a new learning experience, and a case study. Internet High. Educ. **9**(4), 287–301 (2006)
11. Cook, J., Mor, Y., Santos. P.: Three cases of hybridity in learning spaces: towards a design for a Zone of Possibility. Br. J. Educ. Technol. **51**(4), 1155–1167 (2020)
12. Hariadi, B., Sunarto, D., Sudarmaningtyas, P., Jatmiko, B.: Hybrid learning by using brilian applications as one of the learning alternatives to improve learning outcomes in college. Int. J. Emerg. Technol. Learn. (iJET) **14**(10), 34–45 (2019)
13. mmb Institut Homepage, Goertz, L., et al.: Systematik der Lernformen in neuem Gewand. https://www.mmb-institut.de/blog/systematik-der-lernformen-im-neuen-gewand/. Accessed 5 Mar 2021
14. LITERATURE Homepage, Huddart, D.: Hybridity. https://oxfordre.com/literature/view/ https://doi.org/10.1093/acrefore/9780190201098.001.0001/acrefore-9780190201098-e-1124. Accessed 12 Mar 2021
15. Booch, G.: Object Solutions: Managing the Object-Oriented Project. Pearson Education, London (1995)
16. Hariadi, B., Sunarto, D., Sudarmaningtyas, P., Jatmiko, B.: Hybrid learning by using Brilian applications as one of the learning alternatives to improve learning outcomes in college. Int. J. Emerg. Technol. Learn. (iJET) **14**(10) (2019)
17. Johnston, D.L.: Scientists become managers-the 'T'-shaped man. IEEE Eng. Manage. Rev. **6**(3), 67–68 (1978)

18. Gallenkämper, J., Schumann, C.-A., et al.: Ingenieurausbildung für die digitale Transformation. Diskussionspapier zum VDI-Qualitätsdialog. Verein Deutscher Ingenieure. Düsseldorf, 15–19 (2018)
19. Götze, J., Schumann, C.-A. et al.: Cross-linking and multivalent use of knowledge modules for special target and user groups. In: Reynolds, N., Webb, M. (eds.) Proceedings IFIP World Conference on Computers in Education WCCE 2013, pp. 184–193. Nicolaus Copernicus University Press, Torun (2013)
20. British Library: Contemporary Theories of Learning: Learning Theorists ... In Their Own Words. Abingdon (2018)
21. Bates, B.: Learning Theories Simplified. Sage, London (2019)
22. Matthews, L.: Theories of Learning. An Educational Perspective. Larsen and Keller Educatio, New York (2019)

Coherence Between Online Learning and Project-Based Courses in Higher Education

Vilmos Vass(✉) and Ferenc Kiss

Budapest Metropolitan University, Budapest, Hungary
{vvass,fkiss}@metropolitan.hu

Abstract. The context of the presentation is a challenge of a "skill gap" of the latest Future of Jobs Report, emphasizing the importance of some relevant skills for project-based courses in higher education—for instance, active learning and learning strategies, complex problem solving, and critical thinking and analysis. Under COVID-19, higher education has transformed to online learning, which requires renewing the concept of project-based courses. The presentation focuses on innovation of assessment, especially giving some examples using Zoom in order to strengthen collaborative assessment culture under the umbrella of formative assessment. The actual outcomes of the presentation refer to collecting opportunities, experiences and challenges on the feasible coherence between online learning and project-based courses, especially focusing assessment on giving some examples on sharing aims and expectations, using self-assessment, a progression plan and shared assessment criteria on final presentations, team digital portfolios and individual reflective papers. Finally, the conclusion has some dilemmas and questions for discussion.

Keywords: Skill gap · Transformation · Digital mindset · Knowledge management

1 Contextual Background

In the first approach (narrow sense), a skill gap is the difference in the skills required on the job and the actual skills possessed by the employees. A skill gap is related workforce readiness for the challenges of the fourth Industrial Revolution. One of the biggest challenges is to prioritize soft skills in the workplace—for instance, ethics, personal organization and work habits; time management; teamwork and interpersonal communication; anger management; reasoning and problem solving; and managing one's learning. In the second approach (broad sense), a skill gap is based on changing the future-oriented skill set. Half of all employees will need reskilling by 2025, as adoption of technology increases, according to the World Economic Forum's Future of Jobs Report [11]. Behind the growing needs of reskilling and upskilling process, the top 10 job skills of tomorrow are:

1. Analytical thinking and innovation

D. Guralnick et al. (Eds.): TLIC 2021, LNNS 349, pp. 333–338, 2022.
https://doi.org/10.1007/978-3-030-90677-1_32

2. Active learning and learning strategies
3. Complex problem solving
4. Critical thinking and analysis
5. Resilience, stress tolerance, and flexibility
6. Creativity, originality, and initiative
7. Leadership and social influence
8. Reasoning, problem-solving, and ideation
9. Emotional intelligence
10. Technology design and programming

In fact, these future-oriented skills are related to the challenges of the fourth Industrial Revolution. This is the digital-driven process, which prioritizes transformation towards the knowledge-based society highlighting technological changes, such as Artificial Intelligence and Big Data. This process of transformation has enormous impacts on rethinking higher education, especially curriculum development, teaching methodology and assessment function. Regarding curriculum development, collaborative planning comes to the front. Rethinking teaching methodology is based on interactions; accordingly, this process puts great emphasis on project method and problem-based learning and Socratic teaching. Traditional higher education stresses summative assessment. Analytical thinking and innovation, active learning and learning strategies, complex problem-solving, and critical thinking and analysis require diagnostic and formative assessment. Du Preez et al. (2016) [1] suggest that transformation is a complex, open-ended process, which "can refer to changes in institutional structures and culture, as well as to specific elements such as curriculum and academic and student experiences." [9]. This is a transformational learning, where learning is a central point at the individual, team and organizational levels. In other words, the trend is focusing on teaching and learning. No doubt, this process characterizes fluidity, open-endedness and inherent complexity [1]. On the other side of the coin, under the COVID-19 pandemic, higher education has transformed to online learning, which requires renewing the concept of transformational learning. How can online learning transform learning and teaching process and higher education? In order to answer this question, the presentation focuses on the triangle of digital mindset, methodology of online learning and technological innovation.

2 Phenomena on Coherence Between Online Learning and Project-Based Courses

In order to understand changing a traditional to a digital mindset in higher education, we need to analyze some phenomena from the point of required coherence between online learning and project-based courses. The traditional mindset of higher education is based on a huge lecture hall, where hundreds of students listen to the teacher. They take many notes; formerly, they wrote these notes on the paper, and nowadays they type them on the laptop or iPad. This process requires high concentration and memorization. Seminars prefer text analysis with discussion and convergent thinking. Small interactions, questioning and debating take place. The traditional mindset has a behavioristic approach of learning and teaching process, where teachers teach and students listen to the teacher.

The conception of knowledge is mainly declarative and explicit. According to Ten Berge and van Hezewijk, *"traditionally, cognitive psychology has viewed the human mind as a general information-processing device"* [10]. In this sense, the peak of this learning and teaching is recording the information (facts, data) with different levels of understanding and application of knowledge. Expression of declarative knowledge requires directed attention, as opposed to the expression of skills, which is automatic [12]. Obviously, neglecting procedural knowledge cannot answer the above-mentioned challenges of the fourth Industrial Revolution and the future of jobs. The good news is in artificial intelligence research started to find the required balance between declarative and procedural knowledge [10]. It is the first phenomenon on coherence between online learning and project-based courses. In order to understand the second phenomenon, we need to turn back to the Polanyi's concept of knowledge, especially the difference of explicit and tacit knowledge [6, 7]. In a narrow sense, explicit knowledge is easy to articulate, write down, and share. Tacit knowledge, which is based on your personal experience, is more difficult to express. In a broader sense, Polanyi [8] argued that knowing was always an act of a particular individual. The other famous statement that *"we can know more than we can tell"* [7]. Tacit knowledge has experienced-based, individualized characteristics [6, 7]. Finding the coherency between online learning and project-based courses, it is a fundamental statement that tacit and explicit knowledge should not be seen as two separate types of knowledge [2]. As Jasimuddin, Klein and Connell stated: *"all knowledge has both tacit and explicit components"* [4]. No doubt, digitalization, the fourth Industrial Revolution and growing needs for reskilling have changed the traditional mindset of higher education, especial concept of knowledge and learning. COVID-19 has accelerated the above-mentioned process; online learning, especially, has some opportunities in order to rethink curriculum development, teaching methods and assessment functions.

3 Practical Case

Budapest Metropolitan University was the first higher education institution in Hungary that implemented online learning last year in March using Zoom platform (zoom.us) for remote teaching. Changing the traditional mindset, we organized online Teachers' Club for mentoring on lectures, especially using Zoom with some functions:

1. Chat - put comments, questions, fast information from the students' and teachers' points of view, short brainstorming, collecting students' questions, collecting key words, quotes
2. Raise hand – answering the questions, indicating the level of understanding
3. Breakout Rooms – group work with teachers' mentoring
4. Sharing – videos, lectures, music
5. Reactions – giving feedback, emotions

Fundamentally, without changing the traditional mindset moving forward to a digital mindset, lectures have the same characteristics on online as an offline learning environment. A digital mindset is much more than methodological innovation. Digital mindset is an approach that focuses on learning and learners at the cognitive and affective dimensions. The "Art and Science of Teaching" [5] in this sense, means balancing among knowledge, skills and attitudes via the learning and teaching process. Essentially, learn first, teach second. Redesigning project-based courses prioritize collaborative curriculum development using e-mind mapping and MIRO (miro.com) in order to share aims and expectations, and rethink learning outcomes. This process is based on the concept of collaborative professionalism. As Hargreaves and O'Connor state: *"Collaborative professionalism is about how people collaborate more professionally and also how they work as a profession in a more collaborative way"* [3]. Greater collaborative professionalism appears at methodological level, where teachers - students, students - students, teachers - teachers collaboration are the fundamental point. No doubt, project-based courses have collaborative characteristics. However, online learning requires stronger intrinsic motivation at the levels of personalized and cooperative learning. Organizing breakout rooms, working with our students, together giving continuous feedback, is the starting point on this process. Motivating our students for using online collaborative platforms and communication tools can strengthen coherence between project-based courses and online learning. Practically, mapping the students' prior knowledge via brainstorming and e-mind mapping, promoting questioning on question cards, developing critical thinking, communication and debating on place mat are effective learning and teaching methods. Turning back to the top 10 job skills of tomorrow, especially to the first competency areas—analytical thinking and innovation, active learning and learning strategies, complex problem-solving, and project-based courses—is extremely relevant. Complex problem-solving combining with analysis is a traditional concept of project-based learning. Nevertheless, project presentations are based on analysis and conclusion, which are raising some questions and dilemmas. In this sense, it is not necessary to put a point at the end of the sentence. The fundamental key point of the presentation is changing a traditional to digital mindset via using online learning with some teaching methods—for instance, e-mind mapping, breakout rooms, place mat with using collaborative learning platforms. Regarding assessment functions, it is obvious that formative assessment comes to the front. Concept of assessment for learning is a horizontal aspect for a project-based course in higher education—for instance, giving continuous feedback during the learning and teaching process, using e-portfolio following the progression and sharing the evaluation criteria. An e-portfolio is a significant part of self-branding collecting some products and outcomes with reflective analysis [13]. In this paper, the basic innovation turns to formative assessment, especially sharing aims and expectation with the students in the beginning the course. It requires flexible curriculum development. For instance, our curriculum of Social Studies project-based course formulates some aims and learning outcomes:

> *"Students should be able to do collaborative presentations and evaluation on the project work. Students should be able to reflect own project work and reflect to others work. At the end of the course, students should be able to solve problems in a creative way and be able to listen and understand others opinion, make conscious decision."*

In order to reach these aims and outcomes with students, at the first lesson, we need to map their prior knowledge (vision, values and beliefs) about collaborative assessment. We think that changing assessment culture is a fundamental point of innovation.

Sharing assessment criteria is important in this innovative process, as well. In fact, diagnosis of prior knowledge, interest and competency strengths are significant milestones, especially using breakout rooms for introduction via name cards with letters of personal characteristics. Students make self-evaluation about their strengths and weaknesses for project work. Concerning following the progression, assessment focuses on continuous feedback and assessment of project plans, progression—for instance, what we did (3–4 points or sentences), what we are going to do (steps and deadlines), risk factors, and success criteria. Students present their Progress Report and give comments, advice and feedback in order to strengthen collaborative assessment culture. Final presentations are based on sharing assessment criteria: in the content portion, relevance, coherency, consistency are the criteria, while in the presentation, transparency, understanding, and creativity are the criteria. Finally, each project team needs to make their digital portfolio, where they assess team progression on some competencies (communication, collaboration, creativity, decision-making, complex problem solving). At the individual level, they need to write a reflective paper on personal progression. In order to prepare the final assessment level, we use chat function and reactions via Zoom to discuss the tasks, giving continues feedback to the students, while showing scaffolding and best practices on team digital portfolio and individual reflection.

4 Conclusion

Eo ipso, feasible coherence between online learning and project-based courses is a huge challenge in higher education. Some contextual elements can promote this process. In order to give relevant answer to the questions from the digitalization, the fourth Industrial Revolution and the growing needs of upskilling are motivated higher education institutions rethink strategy and practice, as well. As we see, behind this challenging contextual background, it is necessary to conceptualize the vision of knowledge and learning in order to find the balance between declarative and procedural, explicit and tacit knowledge. How can online learning transform learning and teaching processes and higher education? Introducing some examples and cases, these are raising some other questions and dilemmas. How can we change a traditional mindset moving forward to a digital mindset? We would like to point out, this is not an individual progression. It requires collaborative professionalism and changing organizational culture. How can we use the opportunities of online learning effectively? No doubt, intrinsic motivation is a basic point. But collaborative curriculum development, professional mentoring on teaching methods and assessment techniques are necessary on this process. Fundamentally, this learning journey has enormous impact on changing your mindset, teaching methodologies and assessment culture, but mutual trust is a fundamental factor.

References

1. Du Preez, P., Simmonds, S. Verhoef, A.H.: Rethinking and researching transformation in higher education: a meta-study of South African trends', Transform. High. Educ. **1**(1), 1–7 (2016). https://www.researchgate.net/publication/308756590_Rethinking_and_researching_transformation_in_higher_education_A_meta-study_of_South_African_trends. Accessed 4 Jan 2021

2. Hall, R., Andriani, P.: Managing knowledge associated with innovation. J. Bus. Res. 56, 145–52 (2003). https://www.researchgate.net/publication/222242345_Managing_knowledge_associated_with_innovation. Accessed 6 Jan 2021
3. Hargreaves, A., O'Connor, M.T.: Collaborative Professionalism. CORWIN a SAGE Publishing Company, Thousand Oaks, CA (2018)
4. Jasimuddin, S.M., Klein, J.H., Connell, C.: The paradox of using tacit and explicit knowledge: Strategies to face dilemmas. Manage. Dec. **43**(1), 102–112 (2005). https://www.researchgate.net/publication/37146053_The_paradox_of_using_tacit_and_explicit_knowledge_Strategies_to_face_dilemnas. Accessed 5 Jan 2021
5. Marzano, R.J.: The art and science of teaching: a comprehensive framework for effective instruction. ASCD, Alexandria, VA (2007)
6. Polanyi, M.: Personal Knowledge: Towards a Post-Critical Philosophy, 2nd edn. Harper and Row, New York (1964)
7. Polanyi, M.: The Tacit Dimension, Garden City. Doubleday & Company (1966)
8. Polanyi, M.: Knowing and Being. Routledge & Kegan Paul, London (1969)
9. Smith, M.K., Vass, V.: The relationship between internationalisation, creativity and transformation: a case study of higher education in hungary. Transform. High. Educ. **2**, 1–9 (2017)
10. Ten Berge, T., van Hezewijk, R.W.J.V.: Procedural and declarative knowledge: an evolutionary perspective. Theory Psychol. **9**(5), 605–624 (1999). https://www.research.ou.nl/en/publications/procedural-and-declarative-knowledge-an-evolutionary-perspective. Accessed 4 Jan 2021
11. The Future of Jobs Report 2020. World Economic Forum (2020). http://www3.weforum.org/docs/WEF_Future_of_Jobs_2020.pdf. Accessed 4 Mar 2021
12. Tulving, E.: How many memory systems are there? Am. Psychol., **40**(4), 385–398 (1985). http://www.bryanburnham.net/wp-content/uploads/2014/01/Tulving-1984-How-Many-Memory-Systems-Are-There.pdf. Accessed 1 Feb 2021
13. Vass, V., Kiss, F.: The role of competency development in the implementation of portfolio-based education in higher education. In: Auer, M.E., Centea, D. (eds.) Visions and Concepts for Education 4.0. Advances in Intelligent Systems and Computing, vol. 1314, pp. 42–48. Springer, Cham (2021). https://doi.org/10.1007/978-3-030-67209-6_5 . Accessed 4 Jan 2021

ALICE (Adaptive Learning via Interactive, Collaborative and Emotional Approaches) Special Track

Grasping the Shape of Ethical Dilemmas: Towards an Educational Dilemma Generator

Joan Casas-Roma(✉) and Jordi Conesa

Department of Computer Science, Multimedia and Telecommunication, Universitat Oberta de Catalunya, Barcelona, Spain
{jcasasom,jconesac}@uoc.edu

Abstract. The incorporation of data science and artificial intelligence (AI) technologies in different fields and areas provide many benefits in automation, resource consumption, and efficiency. However, the deployment of information systems that can exhibit autonomous behavior—and, thus, that can make decisions that have an effect on their users—may need to be prepared to face some ethical dilemmas. Those dilemmas cannot often be completely covered by following an ethics by design approach, which is based on foreseeing, beforehand, all potentially detrimental scenarios. In order to allow autonomous systems to act ethically and fairly, artificial morality techniques are a promising way to go. Our goal is to use artificial morality systems to support the incorporation of AI in eLearning environments. However, before implementing systems that are able to identify and react to ethically relevant scenarios, there is the need to understand what makes an educational situation be considered ethically relevant and why. In order to gain this knowledge, this work presents a preliminary prototype that automatically generates hypothetical educational situations that may involve (or not) potential ethically relevant decisions: the Educational Dilemma Generator. The work characterizes ethical dilemmas in education, presents the design of the Educational Dilemma Generator, and shows the results of the preliminary tests conducted. The contribution of the research is to create tools to advance in the knowledge about what ethical dilemmas in education are and what characterizes them.

Keywords: Artificial intelligence · Online learning · Ethics · Ethical dilemma

1 Introduction and Motivations

The new disciplinary approach of learning engineering as the merge of breakthrough educational methodologies and technologies based on internet, data science and artificial intelligence (AI) have completely changed the landscape of online education over the last years by creating accessible, reliable, and affordable data-rich powerful learning environments [1]. Particularly, AI-driven technologies have managed to automate pedagogical behaviours that we would deem as “intelligent” within an online education setting. Nevertheless, and just as it happened in other fields where AI tools have also been extensively deployed, there is the risk of those semi-autonomous tools failing to grasp the ethical dimension behind some of their decisions [2–4]. If we want to be able

D. Guralnick et al. (Eds.): TLIC 2021, LNNS 349, pp. 341–351, 2022.
https://doi.org/10.1007/978-3-030-90677-1_33

to confidently use the full potential that AI-driven technologies can bring to a wide variety of fields, we must ensure that these technologies are able to identify and address scenarios bearing an ethical dimension in a satisfactory way. Nowadays, the community is aware of such a need and, therefore, ethics by design (that is, foreseeing, as part of the design process, potential ethically relevant effects that the deployment of such technology might have) is a common practice in new projects. However, and although ethics by design is necessary, it is not enough to guarantee that autonomous systems will always behave in a way that we would deem to be fair and ethically-desirable—specially, as artificial agents become more autonomous, are deployed in more areas, and need to be prepared to face a larger number of potential situations and decisions [5].

Particularly in the field of education, cases such as the ones presented in Ref. [6] show how many decisions can have a profound ethical impact on the people involved in them. As such, these situations usually require taking a holistic perspective and balance educational, personal and social needs in order to make an ethically correct decision. Although the cases depicted in the aforementioned work concern offline educational environments, the increase of online education platforms, in most cases supported by some sort of automated decision procedures, opens up the question about how these ethically relevant scenarios will translate to these more ubiquitous learning environments. Particularly, as the ratio between students and human teachers increases, many day-to-day tasks in the virtual classroom are partly handled by AI-assisted systems (such as the processing and classification of students' messages, or the use of conversational agents to provide support in online learning environments) [7]. In order to ensure that the integration of such autonomous systems in educational settings will not lead to unexpected detrimental decisions, it is paramount to gain a thorough understanding of what makes educational situations ethically relevant.

Gaining a deeper understanding on this will eventually serve a triple purpose: 1) to further refine a conceptual definition of what elements, from the perspective of human educators, are key in determining whether a situation bears a clear ethical dimension with it; 2) to furnish artificial pedagogical agents with some sort of *ethical sensor* [5] that would allow them to understand when a decision is clearly ethically relevant; and 3) to generate a knowledge base suitable to be used by machine learning techniques in order to provide those artificial pedagogical agents with some sort of artificial morality [5, 8–11]. The purpose of this paper is to introduce a preliminary prototype that automatically generates hypothetical situations featuring a decision that may or may not call for an exception to be made on the ground of ethical reasons. These decisions are inspired in the case entitled "The failing eight-grade student," presented in Ref. [6], and generalizes it in order to explore how different variables would affect hypothetical situations similar to that one.

After this short introduction, we present a layered approach to ethical dilemmas in education, which we use as a starting point to the design of our hypothetical scenarios, in Sect. 2. Then, we provide details on the design decisions, user interface, and technologies behind the prototype in Sect. 3. Furthermore, and after having had the chance to run a few preliminary tests on our prototype, we briefly discuss some of the results we have gathered in Sect. 4. Finally, in Sect. 5 we use those insights to foresee further lines of work to improve our Educational Dilemma Generator before rolling it out for further tests on a larger scale.

2 Ethical Dilemmas in Education: A Layered Approach

Ethical dilemmas in education often involve multiple dimensions that might need to be taken into account in order to reach an ethically acceptable decision to the dilemma [5, 12]. A theoretical model to represent the multiple relevant layers that can appear in an ethical dilemma in education was briefly introduced in Ref. [5]. Nevertheless, in the aforementioned work, the model was focused on a case study example that was used throughout that paper. In the present case, we want to devote some space to the core definition of that model, and we use it as a starting point to guide the design of our Educational Dilemma Generator.

As it was argued in Ref. [12], ethical dilemmas may come from either a *resource allocation problem* (i.e., limited resources, such as teacher hours, that must be somehow allocated among a higher number of students), or a *clashing principles problem* (i.e., different choices inevitably lead to upholding certain principles, whilst disregarding other ones). In particular, when focusing on ethical dilemmas in education involving students, there is the risk to focus solely on the students' academic results, without taking into account certain particularities that can make certain cases sensibly different from similar ones. Some works focusing on ethical dilemmas in educational contexts [13–15] already identify the struggle of having to make a decision where a conflict between educational and personal needs might be in conflict. In this sense, and as an example, it is noted in Ref. [15] how acknowledging the personal situation and needs of a student may not always be in accordance with the school's standard rules; this kind of tension, as it happens in other situations featuring a similar conflict, often leads to complex dilemmas where one must bring educational, personal, and social issues into the picture in order to get a holistic perspective on the case and fully understand how the outcomes of a decision might affect the parties involved.

Following these intuitions, we define a layered approach to ethical dilemmas in education that identifies three relevant dimensions surrounding a student:

- *Educational Layer*: Refers to those elements of the learning process that are explicitly taken into account by the educational system and which reflect the student's activities and progress within the classroom. They typically include the student's marks in assessed activities, continuous assessment activities, assignments and homework, the student's past record, etc.
- *Personal Layer*: Refers to the student's personal life and, particularly, to the way this relates to their learning process. In other words, the Personal Layer accounts for the student's goals and expectations, their studying habits, and their effort and motivation, as well as their preferences. Even though this is not explicitly taken into account by the educational system itself, it directly impacts the way the student approaches their learning, and it includes both psychological and emotional factors that affect such learning process.
- *Social Layer*: Refers to the student's social context and the way this context may affect both their Personal and Educational layers. This layer is highly contextual and changing, as it is linked to historical, social, and political realities. It can include,

among many others, geographical and structural inequalities present in different territories, opportunities and limitations associated with certain social and gender roles, the possibility (or the lack) of access to technology, etc.

By capturing the intuitions related to what relevant features one can find in ethical dilemmas, our model seeks to distill and classify those in the aforementioned layers. This can be seen as a step forward towards the systematic identification and analysis of ethical dilemmas in educational settings, and it can help ensure that all relevant features in those situations are accounted for and taken into account, prior to making a decision on what the most ethically desirable course of action would be. It is worth noting how this model can help both human educators gain an understanding of these situations, as well as enabling AI-assisted tools—for instance, in online learning environments, to identify cases that could potentially classify as ethical dilemmas. A schematic representation of these layers is depicted in Fig. 1.

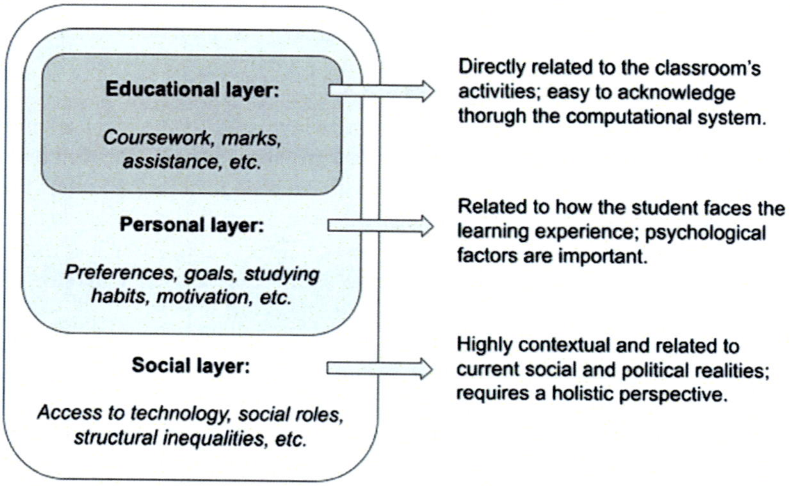

Fig. 1. Understanding the different layers surrounding a student.

We provide, in Ref. [5], an analysis of a particular case study using our multifaceted model for ethical dilemmas in education. In the present work, we instead use the model to identify and distill the specific characteristics that could appear in these dilemmas, and we use those insights to begin prototyping an automated Educational Dilemma Generator.

3 Towards an Educational Dilemma Generator

The first stage of the Educational Dilemma Generator, defined in this paper focus solely on the elements that belong to the Educational Layer—it is worth noting, however, that we do include some minimal demographic information of the hypothetical student involved in the dilemma, as we would like to see, as part of the first preliminary tests, whether this could already highlight certain relationships between this information and the way some human educators would react to that scenario. Our aim is to test, through a series of preliminary trials, whether the elements of the Educational Layer could be, *per se*, enough to generate scenarios that human educators would classify as ethical dilemmas. On future stages of the current project, we will add elements belonging to the Personal and the Social layers to further study how those elements affect the ethical dimension of the generated dilemmas.

3.1 Conceptual Design

Even though ethical dilemmas in education can greatly vary in shape and content, we have chosen to focus our initial development to a pretty specific set of situations. Following the analysis of a case presented in Ref. [6], which we already explored using our multifaceted characterization of dilemmas in education [5], we focus our dilemmas on the same kind of situation: a student has failed an exam, and the user (potentially an educator related to the student or to their institution) has to decide whether an exception should be made in the case of that student with regards of their failed exam. The case presented in Ref. [6] had a very specific set of features, such as the exam being a grade-pass exam, and providing some information regarding the student's Educational, Personal, and Social layers. In this case, we have also included other elements that could likely make a difference to the outcomes of the depicted situation—for example, and among other factors, in our hypothetical situations, we consider different kinds of exams (with a different degree of importance within the educational process), as well as different types of modules and degrees.

Far from providing an exhaustive set of all the potentially relevant variables and values, we have chosen to include the following elements (most of them belonging to the Educational Layer, plus a few of them belonging to the demographic data of the student) in this preliminary test of the Educational Dilemma Generator:

- *Student*:
  - Name: The student's first name.
  - Gender: The student's gender.
  - Ethnicity: The student's ethnicity.
  - Record: The student's performance record, specified as a combination of multiple variables. It can either be a record that consistently shows the same performance over time, or it can show a variation in performance.
  - Effort: The student's effort in the current module, plus potentially some additional details on the student's engagement and shown interest.
- *Exam*:
  - Relevance: Indicates whether the exam is part of a formative, or summative assessment activity, or whether it is the last exam of a module, or even a grade pass exam.

- Mark: The range of the mark achieved by the student. Because of the nature of the dilemma itself, the mark will always be a fail (we have considered the mark to be evaluated in a 0 to 10 scale, being 5 the mark required for a pass), but it can have a variety of ranges featuring a *clear fail* (less, or equal than 3), a *fail* (between 3.1 and 4.4), and an *almost pass* (between 4.5 and 4.9).

- *Module*:

  - Prerequisite: Indicates whether the module is prerequisite to other modules (that is, whether the student not having learned its contents may likely be an impediment for future modules).

- *Course*:

  - Degree: Specifies what sort of degree, within the educational system (that is, primary education, high school, undergraduate, or graduate) the course and the module belong to.

A schematic representation of these elements can be seen in Fig. 2.

The prototype developed will create new random scenarios by generating different combinations of the relevant variables. These scenarios will be presented to teachers by using a web-based interface to get insights on whether the generated scenario poses an ethical dilemma, as well as how the dilemma should be solved and why. An example of one of such scenarios is the following one (we highlight in bold font those elements that are part of the randomly-generated variable values):

> *Alex is a* ***latino female*** *student who recently failed a* ***formative assessment*** *in a module which* ***is a prerequisite module to other modules*** *as part of a* ***high school course****; Alex scored* ***almost a pass (between 4.5 and 4.9)****.*
>
> *The student* ***has shown a good amount of effort, being consistently engaged during the course of this module****. Alex's student record* ***recently has shown outstanding results but, in the past, has shown average results (mostly passes, a few fails)****.*
>
> *According to the standard school rules, Alex should fail the exam. Do you think an exception should be made in this case? Why?*

The aim of our Educational Dilemma Generator is to understand better what the weight of the aforementioned elements are, in relation to that hypothetical situation, being eligible to be considered as an ethical dilemma. Similarly, we also want to understand whether and how different combinations of those variables can make a difference in cases where, say, those elements do not occur all at once, or where other variables may act as inhibitors (for example, is the fact that a module is not a prerequisite to other modules key in determining whether an exception should be made in a failed final exam of such module?).

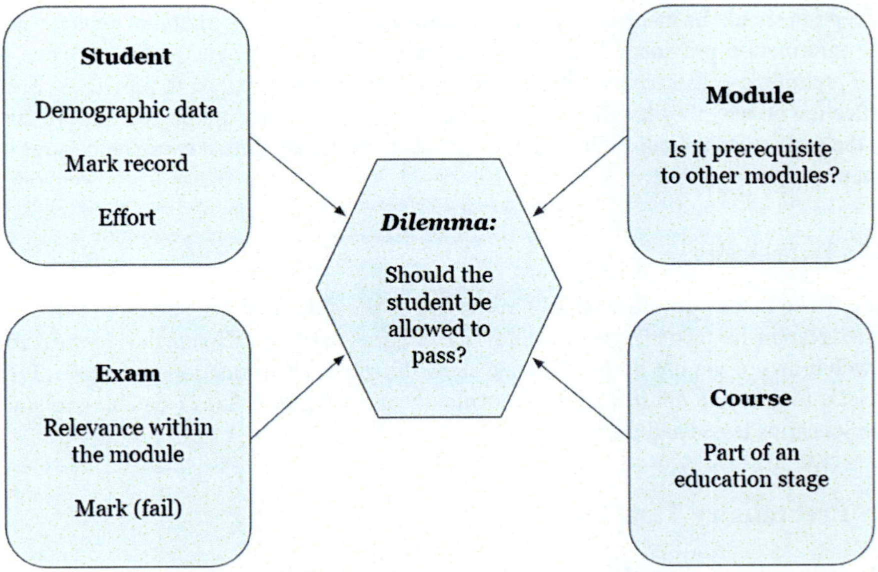

Fig. 2. A conceptual diagram for the dilemmas generated.

3.2 Interface

The prototype runs on a web browser and presents the user with an initial screen that briefly introduces the project and its context, and provides a minimal set of instructions on how to run the tests—plus some information regarding privacy and the data we gather. Once the user clicks the button to start the test, the browser shows five different cases which describe, in a short text, what the hypothetical situation is (see Fig. 3).

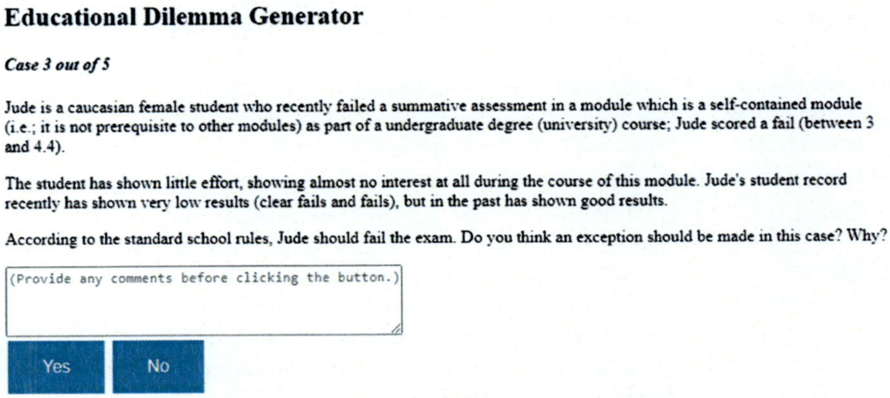

Fig. 3. A screenshot of the prototyped educational dilemma generator.

The user is then told that, according to the standard school rules, the hypothetical student should fail the exam, but the user is then asked whether they think that an exception should be made in this case, according to the particular situation depicted and the information provided. The user can answer with either a "yes" or a "no", and in this preliminary case, we have chosen to include an optional textbox to provide an open reflection on why they have decided to choose one, or the other option. We believe that, in the early stages of our project, the comments provided by the users can help us refine our prototype.

3.3 Technologies

In order to make our test as widely accessible as possible, we have chosen to use web-based responsive technologies that allow the program to run in different devices by using a web browser. We use basic HTML to show the relevant information to the user, PHP 7.3.21, to generate the scenarios automatically and a MySQL 8.0.21 database to store the scenarios tested and the test results.

4 Preliminary Tests and Results

The URL where we initially uploaded our prototype has been sent to a controlled group composed by several eLearning teachers from the field of computer science. We have performed a preliminary set of tests, involving 40 different cases (all of them randomly generated by following the patterns specified in Sect. 3.1); aside from the responses associated to those 40 cases, we have also gathered qualitative feedback regarding the interface of the prototype.

Regarding the users' answers to the cases, we have collated some interesting preliminary insights. Of course, because this is just a preliminary test, and because it is mainly intended to point towards ways of improving it in its next iteration, we do not have yet enough data to draw any solid conclusion; nevertheless, there are some insights that could be interesting to see whether they will appear again in further larger tests. We summarize those comments in the following lines, in no particular order:

- In general, answers suggesting that an exception should be made, with regards to the student failing their exam, mention the student's motivation (effort and engagement) in the module. This is revealing in the sense that this variable already falls within the scope of the Personal Layer (as explained in Sect. 2); this would seem to preliminary suggest that attributes from the Personal Layer are indeed key in this regard.
- Some answers saying that they would not make an exception point to the fact that they would need more information in order to justify making an exception. It is worth noting how, in some of those cases, no information regarding the student's record was available (this is something that we intentionally introduce as a possible variable value for the record). In this sense, it seems that, without having the chance to take into consideration the student's past performance, users did not have enough reasons to support making an exception.
- Some answers saying that they would not make an exception explicitly refer to the "school rules," which we intentionally introduced as part of the test's question (and following the work of Ref. [15], which identifies the tension that sometimes arises

between the *caring climate* and the *formal climate*, and to which the dilemmas depicted in our prototype fit very well).

- The student's attitude is something that has been mentioned in some cases where the hypothetical situation depicts the student as showing little effort, and the user decides not to make an exception. In those cases, the fact of failing the student is suggested as something to foster a change of attitude in the hypothetical student.
- Some of the answers suggest that they would need more information about the student and their case in order to make a decision and that they would like to talk to the student beforehand. This seems to hint that more information on the Personal and Social layers of the student would help make a decision; thus, it may suggest that the Educational layer is not always enough.

Regarding the current state of the prototype's design, its interface, and the way the user interacts with it, we have also received some valuable feedback from the controlled group of users who performed the test. These comments are not related to the prototype's conceptual design or to the content of the cases that it generates, but they rather refer to user interface while performing the tests; thus, it can help us improve it in order to avoid confusion that could compromise the results. We summarize those comments in the following lines (again, in no particular order):

- The prototype currently allows to go back and forth over the cases (through the browser's back and forward buttons); although the cases are kept the same, the user's inputs are recorded independently, which can potentially lead to a confusion in the data we gather regarding the number of cases done in each session. This will need to be amended from a technical point of view.
- Certain phrasings of the variable values should be improved, as they can be confusing for the users. For instance, it should be explicitly mentioned that the hypothetical exam mark is calculated on a scale from 0 to 10, as otherwise the value of the fail mark might be a bit vague. Similarly, the concepts used to distinguish the type of hypothetical exam may not fit all evaluation methods (for instance, those based in continuous assessment activities), so they should be modified accordingly.
- Currently, the test ends somewhat abruptly and redirects to the home page of the prototype. Most users have suggested that it would be good to add a final page showing the overall answers of the test and maybe even including a way to share the test to other people; this would be a good opportunity to spread the test further, once we begin with more large-scale rounds of tests.
- Some of the users have suggested giving the chance, at the end of the test, to modify their decisions; this would allow the test to either gather the first impression the user has about a case, or either to allow further reflection and, once all the cases have been seen, allow the user to put those cases in perspective.

5 Remarks and Future Work

In this paper, we have introduced the design of an automated Educational Dilemma Generator that presents a human user with different hypothetical cases that could be classified as ethical dilemmas in education. The definition of those ethical dilemmas is grounded on our multifaceted characterization of dilemmas in education, which takes into account the Educational, the Personal, and the Social layers surrounding the students. For this first iteration of our prototype, we have chosen to focus mainly on elements from the Educational Layer, plus some extra demographic data of the hypothetical student. This should help us see whether elements from the Educational Layer alone, without getting into the Personal and Social layers, are enough to classify some of our hypothetical situations as ethical dilemmas; furthermore, it would allow us to compare, in future stages of the project, how the addition of elements from the Personal and the Social layers affect those hypothetical cases. We have run a preliminary test of our prototype with a controlled group of eLearning educators in order to gather some initial insights on the content and the design of our prototype; the feedback points to some improvements that can help make our prototype more robust, usable, and reliable.

Once we can start running tests with our prototype on a larger scale, the data gathered will be helpful for three different goals: 1) to further understand what classifies as an ethical dilemma in education, from the perspective of human educators, and what elements are key in distinguishing what cases bear a clear ethical dimension with them; 2) to gather large amounts of human-labelled data that could be used to train *ethical sensors* [5] in order to identify ethically relevant situations; and 3) to gather large amounts of human-labeled data that could be used to train artificial pedagogical agents using techniques from artificial morality (particularly, following either pure bottom-up, or hybrid approaches).

Regarding goal number 2, and as we discuss in Ref. [5], artificial agents taking broader roles in educational settings would need to be embedded with a mechanism allowing them to distinguish which decisions bear a clear ethical dimension with them. For instance, it is not hard to imagine how an automated system assigning the grades of students in MOOCs could stumble upon a scenario similar to the ones depicted in our dilemma generator. Our preliminary results show that some human teachers think that, in the student's best interest, an exception could be made in some cases. Having a large amount of data that distinguish which scenarios are clearly ethically relevant, and that points to the reasons behind why exceptions might be considered, could allow these automated systems to be trained to recognize such situations and raise an alert to involve teachers in their consideration. Similarly, and once the automated system has been furnished with this kind of ethical sensitivity, deploying a layer of moral reasoning into the system could either allow it to make ethically relevant decisions by itself, or act as an ethical advisor to human teachers in the loop: this, which corresponds to goal 3, could potentially be achieved through artificial morality techniques based on some sort of machine learning process, which would require the large amounts of labeled data we aim to gather with our tests. In order to do any of this, however, we would need to run the tests at a much larger scale. Before doing so, nevertheless, there are some parts that we still need to test, iterate and improve.

As future work, we want to start by integrating the feedback received from this first test in order to improve our prototype; both comments regarding the prototype interface, as well as related to the dilemmas generated, will be considered and taken into account in the next iteration of our prototype. Once we have polished the current state of our prototype, adding elements from the Personal and the Social layers would be the next step. By adding those elements, we not only open up the possibility of generating more nuanced hypothetical scenarios, but we can also think about generating paired hypothetical scenarios in which one of them features Personal and Social data, whereas the other does not: we can then perform A/B comparison tests and see what the significance of taking this information into account is, with respect to the way the users would react to the same dilemma.

Acknowledgments. This work has been partially supported by European Commission through the project "colMOOC: Integrating Conversational Agents and Learning Analytics in MOOCs" (588438-EPP-1-2017-1-EL-EPPKA2-KA) and by a UOC postdoctoral stay.

References

1. Dede, C., Richards, J., Saxberg, B. (eds.): Learning Engineering for Online Education: Theoretical Contexts and Design-based Examples. Routledge (2018). https://doi.org/10.4324/9781351186193
2. Angwin, J., Larson, J., Mattu, S., Kirchner, L.: Machine bias. ProPublica (2016)
3. Yapo, A., Weiss, J.: Ethical implications of bias in machine learning. In: Proceedings of the 51st Hawaii International Conference on System Sciences (2018)
4. Favaretto, M., De Clercq, E., Elger, B.S.: Big Data and discrimination: perils, promises and solutions. A systematic review. J. Big Data **6**(1), 1–27 (2019). https://doi.org/10.1186/s40537-019-0177-4
5. Casas-Roma, J., Conesa, J., Caballé, S.: Education, ethical dilemmas and AI: from ethical design to artificial morality. In: Sottilare, R.A., Schwarz, J. (eds.) HCII 2021. LNCS, vol. 12792, pp. 167–182. Springer, Cham (2021). https://doi.org/10.1007/978-3-030-77857-6_11
6. Levinson, M., Fay, J.: Dilemmas of Educational Ethics: Cases and Commentaries. Harvard Education Press (2019)
7. Capuano, N., Caballé, S.: Multi-attribute categorization of MOOC forum posts and applications to conversational agents. In: Barolli, L., Hellinckx, P., Natwichai, J. (eds.) 3PGCIC 2019. LNNS, vol. 96, pp. 505–514. Springer, Cham (2020). https://doi.org/10.1007/978-3-030-33509-0_47
8. Wallach, W., Allen, C.: Moral machines. Oxford University Press (2009)
9. Wallach, W., Allen, C., Smit, I.: Machine morality: bottom-up and top-down approaches for modelling human moral faculties. AI & Soc. **22**(4), 565 (2008)
10. Anderson, M., Anderson, S.L., Armen, C.: An approach to computing ethics. IEEE Intell. Syst. **21**, 56 (2006)
11. Cervantes, J.A., López, S., Rodríguez, L.F., Cervantes, S., Cervantes, F., Ramos, F.: Artificial moral agents: a survey of the current status. Sci. Eng. Ethics **26**(2), 501 (2020)
12. Casas-Roma, J., Conesa, J.: Towards the design of ethically-aware pedagogical conversational agents. In: Barolli, L., Takizawa, M., Yoshihisa, T., Amato, F., Ikeda, M. (eds.) Advances on P2P, Parallel, Grid, Cloud and Internet Computing: Proceedings of the 15th International Conference on P2P, Parallel, Grid, Cloud and Internet Computing (3PGCIC-2020), pp. 188–198. Springer International Publishing, Cham (2021)
13. Pope, N., Green, S.K., Johnson, R.L., Mitchell, M.: Examining teacher ethical dilemmas in classroom assessment. Teach. Teach. Educ. **25**(5), 778 (2009)
14. Ehrich, L.C., Kimber, M., Millwater, J., Cranston, N.: Ethical dilemmas: a model to understand teacher practice. Teach. Teach. Theory Pract. **17**(2), 173 (2011)
15. Shapira-Lishchinsky, O.: Teachers' critical incidents: ethical dilemmas in teaching practice. Teach. Teach. Educ. **27**(3), 648 (2011)

Adaptivity in the Wild: Individualizing Reading Supports in Open Learning Scenarios

Robert P. Dolan(✉) and Kim Ducharme

CAST Inc, Wakefield, MA 01880, USA
bdolan@cast.org

Abstract. Adaptive learning solutions generally assume curricular, instructional, content, and/or learning models to be known in advance. This paper describes an ongoing research project being conducted by the Center for Inclusive Software for Learning (CISL), funded by the U.S. Department of Education, that is developing a suite of open-source tools designed for diverse learners by making digital educational materials—including open educational resources (OER)—accessible, flexible, and engaging. The *Clusive Reader* at the heart of CISL leverages Universal Design for Learning (UDL) principles and includes a number of adaptive features designed to operate without prior knowledge of how the materials support learning goals. These adaptive features include content complexity leveling, inclusion of summaries and highlighted main ideas, dictionary and glossary tools, and just-in-time tips and suggestions regarding preference setting and use of learning tools. Information to support adaptivity heuristics comes from user interactions, an onboarding skills discovery process, and embedded affective, comprehension, and learning goal prompts and game-like vocabulary challenges. Per UDL, adaptivity in *Clusive* is designed to support not only immediate learning goals, but also the building of "expert learners" who independently choose the most effective content, contexts, and supports. As such, adaptivity has been designed as a learning scaffold, with the intention of gradual reduction over time, in two ways. First, adaptivity operates transparently, providing students with insights as to why choices and recommendations are made. Second, students are allowed to override system decisions to promote self-agency. Evaluation of student impressions and efficacy are being conducted as part of the CISL project. This paper provides the research foundations and philosophical approach taken for this effort, describes the current adaptive learning features being designed, and concludes with remarks on next steps and challenges being faced.

Keywords: Adaptive learning · Universal design for learning · Reading supports

1 Introduction

1.1 Need for Reading Supports

A majority of America's students fail to meet proficiency standards for reading of literature and academic texts [16], a situation compounded by recent COVID-19-based

D. Guralnick et al. (Eds.): TLIC 2021, LNNS 349, pp. 352–363, 2022.
https://doi.org/10.1007/978-3-030-90677-1_34

distance learning requirements, especially for at-risk populations. Differentiation and individualization comprise common strategies among educators to address the need for improved reading instruction, and educational technologies play a critical role in making such personalization possible. These include perceptual supports, such as adjustable sizing and spacing of text [6, 22, 27, 41] and adjustable color and contrast [13, 28]; linguistic supports, such as reading aloud through text-to-speech [9, 18, 19]; comprehension and cognitive supports, such as allowing students to highlight [12, 39] and annotate [17, 33] text; access to digital glossaries and dictionaries [11, 20]; and text complexity reduction [10, 31, 32]. Learning technologies frequently feature built-in assessment tools that provide prompt-based and diagnostic feedback on student performance, such as progress tracking and identification of patterns in content acquisition. Additional personalization can come in the form of content area selection and mastery-based advancement, in which the technology directs students to additional resources, lessons, and activities based on completion of prior work and student interest [34].

The challenge with all these supports is ensuring that students understand the benefits that may—or may not—accompany their use. This is compounded by large variation as a function of the learner, content, context, and reading goal [30].

1.2 Promise and Limitation of Adaptive Learning

Adaptive learning solutions—in which students are provided instruction, content, and/or supports as a function of their individual, ongoing learning needs—have shown increasing promise as an effective tool to improve learning. Most success has been achieved within "well-structured domains," those in which models of pedagogy have been well established, such as some STEM areas (e.g., physics and algebra), and for which mastery learning approaches can be defined [3, 29].

Less effectiveness of adaptive learning approaches has been demonstrated in domains where learning progression is harder to define and for which instructional models are cruder. Within reading, the problem is further compounded. There are many reasons a student may be struggling in reading. Even with a diagnosis of dyslexia, for example, there is ongoing debate whether this is a monolithic diagnosis for which a singular interventional approach is warranted, as opposed to a cluster of neurodiverse factors warranting more "agile" intervention approaches.

Furthermore, adaptive learning solutions to support reading that generally assume curricular, instructional, content, and/or learning models are known in advance and, thus, are limited to self-contained learning systems [5]. Lastly, models don't typically include affective components of learning. This is a significant shortcoming, given that affect is a critical, gating component for learning and has been shown to be especially deterministic of reading success.

1.3 Promise and Limitation of Open Education Resources

Openly licensed and free available Open Education Resources (OER) are increasingly being used in K-12 and post-secondary learning. OER provide greater equity for disadvantaged regions globally and for students who are economically disadvantaged. Through their open licensing, they support the "5R Permissions of OER," namely the rights to: (1) retain and make your own copies; (2) reuse in a wide range of ways; (3) revise, adapt, modify, and improve; (4) remix by combining two or more; and (5) redistribute to share your contributions with others (Lumen Learning, 2014). As such, OER offer the potential to create many variants and choices to match the variability and diversity of student needs. This same latitude and variety is generally unavailable with mass-produced, copyright-protected commercial curriculum products. Furthermore, despite their increasing popularity and use, OER still rarely include adaptive learning supports.

1.4 Current Effort

This paper describes an ongoing research project in which we are developing an open-source reading tool that contains adaptive reading supports for diverse learners designed to operate "in the wild" without a priori information about curricular, instructional, content, or learning models. We will explain the research foundations and philosophical approach taken for this effort, and we will describe the current adaptive learning features being designed. The paper concludes with remarks on next steps and challenges being faced.

2 Center for Inclusive Software for Learning

The Center for Inclusive Software for Learning (CISL) has been funded by the U.S. Department of Education to develop a suite of open-source tools to make digital educational materials—including OER—accessible, flexible, and engaging for diverse learners. This effort is predicated on the need for an ecosystem of open source and interoperable software applications available to be implemented in state and local systems, so that equitable opportunities for learning become a reality for all students served in K-12 schools. While accessibility—in the form of compliance with guidelines such as the WCAG 2.1 [37]—is necessary, it is insufficient. It's critical to implement existing and new technologies as a means of providing equitable and improved learning opportunities to ensure that K-12 systems can adequately prepare all students for college and career readiness and are equipped to succeed in postsecondary education or the workforce.

The goal of CISL is three-fold. First, it will provide a body of research to increase the knowledge and understanding of the current challenges, a market scan, and gap analysis of which accessibility features and learning scaffolds are most useful for whom, and under what conditions. This will include descriptions of what features and scaffolds already exist and which barriers to access and learning need to be addressed by the creation of new features and supports. Second, CISL will undertake an iterative co-design process to develop an open-source applications suite combining extant and newly developed software that allows students and end-users to customize their access to digital content, online and off. Third, CISL will create developer's guidelines and an ed tech stakeholder's group to encourage widespread use of the software, its discrete features, and/or its design methodology by OER developers, publishers, and ed tech designers. The ultimate goal of the CISL is to improve access and learning for all learners using the full range of curriculum content and modalities available today and in the foreseeable future.

2.1 Universal Design for Learning

As an approach to instructional differentiation to meet individual students' learning needs [15], the Universal Design for Learning (UDL) framework draws from seminal work in the learning sciences, cognitive psychology, and neuroscience that highlights variability among learners [21, 25, 36, 38]. UDL stipulates that, to address this variability, effective learning opportunities must provide (1) multiple ways to present content, (2) multiple ways for students to interact with content and express what they are learning, and (3) multiple ways to engage students in learning [23, 30]. With this flexibility in instructional goals, assessments, methods, and materials, potential barriers to learning are lowered and opportunities to engage in productive struggle and learn are increased. Awareness of UDL's utility and potential has been growing, as evidenced by references to it in the Individuals with Disabilities Education Act (IDEA), the Higher Education Opportunity Act, and the Every Student Succeeds Act (ESSA), as well as in the 2010 and 2016 U.S. Education Technology Plans [35]. In addition, original work demonstrating the effectiveness of UDL in the area of language arts is now being extended to math and science, a wide range of populations, and to assessments [2, 9].

UDL implementation leverages students' use of scaffolds and supports to promote mastery of standards-based content and learning strategies, and to facilitate engagement in learning. The scaffolds and supports operate by actively guiding students through just-in-time feedback and contextual supports that can be gradually withdrawn as student expertise increases [7, 8, 24, 30].

In addition to supporting students with immediate learning goals, UDL provides a framework for fostering development of "expert learners" with the metacognitive and self-regulated learning skills to be—in their own way—purposeful and motivated, resourceful and knowledgeable, and strategic and goal driven. As such, expert learners can decide for themselves which content, contexts, and supports work best in the moment. Even when curricula, instruction, and materials adhere to UDL principles and provide flexibility, it can be daunting for students to make effective choices. An excellent way to support students in building the metacognitive skills foundational for expert learning is through modeling. This is where effective adaptive learning systems can best operate—and why they must operate transparently. Adaptive learning solutions that indicate why choices and recommendations are made can scaffold students in making effective choices. Allowing students to override system decisions further provides opportunities for self-agency. In the end, students are provided not only with improved opportunities for immediate learning goals, but they're supported in becoming expert learners.

3 CISL Adaptive Learning Approach

Per UDL, adaptivity in the student-facing *Clusive Reader* at the heart of CISL has been designed according to three approaches. First, adaptivity operates transparently, providing students with insights as to why choices and recommendations are made. Second, students can choose whether adaptivity is provided in terms of automated delivery vs. recommendations. Third, even when supports are delivered automatically, students are allowed to override system decisions. Collectively, these approaches promote student self-agency and provide students opportunities to improve metacognitive and self-regulated learning skills [1, 4, 26, 40].

CISL is researching a number of adaptive features designed to operate without prior knowledge of how the materials support learning goals. These adaptive features include content complexity leveling, inclusion of summaries and highlighted main ideas, dictionary and glossary tools, and just-in-time tips and suggestions regarding preference setting and learning tools use. Information to support adaptivity heuristics comes from students' interactions with *Clusive*, an onboarding skills and preferences discovery process, and embedded affective, comprehension, and learning goal prompts and game-like vocabulary challenges. These are described in the following sections.

3.1 Complexity Learning

Clusive adaptively delivers and recommends different versions of texts varying in lexical and syntactic complexity. Currently, this is based on students' reported knowledge of key vocabulary words (Fig. 1), additional word look-ups in the text, and further interactions and ratings of the terms in their word bank.

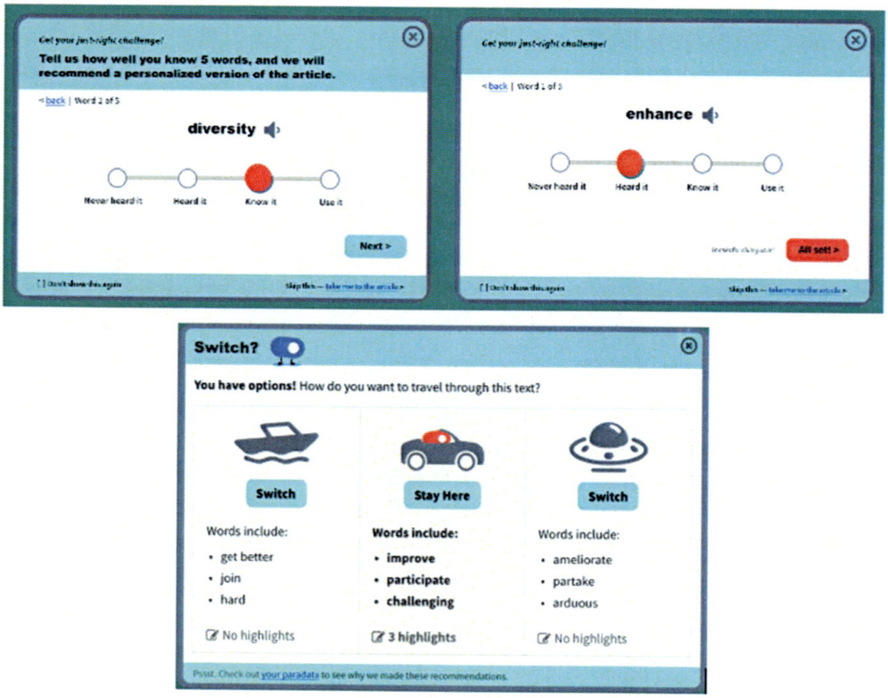

Fig. 1. Initial word rating panels and subsequent text challenge level "switch" panel.

We are exploring other means to reduce text complexity, including outline or summary views, views with the main ideas highlighted, and views with expanded vocabulary support (e.g., simpler synonyms or pictures attached to difficult words).

3.2 Adaptive Onboarding and Learning Tool Adaptivity

Clusive currently provides adaptive onboarding tooltips that identify and suggest features of the application to new users. These are shown not in a fixed sequence or timing, but based on each user's actions and current context. For example, if the student hasn't used the text context menu, they would be prompted to try selecting text, then to highlight a passage, look up a definition, or hear words read aloud (Fig. 2).

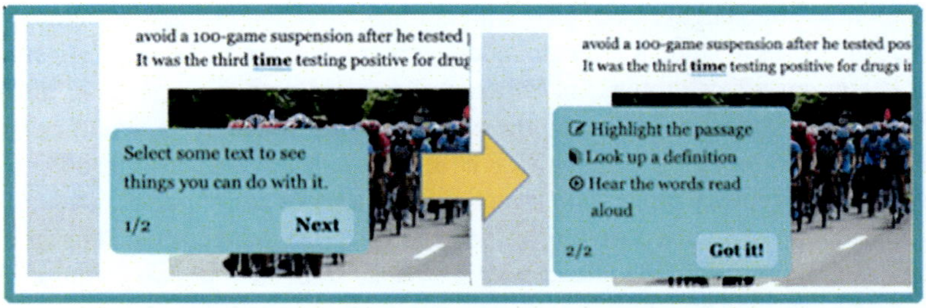

Fig. 2. Onboarding tooltip prompting students to try the text context menu.

Clusive also prompts students to consider using a learning feature (e.g., choosing text level, text-to-speech, highlighting, word look-up) based on whether they have used the feature yet, and if so, how recently (Fig. 3).

Fig. 3. An article showing adaptive learning tooltips.

We are exploring ways to enhance this feature to adaptively prompt students when and how to use learning features based on more sophisticated models of their past interactions.

For example, recommendations for using text-to-speech could be provided to students based on their reading behaviors, frequency of tool use, student-identified reading goals, and the challenge of the text vis-à-vis their reading skills. These recommendations could include not only reminders to use specific supports, but also suggested settings for their use (e.g., speed in the case of text-to-speech).

3.3 Affect-Based Adaptivity

As student affect and engagement play critical roles in reading, we are researching how to include measures of student affect in recommending content and supports to students. We are designing affect prompts in which students evaluate the emotional impact of specific readings during or after reading activities. This information would be used in combination with analysis of their general content choices, as well as information gathered during the discovery process, to improve specificity of recommendations. Figure 4 shows an example of an affective prompt where the student has selected "engaged" and "curious" and is then asked to indicate why. The open response "say why" option shown below will be replaced by a multiple-checkbox response format for better automation.

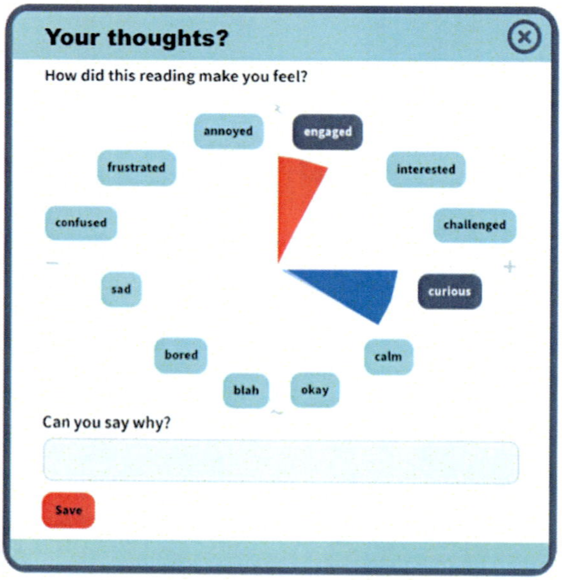

Fig. 4. An affective prompt at the end of an article.

3.4 Comprehension-Based Adaptivity

We are currently designing comprehension checks—ranging from maze to open response questions—for embedding within content as a means to improve student understanding [14]. We are considering the use of student responses (to comprehension probes amenable to automated scoring) as data to support adaptive delivery and recommendation of content and supports. However, we remain skeptical about their ability to provide valid and reliable information unless we are able to obtain pre- and post-reading measures.

3.5 Learning Goals-Based Adaptivity

Learning goals-based adaptivity can occur across or within specific texts. For example, the student might be prompted on the dashboard with, "What goals do you want to set this week?" They may want to work on fluency, read for fun, learn something new, test their knowledge, or just get their reading done. Based on their response, *Clusive* would display and suggest appropriate tools. A student wanting to work on fluency, for example, may be encouraged to use read-aloud, but set to a faster speed. An example of setting content-specific learning goals is shown in Fig. 5. A student wanting to get ready for a class discussion, for example, may see a version of the text with key ideas highlighted, while one wanting to build skills may encounter a vocabulary game that utilizes terms specific to that text.

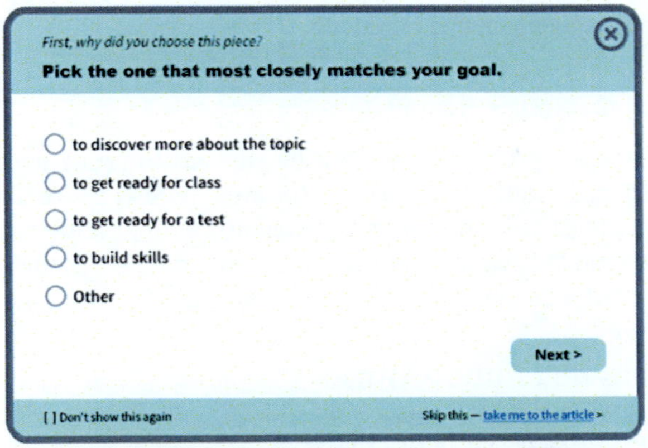

Fig. 5. The goals prompt panel, shown as a student heads into a reading, provides finer-grained information to *Clusive's* adaptive algorithm engine.

3.6 Reading Skills-Based Adaptivity

During the discovery process, we are considering having students conduct a self-check in which they identify their preferences for skill level based on comfort, interest, and motivation. Figure 6 shows an example of an interactive sequence in which students indicate the comfort with varying levels of vocabulary and sentence structure.

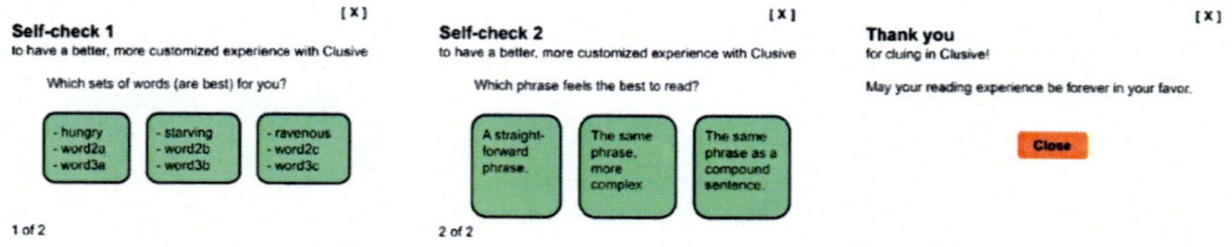

Fig. 6. Example of how students could do a skill-level self-check to have a more customized experience with *Clusive.*

The discovery process could also include more formal measures of students' reading decoding and comprehension skills.

3.7 Adaptive Self-regulation and Time Management Supports

We are currently exploring how *Clusive* could encourage students to set specific goals, such as an amount of time they want to spend reading or a number of pages they would like to read, and then support them in achieving their goals by tracking these metrics and providing encouragement and reminders.

3.8 Contextual Preference Adaptivity

We had considered the possibility of allowing students to create multiple preference configurations for use in different contexts. However, our user research and beta testing has shown us that students prefer to have preference settings at their fingertips, easily and flexibly changeable based on a particular context (e.g., goals, topic, environment). As such, *Clusive* will maintain a single set of preferences for each user.

4 Conclusions and Future Work

Clusive is being soft-launched in the spring of 2021, after which we will be able to evaluate students' impressions and use of features.

Generation of complexity-leveled content is a labor- and skill-intensive process, and as much art as science. Although there is promising research into automated text simplification processes and some tools are available, the inaccuracy and technical limitations of the tools limit their suitability; they are error prone, often changing the meaning of the text along with the reading level. Furthermore, the best performing of the current systems are commercial services and, therefore, not readily incorporable into any open-source software. Nonetheless, we are continuing to refine *Clusive*'s text adaptivity features, with the assumption that viable automated and/or crowdsourced solutions for producing leveled content are on the horizon.

Another challenge is the limited availability of information about students. We are making it a point not to ask too many questions that may be intrusive and allow students to opt out of any questions. Thus, we operate under the constraint of effectively adapting content and supports based on limited information that we can transparently gather with the user's full awareness and consent.

Acknowledgments. This work is being supported under a grant from the U.S. Department of Education, #H327A170002. This manuscript doesn't necessarily represent the policy of the U.S. Department of Education, and endorsement by the Federal Government shouldn't be assumed.

References

1. Azevedo, R., Johnson, A., Chauncey, A., Burkett, C.: Self-regulated learning with MetaTutor: advancing the science of learning with metacognitive tools. In: New Science of Learning, pp. 225–247. Springer, New York (2010). https://doi.org/10.1007/978-1-4419-5716-0_11
2. Basham, J.D., Stahl, S., Zhang, L., Blackorby, J.: Universal design for learning because students are (the) variable. In: Handbook of Research on K-12 Online and Blending Learning Second, pp. 477–508. Carnegie Mellon University: ETC Press (2018)
3. Bloom, B.S.: Learning for mastery. Eval. Comment **1**(2), 1–12 (1968)
4. Capuano, N., Mangione, G.R., Pierri, A., Salerno, S.: Learning goals recommendation for self regulated learning. Int. J. Eng. Educ. **28**(6), 1373–1379 (2012)
5. Capuano, N., Mangione, G.R., Pierri, A., Salerno, S.: Personalization and contextualization of learning experiences based on semantics. Int. J. Emerg. Technol. Learn. **9**(7), 5–14 (2014)
6. Chen, C.J., Keong, M.W.Y., Teh, C.S., Chuah, K.M.: Learners with dyslexia: exploring their experiences with different online reading affordances. Themes Sci. Technol. Educ. **8**(1), 63–79 (2015)
7. Cognition & Technology Group at Vanderbilt: Anchored instruction and situated cognition revisited. Educ. Technol. **33**(3), 52–70 (1993)
8. Collins, A., Brown, J.S., Newman, S.: Cognitive apprenticeship: teaching the craft of reading, writing, and mathematics. In: Resnick, L. (ed.) Knowing, learning and instruction: Essays in honor of Robert Glaser, pp. 453–494. Erlbaum (1989)
9. Dolan, R.P., Hall, T.E., Banerjee, M., Chun, E., Strangman, N.: Applying principles of universal design to test delivery: the effect of computer-based read-aloud on test performance of high school students with learning disabilities. J. Technol. Learn. Assess. **3**(7), 1–32 (2005)
10. Djamasbi, S., Rochford, J., DaBoll-Lavoie, A., Greff, T., Lally, J., McAvoy, K.: Text simplification and user experience. In: Schmorrow, D., Fidopiastis, C. (eds.) Foundations of Augmented Cognition: Neuroergonomics and Operational Neuroscience. AC 2016. Lecture Notes in Computer Science, vol 9744. Springer, Cham (2016). https://doi.org/10.1007/978-3-319-39952-2_28
11. Dziemianko, A.: Dictionary form in decoding, encoding and retention: further insights. ReCALL **29**(3), 335–356 (2017)
12. Fowler, R.L., Barker, A.S.: Effectiveness of highlighting for retention of text material. J. Appl. Psychol. **59**(3), 358–364 (1974)
13. Gradisar, M., Humar, I., Turk, T.: The legibility of colored web page texts. In: 29th International Conference on Information Technology Interfaces, pp. 233–238 (2007)
14. Hall, T.E., Cohen, N., Vue, G., Ganley, P.: Addressing learning disabilities with UDL and technology: strategic reader. Learn. Disabil. Q. **38**(2), 72–83 (2015)
15. Hall, T.E., Vue, G., Strangman, N., Meyer, A.: Differentiated Instruction and Implications for UDL Implementation. National Center on Accessing the General Curriculum (2003)
16. Hussar, B., et al.: The Condition of Education 2020 (NCES 2020-144). U.S. Department of Education, National Center for Education Statistics (2020)
17. Hwang, W.-Y., Wang, C.-Y., Sharples, M.: A study of multimedia annotation of web-based materials. Comput. Educ. **48**(4), 680–699 (2007)

18. Izzo, M.V., Yurick, A., McArrell, B.: Supported eText: effects of text-to-speech on access and achievement for high school students with disabilities. J. Spec. Educ. Technol. **24**(3), 9–20 (2009)
19. Keelor, J., Creaghead, N., Silbert, N., Breit-Smith, A., Horowitz-Kraus, T.: Language, reading, and executive function measures as predictors of comprehension using text-to-speech. Read. Writ. Q. **34**(5), 436–450 (2018)
20. Korat, O., Shamir, A.: Direct and indirect teaching: using e-Books for supporting vocabulary, word reading, and story comprehension for young children. J. Educ. Comput. Res. **46**(2), 135–152 (2012)
21. Luria, A.R.: The Working Brain. Penguin Books (1973)
22. Martelli, M., Di Filippo, G., Spinelli, D., Zoccolotti, P.: Crowding, reading, and developmental dyslexia. J. Vis. **9**(14), 1–18 (2009)
23. Meyer, A., Rose, D.H.: Learning to Read in the Computer Age. Brookline Press (1998)
24. Palincsar, A.S.: Metacognitive strategy instruction. Except. Child. **53**, 118–124 (1986)
25. Piaget, J., Cook, M.: The Origins of Intelligence in Children, vol. 8, no. 5, p. 18. International Universities Press (1952)
26. Pintrich, P.R.: The role of metacognitive knowledge in learning, teaching, and assessing. Theory Pract. **41**(4), 219–225 (2002)
27. Rello, L., Kanvinde, G., Baeza-Yates, R.: Layout guidelines for web text and a web service to improve accessibility for dyslexics. In: Proceedings of the International Cross-Disciplinary Conference on Web Accessibility, pp. 1–9 (2012)
28. Richardson, R.T., Drexler, T.L., Delparte, D.M.: Color and contrast in e-learning design: a review of the literature and recommendations for instructional designers and web developers. J. Online Learn. Teach. **10**(4), 657–670 (2014)
29. Ritter, S., Yudelson, M.V., Fancsali, S.E., Berman, S.R.: How mastery learning works at scale. In: Aleven, V., Kay, J., Haywood, J., Roll, I. (eds.) Proceedings of the 3rd Annual ACM Conference on Learning at Scale (Learning@Scale 2016), pp. 71–79. ACM (2016)
30. Rose, D.H., Meyer, A.: Teaching Every Student in the Digital Age: Universal Design for Learning. ASCD Press (2002)
31. Saggion, H.: Automatic text simplification. Synth. Lect. Hum. Lang. Technol. **10**(1), 1–137 (2017)
32. Siddharthan, A.: A survey of research on text simplification. Int. J. Appl. Linguist. **165**(2), 259–298 (2014)
33. Simpson, M.L., Nist, S.L.: Textbook annotation: an effective and efficient study strategy for college students. J. Read. **34**(2), 122–129 (1990)
34. U.S. Department of Education—Office of Educational Technology: Learning Technology Effectiveness. U.S. Department of Education (2014)
35. U.S. Department of Education—Office of Educational Technology: Reimagining the Role of Technology in Education: 2017 National Education Technology Plan update. U.S. Department of Education (2017)
36. Vygotsky, L.S.: Mind in Society: The Development of Higher Psychological Processes. Harvard University Press (1978)
37. Web Content Accessibility Guidelines Working Group (WCAG-WG): Web Content Accessibility Guidelines 2.2 (2020)
38. Wood, D.J., Bruner, J.S., Ross, G.: The role of tutoring in problem solving. J. Child Psychiatry Psychol. **17**, 89–100 (1976)
39. Yue, C.L., Storm, B.C., Kornell, N., Bjotk, E.: Highlighting and its relation to distributed study and students' metacognitive beliefs. Educ. Psychol. Rev. **27**(1), 69–78 (2015)

40. Zimmerman, B.J.: Becoming a self-regulated learner: an overview. Theory Pract. **41**(2), 64–70 (2002)
41. Zorzia, M., Barbierob, C., Facoetti, A., et al.: Extra-large letter spacing improves reading in dyslexia. Proc. Natl. Acad. Sci. **109**(28), 11455–11459 (2012)

Measuring Teachers' Civic Online Reasoning in a MOOC with Virtual Simulations and Automated Feedback Systems

G. R. Marvez[1(✉)], Joshua Littenberg-Tobias[1], Teresa Ortega[2], Joel Breakstone[2], and Justin Reich[1]

[1] Massachusetts Institute of Technology, Cambridge, MA 02142, USA
gmarvez@mit.edu

[2] Stanford University, Stanford, CA 94305, USA

Abstract. In an increasingly polarized digital landscape, evaluating online information has become a critical media literacy skill. Many individuals have difficulty distinguishing satirical from legitimate news sources [1]. One useful strategy is lateral reading, looking up information about a website or social media account in order to judge its credibility from an outside source [2]. We developed a short lateral reading task where we asked users to evaluate the trustworthiness of the satire account @GOPTeens and explain their response in a short text response. We developed a natural language processing classifier to detect whether users correctly identified the account as a satire account, which would indicate that they employed lateral reading to evaluate the trustworthiness of the account. This classifier examines a very specific case and the NLP classifier was highly accurate with a Macro F1 statistic of 0.96 (Overall Macro F1 = 0.96, Yes F1 = 0.99, No F1 = 0.94). In future work, we will employ the classifier to provide targeted feedback to users and will explore the effects of facilitative versus directive feedback on performance with lateral reading tasks.

Keywords: Natural language processing · Media literacy · Assessment

1 Introduction

As people increasingly consume news content online, the ability to correctly evaluate the credibility of online information is becoming an increasingly important literacy skill [3]. With the increased presence of misinformation online, it is essential that citizens be equipped with the skills to distinguish credible from non-credible sources of information [2, 4, 5]. However, traditional methods for evaluating online information encourage the use of long checklists that include characteristics that have little relationship with the sources' reliability, such as whether it is a.com or.org or whether it contains official-looking logos [4, 6]. Checklist approaches can lead to incorrect conclusions about the reliability of a source. For example, the domain name is not a strong indicator of reliability, since any entity can purchase a.org or.edu URL. Moreover, when users employ

D. Guralnick et al. (Eds.): TLIC 2021, LNNS 349, pp. 364–370, 2022.
https://doi.org/10.1007/978-3-030-90677-1_35

these traditional approaches to evaluate websites' credibility, they are often led astray and erroneously conclude that legitimate-looking sources are credible [6].

To address this need, the Stanford History Education Group (SHEG) created the Civic Online Reasoning (COR) curriculum [7]. The COR curriculum introduced a number of strategies to evaluate online information, including "lateral reading." To engage in lateral reading, a user leaves the website in order to search for what other sources say about the website in question [4]. For example, if someone comes across a Facebook post with a headline from an unfamiliar source, the user can search that source in a new tab to determine if it is a legitimate news organization. This method, which is used by professional fact checkers, yields more reliable and faster assessments of online information than spending time evaluating a source itself [4].

Lateral reading lends itself particularly well to evaluating content that may be satirical in nature. When viewed out of context, it can be difficult to distinguish legitimate sources from satirical ones [1]. For example, a user who saw the following post (see Fig. 1) on social media, "Trump Explains Decision to Cut Meals on Wheels Program: 'They just die anyway,'" (February 11, 2020) might incorrectly assume that the headline, from a legitimate-sounding publication, described a real news event. Others may question the source of the information based on the tone of the headline, which suggests an anti-Trump bias, but assume it is referring to a real event. However, a quick internet search for the Breaking Burgh shows that it's a satirical website [8]. When assessing online information, it is important to be able to distinguish between biased accounts of real events and satirical accounts (see Fig. 1).

Fig. 1. A screenshot of a Facebook post that posted the satirical article from Breaking Burgh.

In this paper, we describe a task that was embedded in an online course on 21st century media skills and how we used that data to construct a classifier to determine participants' reasoning in the second version of the task. We then analyzed the effectiveness of that classifier and explored how it could be adapted to other text classification tasks in large-scale learning settings such as massive open online courses (MOOCs). This method could be useful in providing in-the-moment feedback within digital practice spaces to offer redirection to participants in improvisational teaching scenarios. Natural language processing (NLP) methods in other writing tasks show similar grading to experts and can be used to provide feedback to users [9]. Other research on NLP methods in digital teacher development tasks has shown that these methods can be used to detect confusion in participants' responses to provide them feedback and a moment of reflection [10]. Additionally, these feedback mechanisms can be critical in MOOCs where the larger scale means that participants do not often receive one-on-one attention from instructors [11]. Our design of performance tasks for educators in MOOCs with NLP integration would allow them to receive feedback on their performance immediately.

2 Methodology and Analysis of the Task

To prime participants to evaluate satirical content online using lateral reading skills, we developed a task where users had to evaluate the trustworthiness of a Twitter account as a source of information about gun control. We showed participants a Tweet from a satirical account, @GOPTeens, that stated, "#Teens: RETWEET if you walked out because there weren't ENOUGH #guns at #school!!!" (see Fig. 2). Participants were then asked, "Based on your lateral reading, do you think this account is a trustworthy source of information about the school walkouts?" and to explain their answer in the form of a short response. A correct answer would correctly identify that the Tweet is not trustworthy because the account is a parody account.

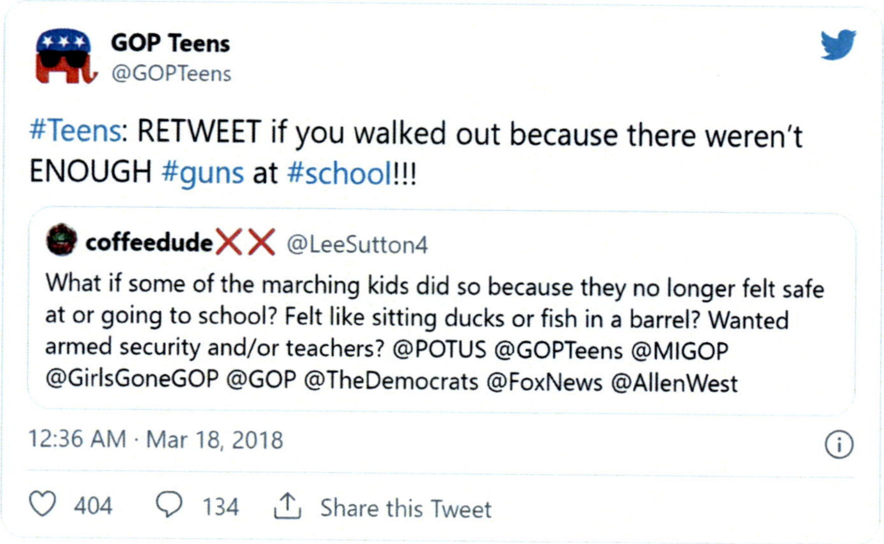

Fig. 2. A screenshot of the Tweet from @GOPTeens that participants saw in the activity.

2.1 Analysis of Participant Responses

We tested this task with 501 unique users who were participating in a MOOC on evaluating online information that used strategies from the COR curriculum. Of the 501 unique users participating, we removed duplicate responses by participants from our dataset, leaving us with 441 valid responses. We then separated the short-answer responses into three categories: those who viewed the Tweet as trustworthy, those who viewed the Tweet as untrustworthy because it was satire, and those who viewed the Tweet as untrustworthy because of political bias in the Tweet. Three researchers qualitatively coded short-answer responses from individuals who labeled the Tweet as untrustworthy to understand the reasoning behind that label; researchers coded responses for reference to two possible explanations of untrustworthiness, satire and political bias (Cohen's Kappa = 0.949). For example, one response coded as related to satire said, "This account is a parody account and is not factual. I found that out by Googling the account name." One response coded as related to political bias said, "This Tweet has political bias so the account can't be trusted for information."

Overall, most participants (73.47%) correctly noted that the account was a parody account based on implementing the lateral reading strategy. Engaging in a brief online search demonstrates that @GOPTeens is a parody account, as it is run by Daniel Kibblesmith, an editor at a website associated with The Onion (another satirical website). The remaining responses did not directly answer the question or expressed confusion with the task.

3 Developing and Deploying the Classifier

In order to provide more directive feedback within the learning environment, we set out to design a text classifier that could identify when a participant failed to determine the Tweet was satire so that we could offer additional support to that participant. Directive feedback offers corrective feedback based on the users' response and is often more effective than general feedback when learners are just starting to learn a new skill [12].

During the task, participants were provided with explanatory video that modeled how an expert would evaluate the credibility of the @GOPTeens account using lateral reading. While we could have offered feedback based on multiple choice responses, limiting the question to pre-constructed options would not have captured participants' full understanding behind the trustworthiness of sources. Developing, training, and deploying a classifier would allow us to provide more robust feedback to participants.

We built our classifier using the coded data reserving approximately 10% (N = 46) as the validation set. We tested several different classifier models on our tagged dataset using two Python packages, Scikit-learn and Natural Language Toolkit. We preprocessed data using multiple Python libraries. We used Beautiful Soup [13] to remove punctuation, NLTK to remove stop words, and Porter Stemmer, a component of NLTK, to pull out word stems [14]. We utilized five stratified shuffled K-folds in our models and analyzed the power of the SVC and MLP models from Scikit [15] (see Table 1) and determined that SVM with all three preprocessing methods provided the best results on the positive class of the data. This resulted in an F-score of .99, precision of .97, and recall of 1.00, in determining that a written response did identify the Tweet as satire in a validation set.

Only one of the responses was miscategorized as a false positive in the validation set for the SVM implementation. "Other sites, unusual names, conservative leaning." was coded by researchers as not including the correct reasoning about satire, but it was marked as containing the correct reasoning by the model. This model was pickled for future use. From there, we deployed this classifier to the back-end server that analyzes participants' written text responses to tasks using the Python script with our classifier. The code for the web socket service implementation of this model is available on GitHub through https://github.com/mit-teaching-systems-lab/dcss-satire-agent.

Table 1. Overall F1, precision, and recall scores for the validation set across different model types and preprocessing methods (Yes: 36, No: 9).

		Macro			Yes			No		
Classifier	Method	F1	P	R	F1	P	R	F1	P	R
SVM	Stemming, stop words	.92	.97	.89	.97	.95	1.00	.88	1.00	.78
	+ punctuation	.96	.99	.94	.99	.97	1.00	.94	1.00	.89
MLP	Stemming, stop words	.89	.91	.88	.96	.95	.97	.82	.88	.78
	+ punctuation	.93	.93	.93	.97	.97	.97	.89	.89	.89

With this classifier model, we then revised the task to include feedback by integrating the classifier into the task to categorize responses. In this task, participants view the Tweet and click whether they believe the account to be a trustworthy source of information about the school walkouts. If they select that it is not trustworthy, they are taken to a screen to explain their answers (those who claimed it was trustworthy receive immediate lateral reading feedback). From the text response slide, the responses are classified as "Yes" or "No" for containing the correct satire reasoning (see Fig. 3). If the classifier returns "Yes," the participant sees positive feedback that they correctly identified the

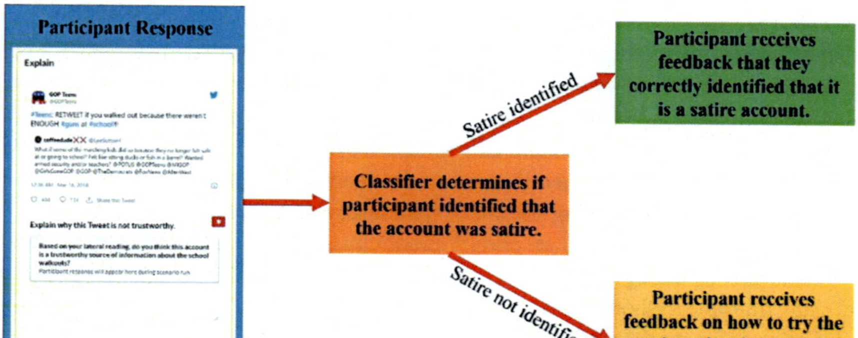

Fig. 3. A representation of the assessment architecture that the classifier produces for participant responses in the task depending on whether their reasoning about why the Tweet was untrustworthy was correct.

Tweet as satire, and if it returns a "No," the participant is shown: "That's not quite it. It looks like you did not correctly identify why this Tweet is untrustworthy. Remember, in lateral reading, it is helpful to open a new tab and type in the name of the organization. Try opening a new tab and searching the term 'GOPTeens' to see what comes up."

4 Discussion

The current implementation of feedback with this lateral reading exercise provides a debrief opportunity to participants that involves watching a video that describes how an expert would perform the task. We anticipate the directive feedback mechanism will be more effective than the current implementation of feedback because it replicates a key aspect of the in-person learning experience that is missing from the online environment—feedback and coaching from an instructor [5]. In upcoming implementations of this design, we plan to integrate questions for participants on whether they believe the classifier correctly identified their written response. For instance, in this satire task, we will ask participants about if the feedback that they received from the classifier was helpful to them and if they felt that the classifier responded appropriately. This will help us refine this preliminary design and iterate using new data.

The creation of classifiers like these will help professional development designers and researchers provide more directive feedback in large-scale environments like MOOCs where the feedback more adequately captures a participant's reasoning. Though our use of a classifier to identify participant responses' pertaining to satire is a small-use case, it provides information on how we can implement more extensive classifiers to teach individuals to evaluate the validity of online information, such as whether a user only trusts a website based on the URL or if a user can discern sponsored content from non-sponsored content. Future research will continue to explore these implementations in MOOCs to allow for personalized feedback and improved assessments that utilize text responses in online spaces.

Acknowledgments. We want to thank Garron Hillaire for technical help in the classifier development, and Sydney Dell and Aria Eppinger for help in coding responses.

References

1. Bowman, N., Cohen, E.: Mental Shortcuts, Emotion, and Social Rewards: The Challenges of Detecting and Resisting Fake News. In Fake News: Understanding Media and Misinformation in the Digital Age. MIT Press (2020)
2. Wineburg, S., McGrew, S.: Lateral reading: reading less and learning more when evaluating digital information. SSRN Electron. J. (2017). https://doi.org/10.2139/ssrn.3048994
3. Kahne, J., Bowyer, B.: Educating for democracy in a partisan age: confronting the challenges of motivated reasoning and misinformation. Am. Educ. Res. J. **54**(1), 3–34 (2017). https://doi.org/10.3102/0002831216679817
4. Wineburg, S., McGrew, S.: Lateral reading and the nature of expertise: reading less and learning more when evaluating digital information. Teach. Coll. Rec. **2019**, 40 (2019)

5. McGrew, S.: Learning to evaluate: an intervention in civic online reasoning. Comput. Educ. **145**, 103711 (2020). https://doi.org/10.1016/j.compedu.2019.103711
6. Breakstone, J., et al.: Students' Civic Online Reasoning: A National Portrait, vol. 49 (2019)
7. McGrew, S., Breakstone, J., Ortega, T., Smith, M., Wineburg, S.: Can Students evaluate online sources? learning from assessments of civic online reasoning. Theory Res. Soc. Educ. **46**(2), 165–193 (2018). https://doi.org/10.1080/00933104.2017.1416320
8. Did Trump Cut Meals on Wheels Because the Elderly "Just Die Anyway"? https://www.snopes.com/fact-check/trump-cut-meals-on-wheels/. Accessed 30 Mar 2021
9. McNamara, D.S., Crossley, S.A., Roscoe, R.: Natural language processing in an intelligent writing strategy tutoring system. Behav. Res. Methods **45**(2), 499–515 (2012). https://doi.org/10.3758/s13428-012-0258-1
10. Hillaire, G., Larke, L., Reich, J.: Detecting confusion in everyday conversations (2020). https://doi.org/10.13140/RG.2.2.31143.50081.
11. McNamara, D.S., Allen, L.K., Crossley, S.A., Dascalu, M., Perret, C.A.: Natural language processing and learning analytics. In: Lang, C., Siemens, G., Wise, A., Gasevic, D. (eds.) Handbook of Learning Analytics, pp. 93–104. Society for Learning Analytics Research (SoLAR) (2017). https://doi.org/10.18608/hla17.008
12. Shute, V.J.: Focus on formative feedback. Rev. Educ. Res. **78**(1), 153–189 (2008)
13. Richardson, L.: Beautiful Soup Documentation. Vol. 84 (2019)
14. Sri, M.: Types of data. In: Practical Natural Language Processing with Python, pp. 1–12. Apress, Berkeley, CA (2021). https://doi.org/10.1007/978-1-4842-6246-7_1
15. Pedregosa, F., e al.: Scikit-learn: Machine learning in python. Machine Learn. Python. **6**

CAERS: A Conversational Agent for Intervention in MOOCs' Learning Processes

Diego Rossi[1](✉), Victor Ströele[1], Regina Braga[1], Santi Caballé[2], Nicola Capuano[3], Fernanda Campos[1], Mário Dantas[1], Luigi Lomasto[4], and Danielle Toti[5]

[1] Federal University of Juiz de Fora, Juiz de Fora, Brazil
{diego.rossi,victor.stroele,mario.dantas}@ice.ufjf.br, {regina.braga,fernanda.campos}@ufjf.br

[2] Open University of Catalonia, Barcelona, Spain
scaballe@uoc.edu

[3] University of Basilicata, Potenza, Italy
nicola.capuano@unibas.it

[4] University of Salermo, Fisciano, Italy
llomasto@unisa.it

[5] Catholic University of the Sacred Heart, Brescia, Italy
daniele.toti@unicatt.it

Abstract. Massive Open Online Courses (MOOCs) make up a teaching modality that aims to reach a large number of students using Virtual Learning Environments. In these courses, the intervention of tutors and teachers is essential to support students in the teaching-learning process, answer questions about their content, and provide engagement for students. However, as these courses have a vast and diverse audience, tutors and teachers find it difficult to monitor them closely and efficiently with prompt interventions. This work proposes an architecture to favor the construction of knowledge for students, tutors, and teachers through autonomous interference and recommendations of educational resources. The architecture is based on a conversational agent and an educational recommendation system. For the training of predictive models and extraction of semantic information, ontology and logical rules were used, together with inference algorithms and machine learning techniques, which act on a dataset with messages exchanged between course forum participants in the humanities, medicine, and education fields. The messages are classified according to the type (question, answer, and opinion) and parameters about feeling, confusion, and urgency. The architecture can infer the moment in which a student needs help and, through a Conversational Recommendation System, provides the student with the opportunity to revise his or her knowledge on the subject. To help in this task, the architecture can provide educational resources via an autonomous agent, contributing to reducing the degree of confusion and urgency identified in the posts. Initial results indicate that integrating technologies and resources, complementing each other, can support the students and help them succeed in their educational training.

Keywords: Massive open online courses · Recommender system · Conversational agent

D. Guralnick et al. (Eds.): TLIC 2021, LNNS 349, pp. 371–382, 2022.
https://doi.org/10.1007/978-3-030-90677-1_36

1 Introduction

With the relevance and prominence of massive open online courses (MOOCs), the debate on new ways to improve this teaching modality's efficiency has gained relevance, as some problems are recurring. MOOCs have expanded and democratized education, since anyone with a computer or smartphone and an Internet connection can access a course in this modality. In most cases, the participant chooses a course free of charge, and, in some cases, there is a small fee for obtaining the certificate [10]. Several major universities produce MOOCs, as well, including Harvard, Stanford, and the Massachusetts Institute of Technology, just to name a few. Several platforms also produce courses of this type, such as Coursera, edX, and Future Learn [9].

Although the expansion of MOOCs is somewhat noticeable, the number of evasions is also very high [11]. School dropout is a problem that can be triggered by several factors, one of which is the difficulty of understanding the content and the lack of support to clarify some questions. However, due to a large number of students for a low number of tutors in this teaching modality [14], it is challenging to be able to carry out a pedagogical intervention on time. Therefore, the use of technology would make it possible to accomplish the task.

In this work, Conversational Agent in an Educational Recommender System (CAERS) is presented. CAERS is a software architecture composed of a conversational agent. It works as follows: when a post in a discussion forum indicates that there is a need for pedagogical intervention, the subject in the message is recognized; then, the educational content is recommended for clarification of questions.

The study followed four main steps: (i) defining an architecture for the conversational recommender system; (ii) using predictive models to automatically detect the need for student intervention; (iii) building an ontology to extract subjects from the student's question in the MOOC forums; and (iv) implementing the system prototype to filter contents corresponding to the student's need.

This work relies upon the recommendation system's overall concepts [12] to design our architecture. The approach proposed includes an infrastructure that automatically recognizes the student intervention necessity and, from a conversational agent, interacts with them to recommend educational resources. The proposed architecture has been evaluated through a usage scenario where messages from a discussion forum were sent, and by using machine learning methods, the system prototype identified the corresponding intervention moment. An ontology extracted the message concepts providing rich data to build the student profile, enabling the recommendation of educational resources to be made.

This paper is structured as follows. In Sect. 2, the related work is briefly presented. In Sect. 3, the CAERS architecture is presented, by detailing each layer with its own features. An evaluation of the proposed architecture and prototype is presented in Sect. 4, where the corresponding experiments are detailed. Finally, Sect. 5 reports the concluding remarks and mentions ideas for future work.

2 Related Work

Several papers discuss the topic of recommending educational resources for students of online courses.

In reference [8], the authors use a very rich data source to tackle the problem of discussion forums in a massive open course environment, which is often not given the desired attention, as the number of students is often higher concerning the number of instructors. This problem can be very limiting, since guided and closely monitored activities by a responsible person seek a lot of effort. A tool has been proposed to label the characteristics of the messages posted on massive course forums to assist this type of interaction. The knowledge is automatically discovered, guiding the instructor in his operations. The information can also be helpful for conversation agents.

The colMOOC platform is proposed in reference [4], where a conversational agent is modeled in a specific domain to mediate between dyads by triggering appropriate interventions in order to facilitate productive dialogue during chat-based collaborative learning activities in massive education, such as MOOCs. The agent's interaction in the chat is based on the Academically Productive Talk, which is a framework for modeling an experienced teacher's interventions during students' dialogues to make them elaborate in the knowledge domain [18].

According to reference [5], students' dependence on poor assistance and adaptation in massive courses leads them to lose motivation and, consequently, drop out of the learning process. As a solution, the authors developed a strategy for adapting activities through a recommendation system. The proposal uses a hybrid recommendation system based on knowledge, supported by ontology, to recommend activities for students in the context of MOOCs.

Considering many questions posted on the forums in massive courses, YouEDU, which helps students, detecting all messages that express content confusion is proposed in reference [6]. The authors trained a set of classifiers to categorize the forum posts in several aspects: feeling, urgency, and other descriptive variables that guide a classifier to detect confusion. Then, the confusing posts are directed to video excerpts from the course.

The authors in reference [7] proposed an agent-based recommendation system that aims to help students overcome their disabilities. The system recommends relevant learning resources to provide support to improve the learning experience. An agent-based cooperative system was designed, where agents act independently and update recommendation data, making the recommender more efficient and enhancing experiences on the learning platform.

The work developed in reference [16] aims to detect and analyze the involvement of course participants in the context of online education, obtaining relevant information related to aspects that indicate student involvement, such as feeling, urgency, and confusion; the probability of evasion of each student is also informed. To accomplish this task, students' posts and comments are considered, using classification algorithms based on machine learning.

In this work, the system architecture proposed uses a combination of techniques to improve the intervention in the students' learning process, detecting their need for help via machine learning techniques and producing the needed help by recommending educational resources through a semantics-based conversational agent.

3 CAERS – Conversational Agent in an Educational Recommender System

As shown in Fig. 1, the CAERS architecture was designed as an architectural structure with seven components: data, extraction, filtering or filter, knowledge model, recommendation or recommender, conversational, and application.

In the **data component**, the information is obtained from external sources that feed the system. This layer has a logical integration between the databases, making it possible to trace the student's profile, details of the student's learning process, objectives, and the possibilities that this training provides.

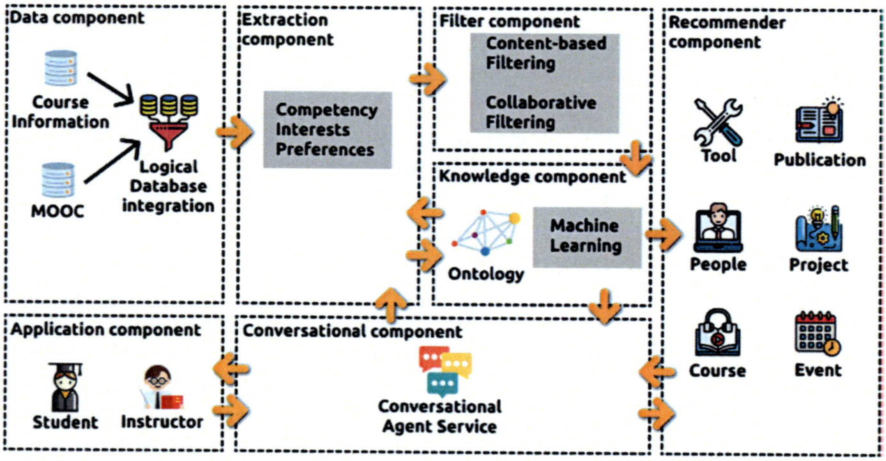

Fig. 1. CAERS architecture.

The **extraction component** is responsible for capturing data from different sources and providing information for both the filter and knowledge models components. The extraction process can be "explicit" and/or "implicit."

The explicit extraction occurs when the system asks the user to fill in their data, promoting their initial profile, which can be updated over time. This filing is done through forms, surveys, and even evaluations of the information presented to him.

The implicit extraction occurs when the information to define the profile is obtained without the user's actions, i.e., it occurs in a passive way, reflecting his behavior in an environment. This layer aims to provide information on skills, preferences, and interests.

The **filtering component** is responsible for implementing two types of filters, which serve to recommend the resources. The first type is content-based filtering, which has, as a principle, the similarity among the recommended items. The basic idea is that if a user likes a particular item, he may also like a similar one. The second type is collaborative filtering, where the principle is to recognize users with similar interests based on positive reviews made by similar users.

In collaborative filtering, after consuming the resources, the users collaborate with evaluations about these resources, indicating whether they are relevant or irrelevant to them. In this way, the system can make new recommendations considering the users' evaluations. Thus, it is possible to recommend resources more adherent to users' preferences considering similar previous evaluations.

The **knowledge component** can bring understanding of the context, providing information for both the conversational agent and the recommendation algorithms. In this component, ontology is used with its rules and knowledge representation. Machine learning techniques are used to create a predictive model [19], where it is possible to recognize the right moment to intervene with recommendations to enrich the student's learning moment.

This component can also adopt techniques responsible for improving the student's profile through ontologies and machine learning. The component techniques are used in conjunction with the ones mentioned above, as they can be applied to enrich user data or objects to be recommended [12].

The **recommendation component** is responsible for selecting open educational resources (OER), which can be obtained from different repositories, e.g., learning objects, linked data, and videos. In an attempt to obtain more relevant resources, a relationship is made between the user's profile and educational resources. This component defines what will be recommended and the priority that each item will have. Tutors and teachers will be able to analyze the available resources, thus indicating which ones are more adherent to the educational context of the course.

The conversational agent is responsible for obtaining information according to the students' forum messages in the **conversational component**. The agent interacts with the student when the machine learning model identifies that it is necessary. In this component, intervention and recommendation needs are detected based on the data extracted from the conversation between the student and the agent. The agent is connected with the course instructor, providing pre-selected educational resources. Hence, the instructor can analyze and determine the best content for the student.

The **application component** is the interface of interaction with the user, responsible for sending and receiving messages through the conversational agent's API, being responsible for improving the usability experience by presenting the agent's responses.

Once the architecture is defined, it is essential to analyze its feasibility to assist students in a virtual learning environment. In the next section, the architecture is evaluated considering the technological components adopted in the development and the behavior of the architecture when using messages posted by students in a MOOC.

4 Evaluation

An experiment was conducted using data from messages posted by students in MOOC discussion forums. A prototype was developed to verify the proposal's feasibility concerning the components proposed in the architecture, the usage of the predictive model, and the ontology. The underlying question of this experiment is: *can the conversational agent identify the student's need for intervention from the forum's posts and carry out an intervention, recommending an educational resource?*

4.1 Environment Setup

A Stanford MOOCPosts dataset[1] with 29.604 messages was used to carry out the experiments, captured from forums of three courses from different areas of knowledge, during 14 months. The data were anonymized to avoid identifying the participants, so it is not possible to tell the number of students who collaborated with these interactions.

Each post has the following attributes: text, opinion (0 or 1), question (0 or 1), answer (0 or 1), feeling (from 1 to 7), confusion (from 1 to 7), urgency (from 1 to 7), course type (education, humanities, and medicine), forum_post_id (unique ID of the respective post), course_display_name (course name), forum_uid (student's unique identifier), created_at (post date), post_type (comment or comment read, the last one is attributed to the post that originated a topic), anonymous (true, if the poster appears to everyone as the anonymous name), anonymous_to_peers (if true, the author of the post will appear as "anonymous" to everyone except the moderator and the instructor), up_count (number of positive votes for the post), comment_thread_id (topic object ID), and readings (the total number of readings recorded in the topic).

Another characteristic is that the phrases were labeled with some attributes that add more knowledge to the text fragments. This work's essential property was the confusion measure, which has a scale from 1 to 7, with level 1 for the least confused and 7 for the most confused. The course that the student was enrolled in also helped in carrying out the work [15].

Based on the confusion attribute, it was possible to train the predictive model using a supervised learning algorithm. Based on other studies [1, 2], logistic regression was used, as it presents good results in the classification of texts.

To use logistic regression, the dataset was divided into two parts, in the proportion of 80/20: 80% for training, and 20% for testing the model. The model was trained using k-fold validation, with $k = 10$, a cross-validation technique to avoid the model overfitting. For the model not to become biased, it was necessary to balance classes, the technique used to equalize was the oversample, which consists of generating new examples, because applying the undersample could reduce the number of examples, causing the loss of important information for the model training.

Total accuracy of 84.3% was obtained. Therefore, based on the trained model, if the need for intervention is detected, the conversational agent will send a message directly to the student.

The conversational agent is fed with terms extracted from the messages, added to facilitate the identification of the topic present in the text. Figure 2 shows all the steps taken to add knowledge to the agent. The process starts with obtaining the dataset with the terms, ending with the term set, called entities.

[1] https://datastage.stanford.edu/StanfordMoocPosts/.

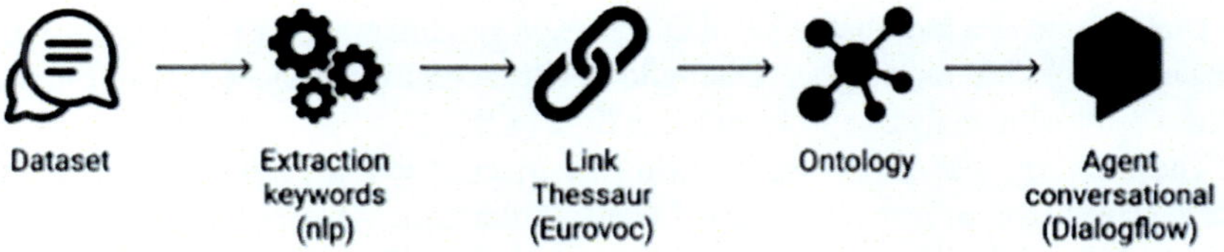

Fig. 2. Workflow to feed the agent.

The first action is to extract keywords from forum messages, and it was necessary to use natural language processing (NLP) techniques to do that. This step was performed using the Python language and the Spacy library to enable keyword identification.

Later, only words that had a match in the Eurovoc thesaurus were considered to avoid identifying terms that did not add relevant information about the domain. In this step, a dataset consisting of 5,933 descriptive terms was built, organized by areas of knowledge [3].

Once the extraction of the relevant terms has been completed, they are loaded in the ontology, shown in Fig. 3. This step was implemented via the Java OWL APIs, meant to create, manipulate, and serialize OWL ontologies. Using inference machine and ontology rules, instances from the ontology were extracted and then loaded in the Dialogflow tool [13].

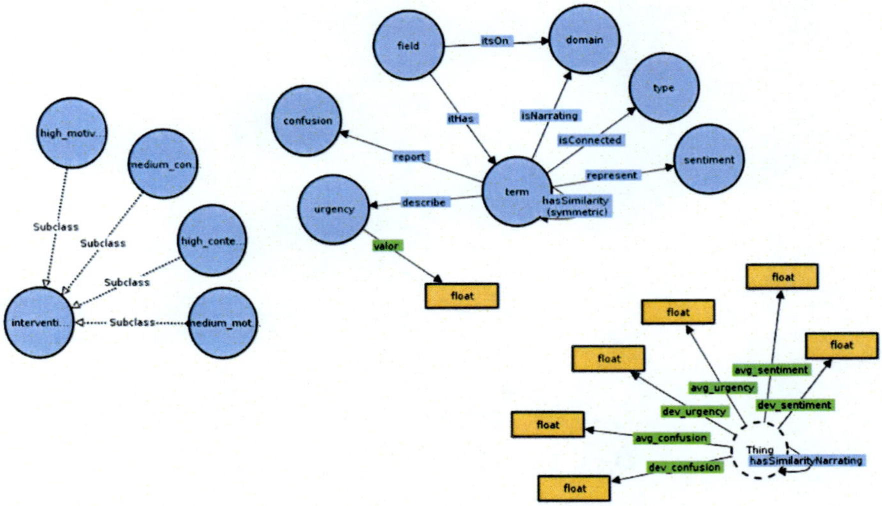

Fig. 3. Visualization of the classes that make up the ontology.

The classes of the ontology shown in Fig. 3 are term, domain, field, type, sentiment, confusion, urgency, intervention, high_motivation, medium_motivation, high_intervention, and medium_motivation. The urgency class has an attribute that represents the urgency value. The ontology has some attributes, which are avg_confusion, dev_confusion, avg_urgency, dev_urgency, avg_sentiment, and dev_sentiment.

Dialogflow[2] is a tool that helps in the creation and integration of conversation interfaces. It is possible to integrate Dialogflow with cognitive services through its APIs, such as sentiment analysis and knowledge base [17].

The prototype was developed following these steps and using Dialogflow as an initial user interface. The conversational agent waits for the student message, and, based on the predictive model decision, the agent can intervene or not. The agent working is described in the next section.

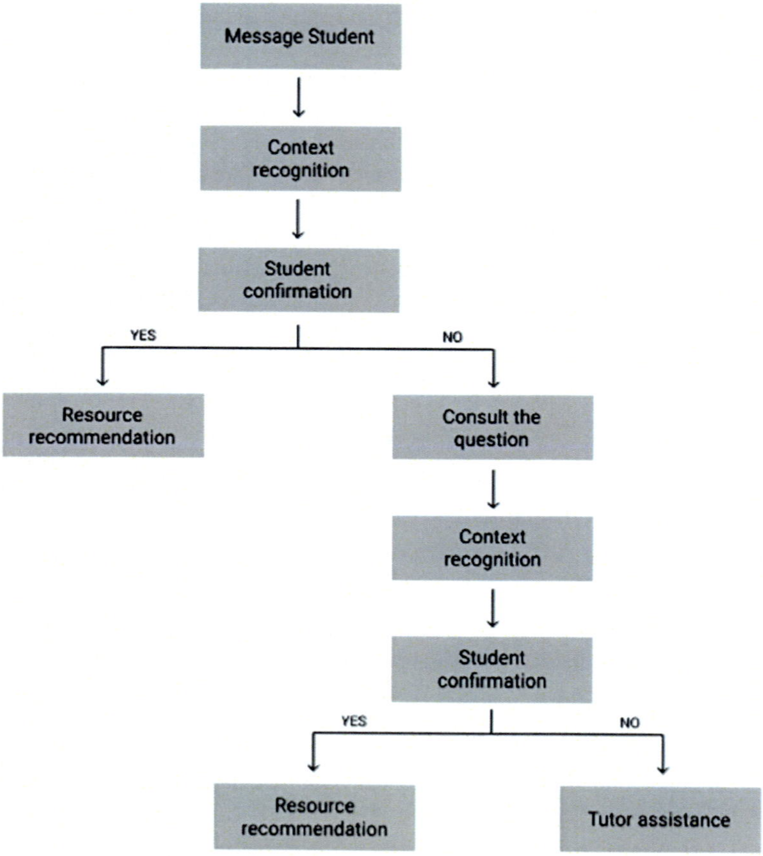

Fig. 4. Diagram of the agent's operation.

4.2 Conversational Agent in Action

In order for the agent to perform the intervention with the user, it was necessary to create a functioning flow represented in Fig. 4. The first step requires the agent to identify the

[2] https://dialogflow.cloud.google.com/#/login.

student message's context, which is enabled by the ontology. Later, the student is asked a question on whether he wants to get more content on the identified topic, and if the suggestion is accepted, a YouTube link is reported. If the student refuses the agent's content suggestion, the user will contribute with more details regarding the message's content. It gives the agent another chance to recognize the message subject and provide the student with useful educational recommendations. However, if the agent fails to identify the student's necessity the second time, the message history is forwarded to the tutor, asking him for a more specific intervention.

In this way, after performing the predictive model's training, the new message sent for the student is classified according to the model. Ten recent messages were selected from the forum and presented to the conversational agent to check its operation, with three of them classified as requiring intervention.

The ontology processed these three messages, and the agent gave an answer for each of them. The students interacted with the agent, saying that they wanted to receive some complimentary educational content. In this first moment, in order to verify the solution feasibility, the agent selected videos from YouTube using the concepts identified by the Ontology in the search string. Once the search was done, the agent provided a link where the student could access the complementary content. Figure 5 shows one of the chats between the conversational agent and the student.

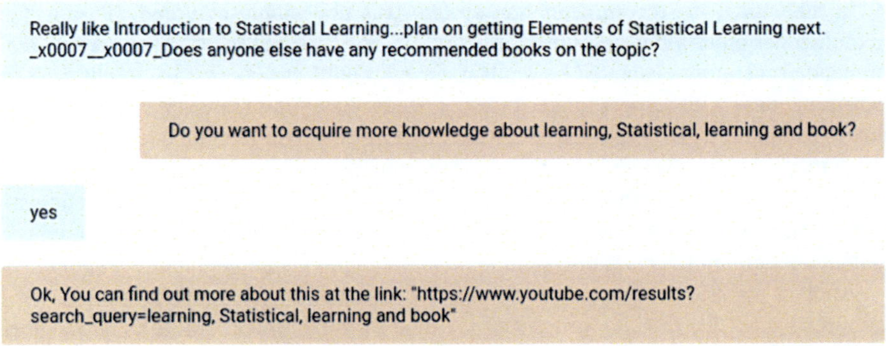

Fig. 5. Operating agent.

If the subject identified by the agent was not relevant to the student, the agent would ask for more details, allowing a new content identification from the ontology. If the student deems the recommendations not interesting, this conversation's history is forwarded to the tutor responsible for the course, requesting help for this intervention.

4.3 Observed Evidence

The experiment demonstrated the feasibility of the proposed architecture, the concepts and technologies involved, the use of the ontology, and the predictive model. It was also feasible to integrate a recommendation system with a conversational agent, covering a complete cycle of questions in a discussion forum within a MOOC.

Concerning the ontology, the conversational agent was able to identify the content of messages posted by students on the forum. On the other hand, the predictive model achieved an accuracy of 84.3%, allowing the agent to identify the messages in which immediate intervention was needed.

The conversational agent's action flow allows educational resources to be recommended directly to the student who presented the question. However, more complex cases, where the agent cannot identify the content of the question correctly, are forwarded to the instructors of the course.

5 Final Remarks and Feature Works

This paper presented CAERS, an architecture for a recommendation system with conversational agent interventions, which acts according to the students' messages posted in MOOC forums. Several works in scientific literature seek to solve similar problems for massive courses; the novelty of this work lies in the union of two powerful techniques to assist in the pedagogical intervention, i.e., recommendation systems and conversational agents.

References [4] and [8] were used as a basis for this paper for recommending educational content with a conversational agent. Given that these works had explored forum and chat posts in MOOCs, they contributed to the development of this research proposal.

In this study, the recommendation system was not widely explored, limiting the recommendation to video resources only, on the YouTube platform. The contents recommended were not evaluated by students, even though it is necessary to check their appropriateness with respect to the student's needs. As a future development, there is a plan to expand the prototype and turn it into a more powerful recommendation system, which may consider the user's profile and recommend open educational resources — the latter even more fitting to the students' needs. We intend to use the teacher's didactic plan to improve resource adherence to the students' needs and avoid bringing inappropriate content concerning the class program.

Another point to be addressed in future work is the delay that must be applied to the agent response. The learning process needs time for reflection and time to allow other students to answer the doubts, providing the generation of a community of interest in the course. Responding to students' questions too early may negatively affect their learning or the student's community.

Regardless, it is the author's belief that this work may contribute to the solution of problems related to the intervention in the learning process, providing clarification of questions in MOOC virtual environments. This contribution may positively influence the students' experience of massive courses and avoid student dropout.

Acknowledgments. This work was carried out with the support of the Coordination of Improvement of Higher Education Personnel - Brazil (CAPES) - Financing Code 001, Federal University of Juiz de Fora (UFJF), CNPq, and FAPEMIG.

References

1. Pranckevičius, T., Marcinkevičius, V.: Comparison of naive bayes, random forest, decision tree, support vector machines, and logistic regression classifiers for text reviews classification. Baltic J. Mod. Comput. **5**(2), 221 (2017)
2. Shah, K., et al.: A comparative analysis of logistic regression, random forest and KNN models for the text classification. Augmented Hum. Res. **5**(1), 1–16 (2020)
3. Steinberger, R., Pouliquen, B., Hagman, J.: Cross-lingual document similarity calculation using the multilingual thesaurus EUROVOC. In: Gelbukh, A. (ed.) Computational Linguistics and Intelligent Text Processing, CICLing 2002. LNCS, vol. 2276, pp. 415–424. Springer, Heidelberg (2002). https://doi.org/10.1007/3-540-45715-1_44
4. Demetriadis, S., et al.: Conversational agents as group-teacher interaction mediators in MOOCs. In: 2018 Learning With MOOCS (LWMOOCS), pp. 43–46. IEEE (2018)
5. Harrathi, M., Touzani, N., Braham, R.: A hybrid knowledge-based approach for recommending massive learning activities. In: 2017 IEEE/ACS 14th International Conference on Computer Systems and Applications (AICCSA), pp. 49–54. IEEE (2017)
6. Agrawal, A., et al.: YouEDU: addressing confusion in MOOC discussion forums by recommending instructional video clips (2015)
7. Brigui-chtioul, I., Caillou, P., Negre, E.: Intelligent digital learning: agent-based recommender system. In: Proceedings of the 9th International Conference on Machine Learning and Computing, pp. 71–76 (2017)
8. Capuano, N., Caballé, S.: Multi-attribute categorization of MOOC forum posts and applications to conversational agents. In: Barolli, L., Hellinckx, P., Natwichai, J. (eds.) Advances on P2P, Parallel, Grid, Cloud and Internet Computing, 3PGCIC 2019. LNCS, vol. 96, pp. 505–514. Springer, Cham (2019). Springer, Cham (2019). https://doi.org/10.1007/978-3-030-33509-0_47
9. Foley, K., et al.: Massive open online courses (MOOC) evaluation methods: protocol for a systematic review. JMIR Res. Protocols **8**(3), e12087 (2019)
10. Hew, K.F., Cheung, W.S.: Students' andtutor' use of massive open online courses (MOOCs): motivations and challenges. Educ. Res. Rev. **12**, 45–58 (2014)
11. Dalipi, F., Imran, A.S., Kastrati, Z.: MOOC dropout prediction using machine learning techniques: review and research challenges. In: 2018 IEEE Global Engineering Education Conference (EDUCON), pp. 1007–1014. IEEE (2018)
12. Pereira, C.K., Campos, F., Ströele, V., David, J.M.N., Braga, R.: BROAD-RSI – educational recommender system using social networks interactions and linked data. J. Internet Serv. Appl. **9**(1), 1–28 (2018). https://doi.org/10.1186/s13174-018-0076-5
13. Santoso, H.A., et al.: Dinus Intelligent Assistance (DINA) chatbot for university admission services. In: 2018 International Seminar on Application for Technology of Information and Communication, pp. 417–423. IEEE (2018)
14. Capuano, N., Caballe, S.: Towards adaptive peer assessment for MOOCs. In: Proceedings of the 10th International Conference on P2P, Parallel, Grid, Cloud and Internet Computing (3PGCIC-2015), Krakow, Poland, pp. 64–69 (2015)
15. Capuano, N., Caballé, S., Conesa, J., Greco, A.: Attention-based hierarchical recurrent neural networks for MOOC forum posts analysis. J. Ambient Intell. Hum. Comput. **12**(11), 9977–9989 (2020)
16. Toti, D., et al.: Detection of student engagement in e-learning systems based on semantic analysis and machine learning. In: Barolli, L., Takizawa, M., Yoshihisa, T., Amato, F., Ikeda, M. (eds.) Advances on P2P, Parallel, Grid, Cloud and Internet Computing. 3PGCIC 2020. Lecture Notes in Networks and Systems, vol. 158, pp. 211–223. Springer, Cham (2020). https://doi.org/10.1007/978-3-030-61105-7_21

17. Sabharwal, N., Agrawal, Amit.: Introduction to Google dialogflow. In: Cognitive Virtual Assistants Using Google Dialogflow, pp. 13–54. Apress, Berkeley (2020)
18. Tegos, S., Demetriadis, S., Karakostas, A.: Promoting academically productive talk with conversational agent interventions in collaborative learning settings. Comput. Educ. **87**, 309–325 (2015)
19. Neves, F., et al.: Heath-PRIOR: an intelligent ensemble architecture to identify risk cases in healthcare. IEEE Access **8**, 217150–217168 (2020). https://doi.org/10.1109/ACCESS.2020.3042342

A Tool for Evaluating the Quality of Online Teaching

Antonio Sarasa-Cabezuelo[1] (✉) and Santi Caballé Llobet[2]

[1] Universidad Complutense de Madrid, 28040 Madrid, Spain
asarasa@ucm.es

[2] Universidad Oberta de Catalunya, 28040 Barcelona, Spain
scaballe@uoc.edu

Abstract. The evaluation of the quality of teaching is a critical aspect of an educational system, since it is the basis for knowing which elements work well and which ones work poorly. Thus, through the results of the evaluation, actions can be articulated to improve the system. The evaluation methods are quite developed in face-to-face education, but not so in the case of online education. This article presents a proposal for an evaluation model for online education based on the analysis of the activity developed by students in the context of a virtual campus based on Moodle. This model has been implemented through a web application that allows managing the evaluation of the subjects. For this, two files extracted from Moodle; the grader report and the log of the activity of the virtual subject to be evaluated are taken as input. From them, the quality criteria defined in the model are automatically calculated. Likewise, the system allows users to manually evaluate other types of criteria that can be dynamically configured. In this sense, another feature of the tool is its ability to configure and adapt to different versions of Moodle and to new quality criteria.

Keywords: eLearning · Quality assessment tool · Moodle · LMS

1 Introduction

The evaluation of the quality of teaching is a critical task in an educational system, since it allows us to know those aspects that are working properly and those that cause problems. Quality assessment systems in the field of face-to-face teaching [8] have been widely developed in the last decades. Normally, this evaluation is carried out based on the analysis of different elements that intervene in the teaching process, such as the academic results of the students, the type and quality of the materials used in the classes, the proposed contents, the teaching methodology, or the organization of matter. In recent years, online education has spread to most educational levels, and the need to assess the quality of teaching in this new field has arisen. Although the online context tries to simulate the elements that are used in face-to-face teaching, there are important differences and restrictions that are complex to simulate [2]. For example, the non-verbal language that can be used in a face-to-face environment is not easy to simulate

D. Guralnick et al. (Eds.): TLIC 2021, LNNS 349, pp. 383–395, 2022.
https://doi.org/10.1007/978-3-030-90677-1_37

in an online environment. In addition, there are other generalized phenomena, such as less student intervention in classes or less predisposition of the teacher to teach dynamic classes with active student participation. Likewise, there are also changes in the materials used, since they are generally all electronic, and the development of the classes is carried out entirely in the context of a virtual campus with the tools that these provide. These conditioning factors directly affect the way in which the quality of online teaching [5] should be evaluated and that establishes differences with respect to the evaluation carried out in face-to-face teaching [4].

In recent years, several evaluation initiatives have been developed for different aspects of online teaching [10], such as specifications for evaluating the quality of materials used [3] in the online environment, evaluating the quality of classes taught online, or a more generic and comprehensive type of evaluation of online training [1]. These evaluation systems are based on the definition of different quality criteria whose value can be obtained by measuring certain aspects of the online teaching implemented [9]. The values obtained for these criteria act as an indication of the quality achieved. In general, they are based on filling in and answering sets of questions about various dimensions of the implemented system [7]. The main problem with these specifications is that they are not widely used, since in general, there are no applications that implement them and, in many cases, they require their adaptation to the specificity of the educational context in which they are to be applied.

In this article, a model for evaluating the quality of online teaching that is developed in a virtual campus based on Moodle [6] is presented. In addition, a web application that implements the model is described. This model has been developed based on the experience in the use of the virtual campus and on some automatic assessment models [11]. For this, two types of evaluations have been defined. An automatic evaluation [11] based on a set of criteria whose assessment is obtained from the analysis of a set of data obtained from the virtual campus, as well as a dynamic manual evaluation that consists of answering a set of questions about the educational process implemented. This evaluation can be configured based on a text document that specifies the evaluation questions and a particular PHP script to evaluate that set of questions that defines the qualification ranges of each question and how the values obtained are added. In this way, the implemented tool can be adapted to different subjects and ways of implementing online teaching.

Likewise, the tool provides different graphic ways to show the results obtained, as well as the possibility of obtaining a PDF document with a report that summarizes the evaluation carried out. The application manages both assignments and teachers and evaluators. There is also the profile of the administrator who is in charge of configuring the tool.

The structure of the paper is as follows. In Sect. 2 a model for evaluating the quality of online teaching will be described. Next in Sect. 3 the implementation of the application will be presented. Finally, in Sect. 4, a set of conclusions and lines of future work will be planted.

2 Model for Evaluating the Quality of Online Teaching

The described application implements a model for evaluating the quality of the teaching of online subjects based on 5 criteria that are automatically calculated from the information obtained from the activity developed in the subject displayed in Moodle. For each criterion, one of the following values is obtained:

- 0: Does not have
- 1: Yes, it has partially
- 2: Yes, it has and it is adequate
- NV: not verifiable

The information used to calculate the criteria values is obtained from two files retrieved from Moodle:

- Grade report. The file must contain the following information: "Name", "Surname", "ID", "USER_ID", "DNI/NIF", "Email address", "Procedure course", "Did you pass the continuous assessment?", "Final exam tasks", and "Total of the course".
  All these fields, except the field "Did you pass the continuous assessment?" appear in the Moodle grade report. The field, "Did you pass the continuous assessment?" indicates whether the student passed the subject through continuous assessment in case of having used this assessment method. It can have two values: "Yes" in the case that the student has passed the continuous assessment and "No" in the opposite case.
- Log of the activity developed by the students in the LMS.

The criteria used to carry out the evaluation are:

- **Criterion 1: Student participation (active or passive) in the different activities**

The Log file is taken as input, and the value of the criterion is calculated from the values of three indicators: activity in the evaluation, activity in the forum, and connection activity. Each of these indicators is associated with a weight according to its importance. The calculation of the indicators is carried out with the algorithms shown below:

1. Algorithm for calculating the indicator "Evaluation activity"
   The objective of this algorithm is to calculate a value that represents the participation of the students in the evaluation activities defined in the course: assignments, quizzes, and lessons. The calculation of the indicator is carried out by calculating a value of the participation of each student in this type of activities as the percentage of activities of each type that the student has carried out. Next, the arithmetic mean of the values obtained for each student will be taken as the total value of the indicator. If the values obtained with the mean were smaller than 0.2, then it is not representative and the median of the values obtained for each student will then be taken. The algorithm is described below:

Let A be the number of "assignments" defined in the course.
Let B be the number of "quizzes" defined in the course
Let C be the number of "lessons" defined in the course
For each student, it is calculated:
Let P be the number of "assignments" done by the student
The ratio V1 = P / A is calculated
Let U be the number of "quizzes" carried out by the student
The ratio V2 = U / B is calculated
Let T be the number of "lessons" visited by the student
The ratio V3 = T / C is calculated
The value of the indicator for the student is calculated as:
Value_indicator = (V1 + V2 + V3) / 3
Let T be the arithmetic mean of the indicator values for each student.
Let M be the median of the indicator values for each student.
If T <0.2, then the value T is returned as the indicator value.
Otherwise, the value of M is returned as the indicator value.

2. Algorithm for calculating the indicator "Forum activity"
   The objective of this algorithm is to calculate a value that represents the participation of the students in the course forums both by reading forum entries and by sending new entries. For this, the participation of each student in each type of activity (reading and sending entries) is calculated as the percentages of each type of activity that the student has carried out with respect to the total of activities carried out of each type. These values are multiplied by weights that represent the importance of each activity (the weights are established by the teacher), and the resulting values are added. Next, the arithmetic mean of the values obtained for each student will be taken as the total value of the indicator. If the value obtained with the mean is less than 0.2, then it is not representative and the median of the values obtained for each student will be taken. The algorithm is described below:

Let A be the total number of posts done in the course forums.
Let B be the value of the weight of the importance of the publications that have been sent to the forums.
Let C be the value of the weight of the importance of the publications that have been read in the forums
For each student, it is calculated:
Let D be the number of messages read from the forums by the student.
The ratio V1 = D / A is calculated.
Let E be the number of messages posted to the forums by the student.
The ratio V2 = E / A is calculated.
The value of the indicator is calculated for the student as:
*Value_indicator = C * V1 + B * V2.*
Let T be the arithmetic mean of the indicator values for each student.
Let M be the median of the indicator values for each student.
If T <0.2, then the value T is returned as the indicator value
Otherwise, the value of M is returned as the indicator value.

3. Algorithm for calculating the indicator "connection activity"
The objective of this algorithm is to calculate a value that represents the degree of connectivity of the students to the virtual course. First, a minimum value of weekly connections and the number of weeks the course lasts is established. The indicator value is then calculated for each student. For this, the connections made per week to the course by each student are calculated. If the calculated value is greater than the established minimum connectivity value, then 1 is returned as the indicator value for the student. Otherwise, the ratio between the connections done by the student and the minimum number of connections per week is returned as the indicator value for the student. Next, the arithmetic mean of the values obtained for each student will be taken as the total value of the indicator. If the value obtained with the mean is less than 0.2, then it is not representative and the median of the values obtained for each student will be taken. The algorithm is described below:

Let A be the number of minimum weekly connections.
Let B be the number of weeks.
For each student, it is calculated:
Let C be the number of connections made by the student during the course.
The ratio V1 = C / B is calculated.
If V1> A, then return 1.
Otherwise, return V1 / A.
Let T be the arithmetic mean of the indicator values for each student.
Let M be the median of the indicator values for each student.
If T <0.2, then the value T is returned as the indicator value
Otherwise, the value of M is returned as the indicator value.

When the values for the 3 indicators have been calculated, a value is calculated that adds them. This value is obtained by multiplying each value for each indicator by a weight that represents the importance of the indicator in the criterion and adding the values obtained. Next, depending on the value obtained, the total value of the criterion is established:

- If value obtained is between 1 and 0.60, then score = 2
- If value obtained is between 0.59 and 0.40, then score = 1
- If value obtained is less than 0.40, then score = 0

- **Criterion 2: Percentage of students presented to the final exam of the subject**
  The grade report file is processed, searching for the name of the final exam and calculating the percentage of students presented:

  - If percentage is between 1 and 0.60, then score = 2
  - If percentage is between 0.59 and 0.40, then score = 1
  - If percentage is less than 0.40, then score = 0

- **Criterion 3: Percentage of students who pass the continuous assessment**
  The grade report file is processed by searching for the Boolean "continuous" to verify if the student has passed the continuous assessment, calculating the percentage of students passed the continuous assessment.

  - If percentage is between 1 and 0.60, then score = 2
  - If percentage is between 0.59 and 0.40, then score = 1
  - If percentage is less than 0.40, then score = 0
  - Not verifiable if the grade report does not provide this information.

- **Criterion 4: Percentage of students who pass the course**
  The grade report file is processed by searching for the name of the column that represents the final grade and calculating the percentage of students passed.

  - If percentage is between 1 and 0.60, then score = 2
  - If percentage is between 0.59 and 0.40, then score = 1
  - If percentage is less than 0.40, then score = 0

- **Criterion 5: Statistics**
  In this criterion, the way in which students have passed is analyzed:

  - If they have passed the continuous assessment (criterion score 3 = 2) and also the final exam (criterion 4 = 2), then it is scored with the value 2.
  - If the students have not taken continuous assessment (criterion score 3 = NV) and have passed the final exam (criterion 4 = 2), then they are scored with the value 2.
  - If a value greater than 0 has been obtained in the criterion on the continuous evaluation and if in the criterion on the final exam, the value 1 has been obtained, then the performance has not been so good and is scored with a 1.

- In any other case, it is considered that the use has been bad and is scored with the value 0.

3 Implementation

The quality model described above has been implemented by means of an application that calculates the criteria and allows the results to be viewed and downloaded. In the application, 3 types of actors have been defined:

- Teacher. This user can manage the evaluations of the online subjects assigned. For each subject evaluated, you can view the evaluation or download an evaluation report. In addition, you can delete any of the assigned subjects.
- Evaluator. This user can manage the subjects that have been assigned to him to evaluate: evaluate a subject, continue the evaluation of a subject, or consult the subjects that he has evaluated
- Administrator. It manages the users, the subjects that must be evaluated, the subject evaluations carried out, and the configuration of the evaluation process of the tool.

In order to access the functionality, it is necessary to be registered and have an account (Fig. 1). Thus, when the personal account is accessed, regardless of the role, a page with 3 different regions is displayed:

- The content of the application appears in the Central Panel depending on where the user is.
- The upper panel shows the profile image and also contains a menu in which you can configure the profile and exit the application.
- In the Left Side Menu, there are several options, and these vary depending on the role with which it has been accessed.

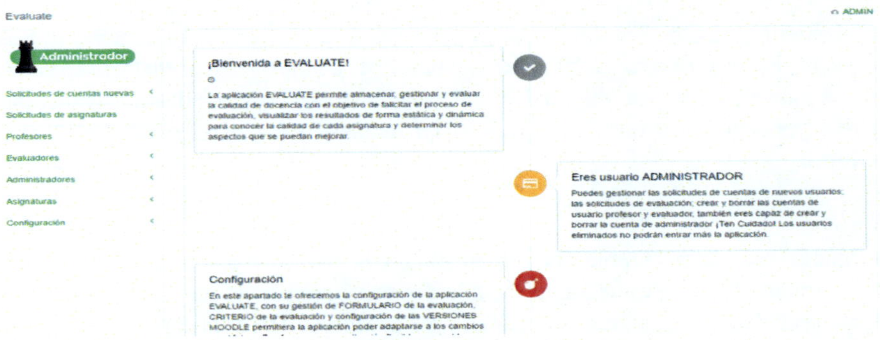

Fig. 1. User account page.

Next, the functionality referred to the evaluation process will be described.

3.1 Organization of the Evaluation

The tool allows two types of evaluations to be carried out (Fig. 2): manual and automatic.

- Manual: It consists of filling in a form with a set of questions that evaluate various aspects of the online subject. The structure and content of the evaluation is configured by the administrator. In this case, the evaluator will have to answer each question on the form based on the contents of the subject that he is evaluating.
- Automatic: It consists of calculating the 5 criteria described in the previous section. For this, the evaluator will have to load, for each subject, the log and the bulletin of grades obtained from the subject displayed on Moodle. You must also indicate the version of Moodle used.

Fig. 2. Ways to evaluate.

3.2 Configuration of the Evaluation

The configuration of the evaluation is carried out by the administrator and consists of defining the following aspects:

- Manual evaluation. The questions in the manual assessment are not static and are established from a configuration file that describes all the questions that will be shown to the assessor. This file is uploaded by the administrator in the application. In this way, the application uses this file (Fig. 3) to dynamically generate the manual evaluation form.

- Manual evaluation criteria. The administrator will be able to upload a txt file in which the criteria for the evaluation of the questions that are part of the manual evaluation are specified, such as the score range of the questions. The application will interpret the content of the file and adapt the form that is shown to the evaluator.
- Automatic evaluation. The structure of the files that come from Moodle and are used in the automatic evaluation vary in each version of the LMS. Thus, to guarantee compatibility with any version of Moodle, the application allows the user to establish

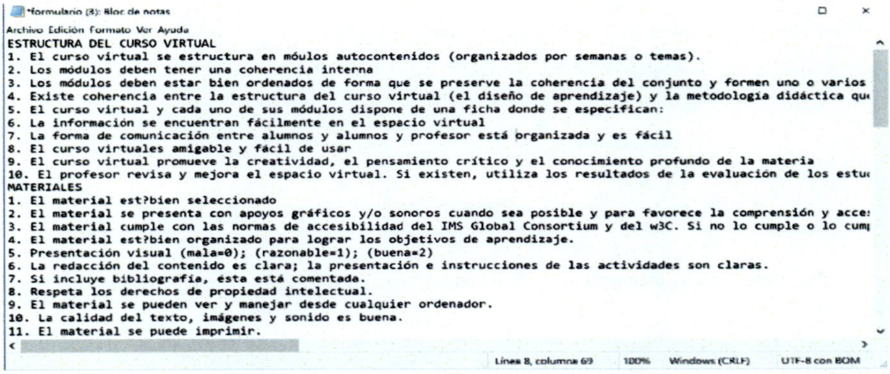

```
*formulario (8): Bloc de notas
Archivo Edición Formato Ver Ayuda
ESTRUCTURA DEL CURSO VIRTUAL
1. El curso virtual se estructura en móulos autocontenidos (organizados por semanas o temas).
2. Los módulos deben tener una coherencia interna
3. Los módulos deben estar bien ordenados de forma que se preserve la coherencia del conjunto y formen uno o varios
4. Existe coherencia entre la estructura del curso virtual (el diseño de aprendizaje) y la metodología didáctica qu
5. El curso virtual y cada uno de sus módulos dispone de una ficha donde se especifican:
6. La información se encuentran fácilmente en el espacio virtual
7. La forma de comunicación entre alumnos y alumnos y profesor está organizada y es fácil
8. El curso virtuales amigable y fácil de usar
9. El curso virtual promueve la creatividad, el pensamiento crítico y el conocimiento profundo de la materia
10. El profesor revisa y mejora el espacio virtual. Si existen, utiliza los resultados de la evaluación de los estu
MATERIALES
1. El material est?bien seleccionado
2. El material se presenta con apoyos gráficos y/o sonoros cuando sea posible y para favorece la comprensión y acce
3. El material cumple con las normas de accesibilidad del IMS Global Consortium y del w3C. Si no lo cumple o lo cum
4. El material est?bien organizado para lograr los objetivos de aprendizaje.
5. Presentación visual (mala=0); (razonable=1); (buena=2)
6. La redacción del contenido es clara; la presentación e instrucciones de las actividades son claras.
7. Si incluye bibliografía, ésta está comentada.
8. Respeta los derechos de propiedad intelectual.
9. El material se pueden ver y manejar desde cualquier ordenador.
10. La calidad del texto, imágenes y sonido es buena.
11. El material se puede imprimir.
Línea 8, columna 69    100%    Windows (CRLF)    UTF-8 con BOM
```

Fig. 3. Manual evaluation configuration file.

the way in which the files of each version of Moodle should be evaluated. To do this, the administrator will upload a file to the system that will contain a PHP script (Fig. 4) that is capable of evaluating a grade report and a log of a specific version of Moodle. The application is capable of managing more than one version, so that every time an automatic evaluation file is uploaded, which version of Moodle it corresponds to must be indicated. Likewise, the administrator can modify the uploaded files or delete supported Moodle versions.

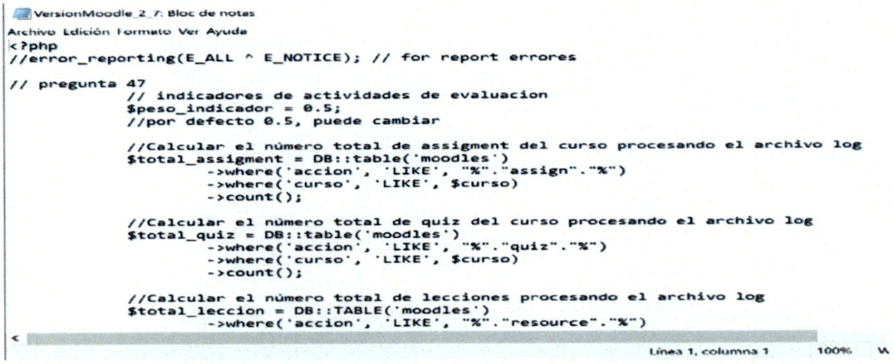

```
VersionMoodle_2_7: Bloc de notas
Archivo Edición Formato Ver Ayuda
<?php
//error_reporting(E_ALL ^ E_NOTICE); // for report errores

// pregunta 47
            // indicadores de actividades de evaluacion
            $peso_indicador = 0.5;
            //por defecto 0.5, puede cambiar

            //Calcular el número total de assigment del curso procesando el archivo log
            $total_assigment = DB::table('moodles')
                    ->where('accion', 'LIKE', "%"."assign"."%")
                    ->where('curso', 'LIKE', $curso)
                    ->count();

            //Calcular el número total de quiz del curso procesando el archivo log
            $total_quiz = DB::table('moodles')
                    ->where('accion', 'LIKE', "%"."quiz"."%")
                    ->where('curso', 'LIKE', $curso)
                    ->count();

            //Calcular el número total de lecciones procesando el archivo log
            $total_leccion = DB::TABLE('moodles')
                    ->where('accion', 'LIKE', "%"."resource"."%")
Línea 1, columna 1    100%
```

Fig. 4. Moodle version 2.7 file evaluation script.

3.3 Ways to View the Results

There are 3 ways to view the results of the evaluations carried out on a subject (Fig. 5): Static, Dynamics and By consultation.

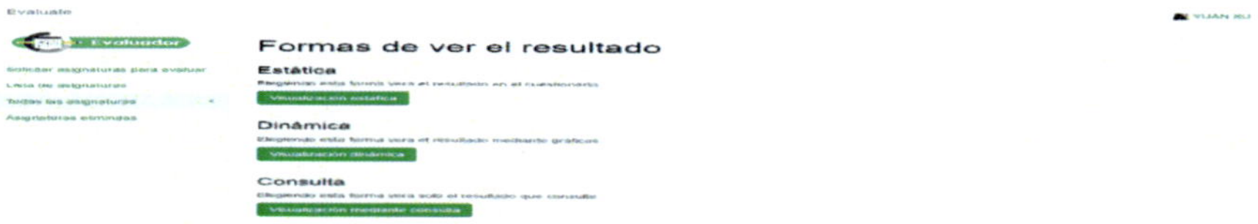

Fig. 5. Ways to view the results.

If the static display is selected, the result will be shown in the questionnaire itself (Fig. 6).

Fig. 6. Evaluation results.

If a dynamic display is chosen, the result will be displayed in the form of graphs. There are two options (Fig. 7a): see the overall result of the evaluation and see the result of a specific section.

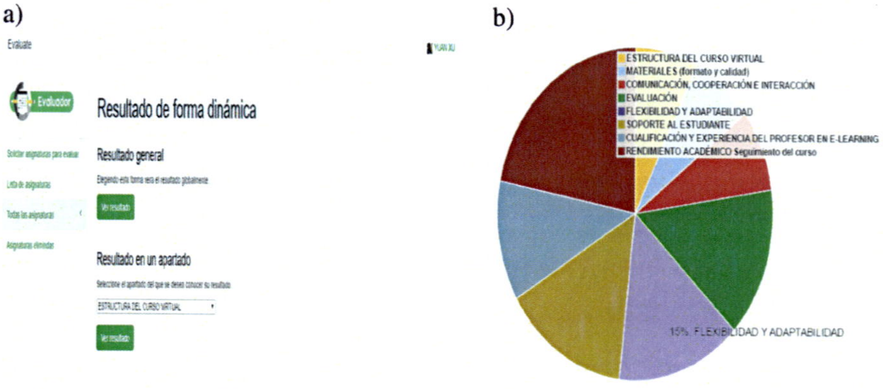

Fig. 7. a) Dynamic display b) Second graph.

In any of the options, four types of graphs are shown:

- The first graph shows the total score obtained in all the sections and those that we need to complete 100% in case we have not done so.
- The second graph (Fig. 7b) shows the score as a percentage that has been obtained in each item. The maximum score that can be obtained in each section is 100%.
- The third graph shows the same content as the second graph but using a pie chart.
- The fourth graph shows the total score obtained in each section.

If you select visualization by query (Fig. 8), a list will appear to be able to select the section of the form that you want to consult, and later, the questions in that section will appear, and you will be able to select those for which you want to know the result.

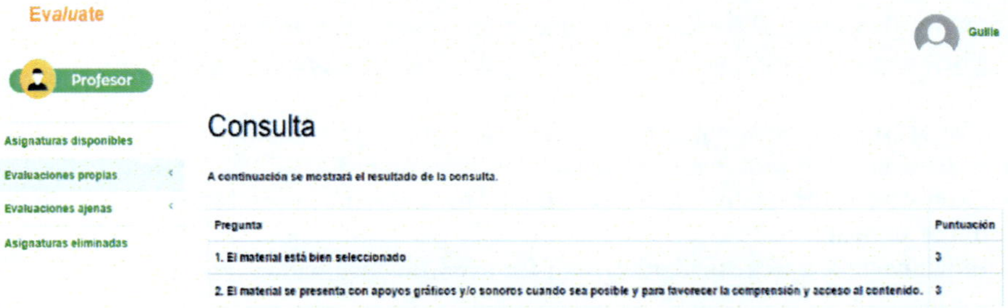

Fig. 8. Visualization by query.

3.4 Evaluation Summary Document

The tool provides both the teacher and the evaluator the possibility of downloading a PDF document that contains the evaluation carried out (Fig. 9).

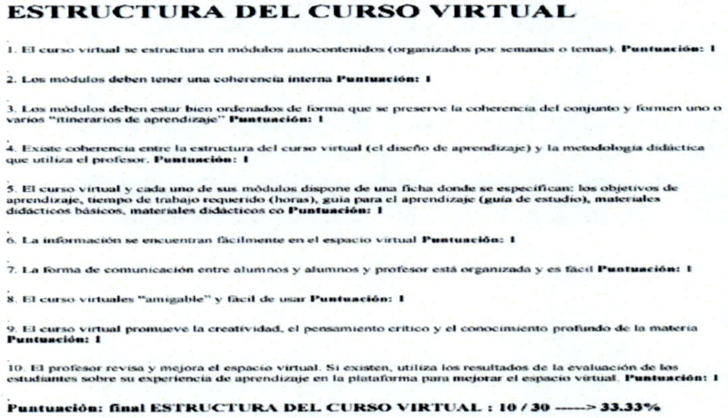

ESTRUCTURA DEL CURSO VIRTUAL

1. El curso virtual se estructura en módulos autocontenidos (organizados por semanas o temas). **Puntuación: 1**

2. Los módulos deben tener una coherencia interna **Puntuación: 1**

3. Los módulos deben estar bien ordenados de forma que se preserve la coherencia del conjunto y formen uno o varios "itinerarios de aprendizaje" **Puntuación: 1**

4. Existe coherencia entre la estructura del curso virtual (el diseño de aprendizaje) y la metodología didáctica que utiliza el profesor. **Puntuación: 1**

5. El curso virtual y cada uno de sus módulos dispone de una ficha donde se especifican: los objetivos de aprendizaje, tiempo de trabajo requerido (horas), guía para el aprendizaje (guía de estudio), materiales didácticos básicos, materiales didácticos co **Puntuación: 1**

6. La información se encuentran fácilmente en el espacio virtual **Puntuación: 1**

7. La forma de comunicación entre alumnos y alumnos y profesor está organizada y es fácil **Puntuación: 1**

8. El curso virtuales "amigable" y fácil de usar **Puntuación: 1**

9. El curso virtual promueve la creatividad, el pensamiento crítico y el conocimiento profundo de la materia **Puntuación: 1**

10. El profesor revisa y mejora el espacio virtual. Si existen, utiliza los resultados de la evaluación de los estudiantes sobre su experiencia de aprendizaje en la plataforma para mejorar el espacio virtual. **Puntuación: 1**

Puntuación: final ESTRUCTURA DEL CURSO VIRTUAL : 10 / 30 -----> 33.33%

Fig. 9. Evaluation summary document.

4 Conclusions and Future Work

In this work, a proposal for a model for evaluating the quality of online teaching has been presented in the context of a Moodle-type LMS. Two types of evaluations have been defined, a manual one that consists of answering a set of questions about the course activity and an automatic one that is based on analyzing the Moodle log file and the grade report. The results of the evaluation can be displayed graphically, and it is also possible to download a document that summarizes the evaluation carried out. The tool allows managing the evaluated subjects and three types of users: teachers, evaluators, and administrator. Furthermore, another feature of the tool is its ability to adapt the evaluation to different versions of Moodle. Finally, one of the advantages of the proposed model compared to other alternatives is that it offers an application that implements it and, secondly, it is a neutral model to the educational context to which it is applied. This tool is a first approximation; however, the following lines of work can be defined:

- Use of Big Data techniques to perform a deeper analysis of the data.
- Extension of the tool to admit information from other LMSs other than Moodle.
- Extension of the quality assessment model to add a dimension based on the opinion of the students.
- Implementation of a standard system for exporting evaluation results in xml or json format.
- Addition to the evaluation of a set of recommendations on how to improve quality.
- Differentiation of the kind of activities/modules in the criterion calculation (summative/formative, mandatory/voluntary), and consideration of different values according to each of them. It is different to avoid reading a mandatory module, as opposed to a voluntary one.
- Allowance of the parametrization of the criteria 2 to 4 in order to represent the differences in different subjects. Some subjects, such as physics, may have finalization or pass rates below 50%.
- Description of the evaluation questionnaires using IMS/QTI.

Use of other types of graphs in order to represent the values of quality aspects of a given element.

Acknowledgments. I would like to thank Guillermo Duque, Yuan Xu Jiang, and Ana María Fernández-Pampillón Cesteros for their participation in carrying out this work. This work has been partially supported by European Commission through the project "colMOOC: Integrating Conversational Agents and Learning Analytics in MOOCs" (588438-EPP-1-2017-1-EL-EPPKA2-KA).

References

1. De Medio, C., Limongelli, C., Sciarrone, F., Temperini, M.: MoodleREC: a recommendation system for creating courses using the moodle e-learning platform. Comput. Hum. Behav. **104**, 106168 (2020)
2. Ehlers, U., Goertz, L., Hildebrandt, B., Pawlowski, J.M.: Quality in e-learning. Voc. Train. Berlin-Cedefop **29**, 3–15 (2004)

3. Fernández-Pampillón Cesteros, A.M.: UNE 71362: calidad de los materiales educativos digitales. Revista AENOR (329) (2017)
4. Gutiérrez-Pérez, B.M., Martín-García, A.V.: Evaluation of quality in blended learning training. In: Martín-García, A.V. (ed.) Blended Learning: Convergence between Technology and Pedagogy. LNNS, vol. 126, pp. 91–111. Springer, Cham (2020). https://doi.org/10.1007/978-3-030-45781-5_5
5. Hadullo, K., Oboko, R., Omwenga, E.: A model for evaluating e-learning systems quality in higher education in developing countries. Int. J. Educ. Dev. ICT **13**(2), 185 (2017)
6. Jiang, L., Wang, X.: Optimization of online teaching quality evaluation model based on hierarchical PSO-BP neural network. Complexity (2020)
7. Littenberg-Tobias, J., Reich, J.: Evaluating access, quality, and equity in online learning: a case study of a MOOC-based blended professional degree program. Internet High. Educ. **47**, 100759 (2020)
8. Martínez-Caro, E., Cegarra-Navarro, J.G., Cepeda-Carrión, G.: An application of the performance-evaluation model for e-learning quality in higher education. Total Qual. Manag. Bus. Excell. **26**(5–6), 632–647 (2015)
9. Muñoz, C., García-Peñalvo, F.J., Morales, E.M., Conde, M.Á., Seoane, A.M.: Improving learning object quality: moodle HEODAR implementation. Int. J. Dist. Educ. Technol. (IJDET) **10**(4), 1–16 (2012)
10. Ortiz-López, A., Olmos-Migueláñez, S., Sánchez-Prieto, J.C.: e-Learning quality assessment in higher education: a mapping study. In: Eighth International Conference on Technological Ecosystems for Enhancing Multiculturality, pp. 833–838, October 2020
11. Wu, D., Wu, L., Palmer, A., Zhou, P.: Automatic evaluation of online learning interaction content using domain concepts. The Electronic Library (2020)

AI-Based Multilingual Interactive Exam Preparation

Tim Schlippe(✉) and Jörg Sawatzki

IU International University of Applied Sciences, Bad Honnef, Germany
tim.schlippe@iu.org

Abstract. Our previous analysis on 26 languages which represent over 2.9 billion speakers and 8 language families demonstrated that cross-lingual automatic short answer grading allows students to write answers in exams in their native language and graders to rely on the scores of the system [1]. With lower deviations than 14% (0.72 points out of 5 points) on the corpus of the short answer grading data set of the University of North Texas [2], our natural language processing models show better performances compared to the human grader variability (0.75 points, 15%). In this paper we describe our latest analysis of the integration and application of a multilingual model in interactive training programs to optimally prepare students for exams. We present a multilingual interactive conversational artificial intelligence tutoring system for exam preparation. Our approach leverages and combines learning analytics, crowdsourcing and gamification to automatically allow us to evaluate and adapt the system as well as to motivate students and increase their learning experience. In order to have an optimal learning effect and enhance the user experience, we also tackle the challenge of explainability with the help of keyword extraction and highlighting techniques. Our system is based on Telegram since it can be easily integrated into massive open online courses and other online study systems and has already more than 400 million users worldwide [3].

Keywords: Artificial intelligence in education · Cross-lingual short answer grading · Conversational AI · Keyword extraction · Natural language processing

1 Introduction

Access to education is one of people's most important assets and ensuring inclusive and equitable quality education is goal 4 of the United Nation's Sustainable Development Goals [4]. Massive open online courses and other online study opportunities are providing easier access to education for more and more people around the world. However, one big challenge is still the language barrier: Most courses are offered in English, but only 16% of the world population speaks English [5]. To reach the rest of the people with online study opportunities, courses would need to better support more languages. The linguistic challenge is especially evident in written exams, which are usually not provided in the student's native language. To overcome these inequities, we present and analyze a multilingual interactive conversational artificial intelligence (AI) tutoring system for exam preparation (*multilingual exam trainer*).

D. Guralnick et al. (Eds.): TLIC 2021, LNNS 349, pp. 396–408, 2022.
https://doi.org/10.1007/978-3-030-90677-1_38

Our system is based on a Multilingual Bidirectional Encoder Representations from Transformers model (M-BERT) [6] and is able to fairly score free-text answers in 26 languages in a fully-automatic way (*en, ceb, sv, de, fr, nl, ru, it, es, pl, vi, ja, zh, ar, uk, pt, fa, ca, sr, id, no, ko, fi, hu, cs, sh*) [1]. Thus, foreign students have the possibility to write answers in their native language during the exam preparation. Since our multilingual natural language processing (NLP) model has been pre-trained with a total of 104 languages, our exam trainer can be easily extended with new languages.

Figure 1 illustrates the concept of our multilingual exam trainer: Iteratively, an exam question is displayed to the user. The user enters the answer (*student answer*) using our chatbot interface. Then the student answer is processed with two AI models—the *multilingual automatic short answer grading (ASAG) model* and the *keyword matching model*—which deliver quantitative feedback in terms of a *score* and qualitative feedback by displaying the *model answer* and highlighting the *keywords* matching the student answer and the model answer.

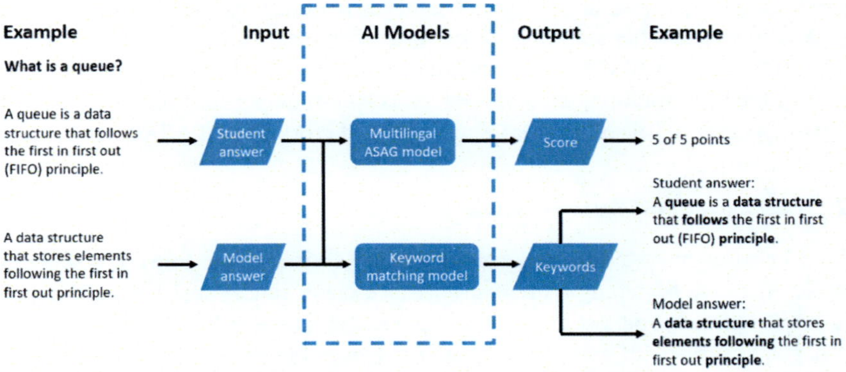

Fig. 1. Concept of the AI-based Interactive Exam Preparation.

To evaluate our approach, we conducted a study where students, former students, and people who enjoy continuing education used our implementation and then completed a questionnaire.

In the next section, we present the latest approaches of other researchers for the components of our multilingual exam trainer. Section 3 describes our specific implementation. Section 4 delineates our experimental setup. Our experiments and results are outlined in Sect. 5. We conclude our work in Sect. 6 and suggest further steps.

2 Related Work

The research area "AI in Education" addresses the application and evaluation of Artificial Intelligence (AI) methods in the context of education and training [7]. One of the main focuses of this research is to analyze and improve teaching and learning processes with NLP models. In the following sections, we describe the use of NLP components in related work for multilingual NLP, ASAG, conversational AI, and keyword extraction to address the challenges of our system.

2.1 Multilingual Natural Language Processing Models

To allow users of our system to answer the exam questions in their native language, we used a multilingual NLP model and adapted it to the task of ASAG. Multilingual NLP models are provided by multiple institutions, e.g., M-BERT [6], RoBERTa [8], or XLM-R [9]. They have the benefit that they can be adapted to a certain task with task-specific labeled text data in 1 or more languages (*transfer learning*) and then perform this learned task in other languages (*cross-lingual transfer*) [10].

To give the users of our system qualitative feedback on their answers, we used M-BERT as the basic multilingual model which is "pre-trained from monolingual corpora in 104 languages" [10] and adapted it to the task of cross-lingual ASAG.

2.2 Automatic Short Answer Grading

ASAG helps us provide feedback on the student answer in the form of a score. The field of ASAG is becoming more relevant since many educational institutions–public and private–already conduct their courses and examinations online [1, 13].

A good overview of approaches in ASAG before the deep learning era is given in [11]. [12] and [13] investigate and compare state-of-the-art deep learning techniques for ASAG. [12] demonstrate that systems based on BERT performed best for English and German and that their multilingual RoBERTa model [8] shows a stronger generalization across languages on English and German.

We extend ASAG to 26 languages and use the smaller M-BERT [10] model to conduct a larger study concerning the cross-lingual transfer [1].

2.3 Conversational AI

For the interaction with the users, we used a conversational AI that takes the input from the users and sends messages based on a dialog flow. The messages of the conversational AI contain the exam question, the student answer score, the model answer with highlighted keywords, information about the progress and motivations.

Conversational assistants in education enable learners to access data and services and exchange information by simulating human-like conversations in the form of a natural language dialogue on a given topic [14]. There are various technologies, frameworks, and services for building a conversational AI, such as Rasa [15], Google Dialogflow [16] or Telegram [17].

Our conversational AI is based on Telegram [17] since it can be easily integrated into massive open online courses and other online study systems and has already more than 400 million users worldwide [3]. However, it can be ported to other chatbot technologies as well. To provide our conversational AI's messages in the students' native languages, we translated them into our 26 languages using Google's Neural Machine Translation System [18]. An overview of the system's BLEU scores over languages is given in [18, 19]. We did not post-correct the translations, as we wanted to check whether our system from scratch already delivers a good user interface in different languages.

2.4 Keyword Extraction and Semantic Similarity

To explain our users the difference between student answer and model answer, we highlight the keywords and their synonyms which are contained in both the student answer and the model answer. This combines two tasks: Keyword extraction and semantic similarity.

Good overviews of automatic keyword and keyphrase extraction are provided in [20] and [21]. A survey of the evolution in semantic similarity is given in [22]. The latest trend for both tasks is to embed the words into a semantic vector space thus working with word embeddings since the semantically similar words are located nearby in vector space.

In our system SpaCy [23] is used to exclude stop words, convert the remaining words into vectors and compute the word similarities.

3 AI-based Interactive Exam Preparation

In this section we describe what components our multilingual exam trainer consists of and how they were implemented.

3.1 Dialog Flow of our Conversational AI

Figure 2 shows the dialog flow of our exam trainer with the following steps:

1. The user activates the conversational AI with the /start command.
2. The conversational AI welcomes the user and presents a list of 26 languages to select from.
3. The conversational AI asks the user a question in the selected language.
4. The user types the answer (any of the 104 languages used in M-BERT is possible).
5. The conversational AI gives feedback in terms of a score and highlights similarities between student and model answer.
6. If the total points collected are equal or greater than THRESHOLD, the goal is reached and the game ends.
7. Otherwise, the user is presented with another student answer that he or she needs to score, considering the given model answer.
8. Proceed with step 3.

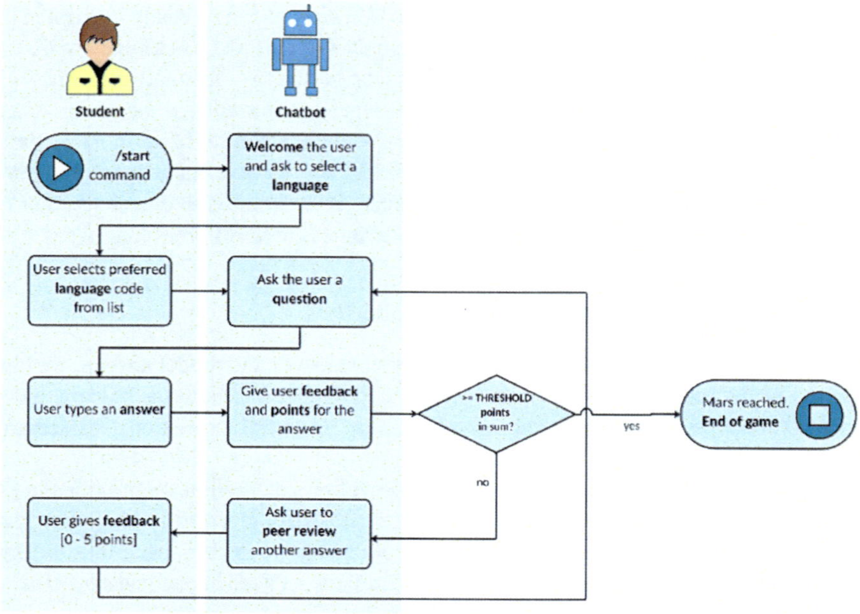

Fig. 2. Dialog Flow of the AI-based Interactive Exam Preparation.

3.2 Gamification and Motivation

Users have the motivation to use our multilingual exam trainer to improve answering open exam questions. However, studies have shown that gamification creates another incentive in learning [24]. To give the users of our system this further incentive, we came up with the following gamification approach: Users are in space and have the goal to fly with their spaceship from Earth to Mars. To get closer to Mars with the spaceship, the users have to answer the displayed exam questions. The points for the answers are converted into kilometers. With better answers, the users get more points and get to Mars faster. Based on the achievement in the student answer, the user is praised and motivated by certain phrases, e.g., "Awesome, that gives us fuel for 3 million more kilometers" and with information of the distance to go. Figure 3 illustrates our gamification in the conversation between a Dutch user and our conversational AI.

3.3 Quantitative Feedback: Multilingual Automatic Short Answer Grading

The AI model which processes the student answers in their native language and delivers the user with quantitative feedback in terms of a score, is based on M-BERT. The model was downloaded and fine-tuned through the Transformers library. We trained 6 epochs with a batch size of 8 using the AdamW optimizer with an initial learning rate of 0.00004. We supplemented each fine-tuned BERT model with a linear regression layer that outputs a prediction of the *score* given an answer. The model expects the model answer and the student answer as input.

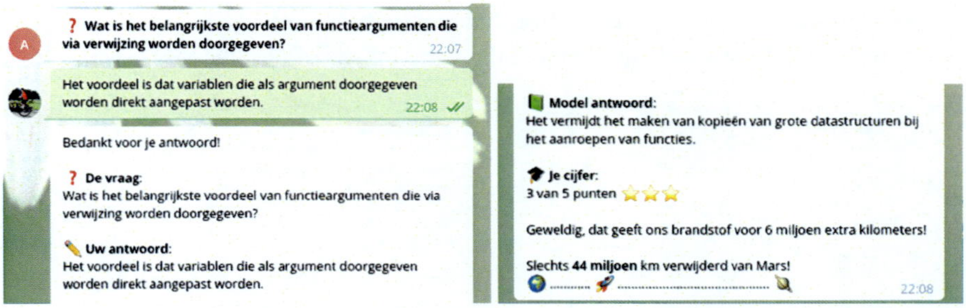

Fig. 3. Conversation with Greeting, Language Selection, Exam Question, Student Answer, Scoring, Model Answer and Motivation.

The ASAG data set of the University of North Texas [2] provided the exam questions, model answers and training data for fine-tuning M-BERT. It contains 87 questions with corresponding model answers and, on average, 28.1 manually graded answers per question about the topic *Data Structures* from undergraduate studies. 80% of the ASAG data set was used for fine-tuning M-BERT. 20% of the data set was used for the exam questions and model answers displayed in our multilingual exam trainer.

After fine-tuning with this original English ASAG training data set, our model would be able to receive a model answer together with a student answer in 1 of the other 103 languages and return a score in terms of points — without the need of fine-tuning with ASAG data in the other languages (*cross-lingual transfer*). However, since we figured out that adding translations of the ASAG data in more languages even improves fine-tuning, we added translations in the 5 languages German, Dutch, Finnish, Japanese, and Chinese [1]. With Mean Absolute Errors between 0.41 and 0.72 points out of 5 points in our analysis of the 26 covered languages, our model has even less discrepancy than the 2 graders which graded the ASAG corpus of the University of North Texas with a discrepancy of 0.75 points [2].

To provide the exam questions and the model answers in our multilingual exam trainer in 26 languages and to get the translations in the 5 listed languages for fine-tuning M-BERT, we used Google's Neural Machine Translation System [18]. Google's Neural Machine Translation System is also used by other researchers who experiment with multilingual NLP models since it comes close to the performance of professional translators [19].

3.4 Qualitative Feedback: Keyword Extraction and Highlighting

Figure 4 shows the keyword highlighting in a snippet of the conversation between the user and the chatbot. For simplicity, we have implemented keyword extraction and highlighting for English only in our prototype. Porting the method to other languages is possible using word vectors and a distance measure.

Our algorithm for keyword extraction and highlighting is shown in Fig. 5. Given are the word tokens of the model answer and the word tokens of the student answer.

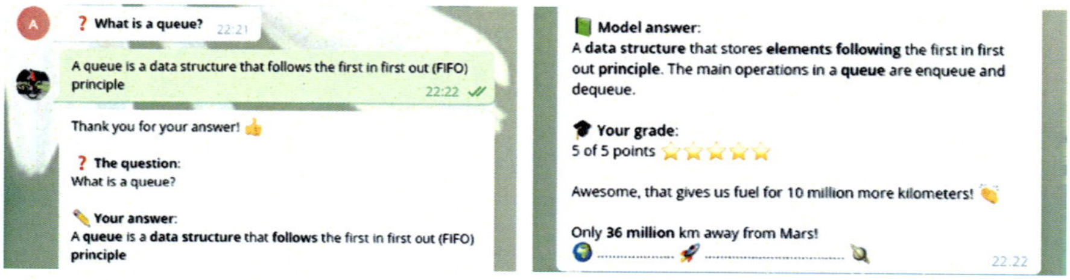

Fig. 4. Conversation with Keyword Highlighting.

```
# Iterate through all tokens in model answer
for model_token in model_answer:
  # Process only tokens not in stop word list and alphanumeric
  if not model_token.is_stop and model_token.is_alpha:
    # Iterate through tokens in the student's answer
    for answer_token in student_answer:
      # If answer token is not a stop word and alphanumeric:
      # Highlight tokens if
      # their vectors' cosine similarity exceeds given threshold
      if not answer_token.is_stop and model_token.is_alpha and \
            model_token.similarity(answer_token) > THRESHOLD:
        highlight(model_token)
        highlight(answer_token)
```

Fig. 5. Algorithm for Keyword Extraction and Highlighting.

We iterate over the word tokens in the model answer and over the word tokens in the student answer and remove stop words and word tokens that do not contain alphanumeric characters. Each remaining word token of the model answer and the student answer is converted into a word vector. Then the word vectors of the model answer are compared with the word vectors of the student answer. If the similarity between two word vectors is lower than a threshold, we consider them as synonyms.

3.5 Crowdsourcing and Peer-Reviewing

In order to continuously improve our multilingual ASAG model with high-quality human labeled training data in a crowdsourcing approach, the user also has the task of scoring another student's answer as part of the game (step 7 in Sect. 3.1). Studies such as [25] have shown that peer-based proofreading is as effective as a professional proofreader. Consequently, the same student answer is demonstrated to different users. This peer-review process makes it possible to detect and filter outliers that would have a negative impact on the model. However, this process also has another advantage for the user: The student deals with the question again, but this time from a different perspective.

4 Experimental Setup

In this section, we describe the structure of our questionnaire and the participants.

4.1 Questionnaire Design

To evaluate our approach, we conducted a study where students, former students, and people who enjoy continuing education first tried our exam trainer and then completed a questionnaire. The study was conducted on a subset of the possible languages and examined 5 different aspects: Learning experience, user experience, motivation, quality of NLP models, and gamification. Our questionnaire contains the following 4 parts:

1. General questions about the scenario of a multilingual interactive conversational AI tutoring system for exam preparation.
2. Specific questions concerning our implementation.
3. Specific questions concerning extensions and improvements.
4. Personal questions (profile and demographic information)

To obtain detailed results, we asked for a score range where it makes sense. The score range follows the rules of a forced choice Likert scale, which ranges from 1 (*strongly disagree*) to 5 (*strongly agree*).

4.2 Participants

51 people from 6 countries filled out our questionnaire, giving us a first impression of the quality and impact of our system. Most were students from the University of Osnabrück, IU International University of Applied Sciences, Karlsruhe Institute of Technology, and Karlsruhe University of Applied Sciences. These people tested our exam trainer in German (64.7%), English (21.6%), Dutch (3.9%), Italian (3.9%), French (3.9%), and Spanish (2.0%).

5 Experiments and Results

As described, our study examined 5 different aspects: Learning experience, user experience, motivation, quality of NLP models, and gamification.

5.1 Learning Experience

Figure 6 shows that participants responded positively to questions about improving the learning experience, meaningfulness, use, and helping fellow students — both in general and for our implementation. The majority also believe that our implementation can accelerate the learning process and that scoring other students' answers is helpful. There is a more divided opinion on the questions whether the exam trainer is good to get familiar with the subject in the native language first, when the actual exam is in English anyway, and whether it can help elderly people to study online. The difference in distribution for the last question about support for elderly people shows that most participants generally rate it as "neutral," while our system scores a bit lower. This feedback plus comments from the participants on this topic lead us to believe that it is possible to optimize such a system in cooperation with elderly people.

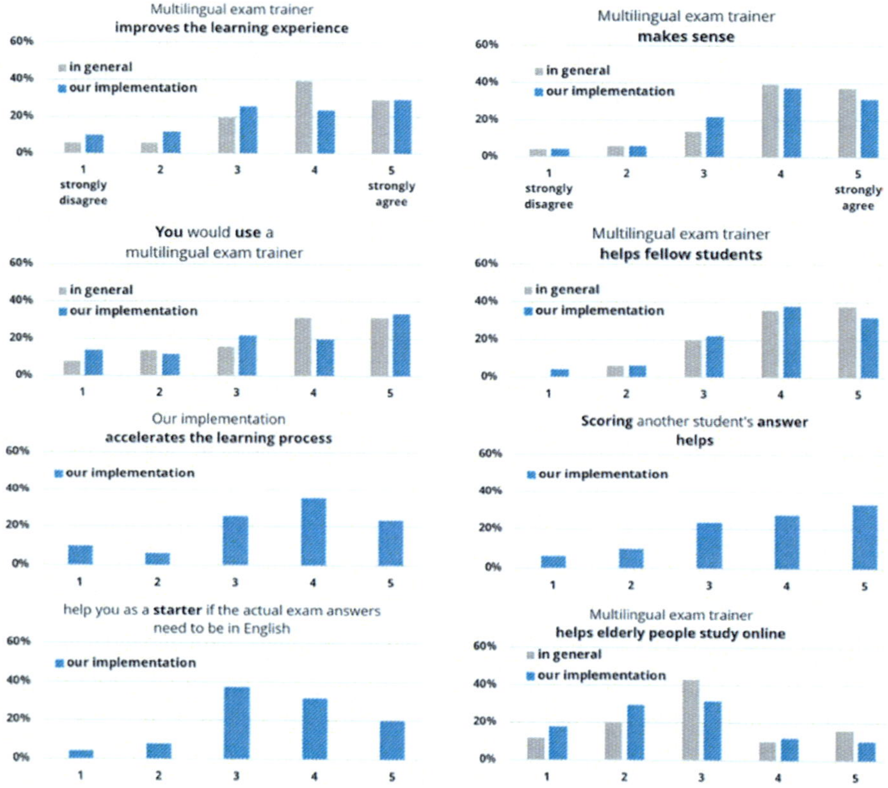

Fig. 6. Learning Experience.

5.2 Motivation

Figure 7 shows a tendency for such an exam trainer in general and for our implementation to motivate people to prepare more for exams.

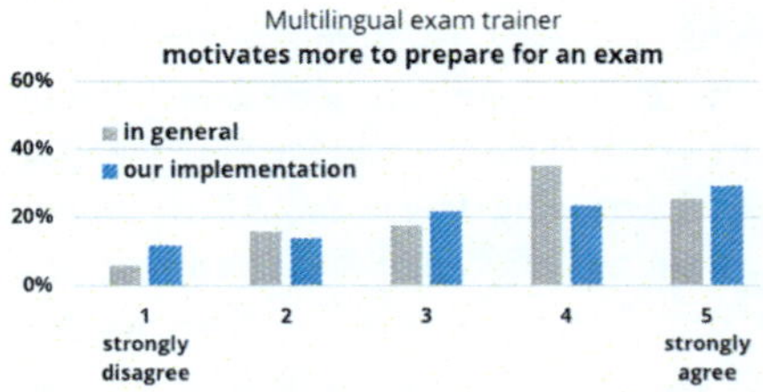

Fig. 7. Motivation.

5.3 User Experience

Figure 8 indicates that the clear majority of participants find that our interface is easy to use and that operating our exam trainer is fun.

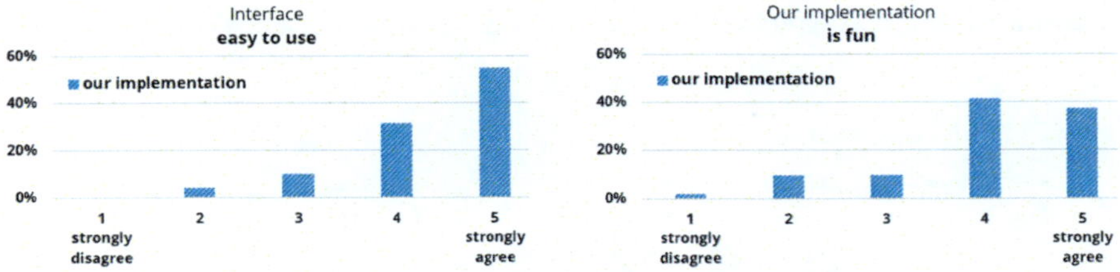

Fig. 8. User Experience.

5.4 Quality of Natural Language Processing Models

Figure 9 shows that the clear majority of participants rate the machine-translated questions as linguistically correct and understandable. This shows that post-correcting the

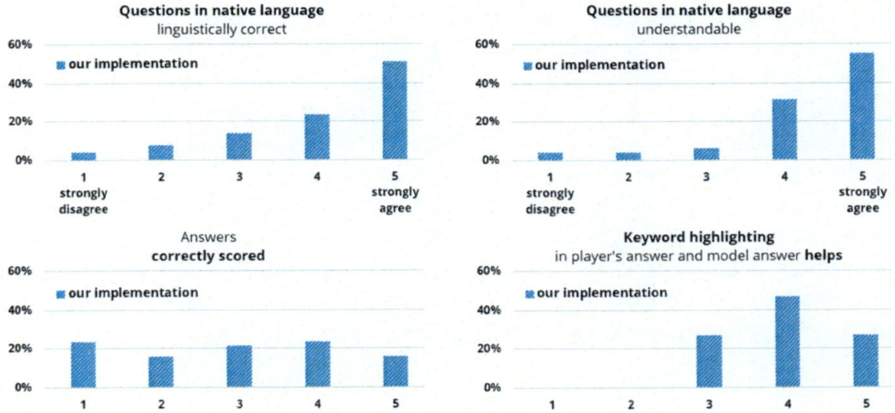

Fig. 9. Quality of NLP models.

translations seems not to be necessary. The scoring with the help of the ASAG was only rated average. Through the users' comments, we learned that many users had randomly entered words as answers and received points for this. This was because these random answers did not appear in the training data of our ASAG model and therefore could not be scored correctly. The training data were taken from exams, where usually no student dares to enter "I don't know" as an answer. Here we see potential for improvement through the evaluation of other students and through simple rules that score such entries with 0 points. Our explainability approach with keyword highlighting was well rated. However, we did not get as much feedback on it because it was only implemented in English.

5.5 Gamification

Figure 10 illustrates the clear majority of participants who like the story and the theme of the game. This demonstrates that even with a simple story — like the trip to Mars — and without special graphical features, a good gamification can be created.

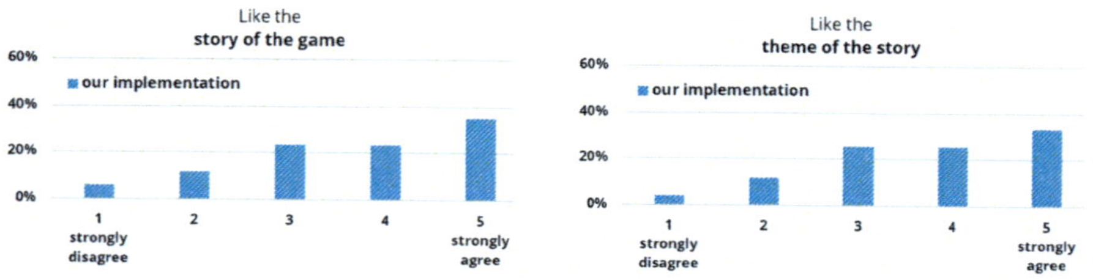

Fig. 10. Gamification.

6 Conclusion and Future Work

We presented a multilingual interactive conversational AI tutoring system for exam preparation that combines multilingual NLP components, ASAG, conversational AI, keyword extraction, learning analytics, crowdsourcing, and gamification. With this multilingual exam trainer, we received positive feedback in a survey regarding learning experience, user experience, motivation, quality of NLP models, and gamification. The results of our survey support our proof-of-concept where users have tested 6 languages so far. Future work may include the extension to other languages. In addition, we would like to further address the issue of explainability to provide even better support to the users of our multilingual exam trainer. To optimize the dialog, it could be investigated how to create a more emotional dialog in written form — e.g., by visualizing voice characteristics and emotions in the textual representation [26, 27].

References

1. Schlippe, T., Sawatzki, J.: Cross-Lingual Automatic Short Answer Grading. In: AIET, Wuhan, China (2021)
2. Mohler, M., Bunescu, R., Mihalcea, R.: Learning to Grade Short Answer Questions using Semantic Similarity Measures and Dependency Graph Alignments. In: ACL-HLT, Portland, Oregon, USA (2011)
3. Statista: Number of monthly active Telegram users worldwide from March 2014 to April 2020 (2021). https://www.statista.com/statistics/234038/telegram-messenger-mau-users
4. United Nations: Sustainable Development Goals: 17 Goals to Transform our World (2021). https://www.un.org/sustainabledevelopment/sustainable-development-goals
5. Statista: The Most Spoken Languages Worldwide in 2019 (2020). https://www.statista.com/statistics/266808/the-most-spoken-languages-worldwide
6. Devlin, J., Chang, M.W., Lee, K., Toutanova, K.: BERT: Pre-training of Deep Bidirectional Transformers for Language Understanding. In: NAACL-HLT, Minneapolis, Minnesota (2019)
7. Libbrecht, P., Declerck, T., Schlippe, T., Mandl, T., Schiffner, D.: NLP for Student and Teacher: Concept for an AI based Information Literacy Tutoring System. In: CIKM. Galway, Ireland (2020)
8. Liu, Y., et al.: RoBERTa: A Robustly Optimized BERT Pretraining Approach (2019). https://arxiv.org/abs/1907.11692
9. Conneau, A., et al.: Unsupervised Cross-lingual Representation Learning at Scale (2018). https://arxiv.org/abs/1911.02116
10. Pires, T., Schlinger, E., Garrette, D.: How Multilingual is Multilingual BERT? In: ACL. Florence, Italy (2019)
11. Burrows, S., Gurevych, I., Stein, B.: The Eras and Trends of Automatic Short Answer Grading. Int. J. Artif. Intell. Educ. **25**(1), 60–117 (2014). https://doi.org/10.1007/s40593-014-0026-8
12. Camus, L., Filighera, A.: Investigating Transformers for Automatic Short Answer Grading. AIED. Cyberspace (2020)
13. Sawatzki, J., Schlippe, T., Benner-Wickner, M.: Deep Learning Techniques for Automatic Short Answer Grading: Predicting Scores for English and German Answers. In: AIET, Wuhan, China (2021)
14. Wölfel, M.: Towards the Automatic Generation of Pedagogical Conversational Agents from Lecture Slides. In: Fu, W., Xu, Y., Wang, S.-H., Zhang, Y. (eds.) Multimedia Technology and Enhanced Learning. Lecture Notes of the Institute for Computer Sciences, Social Informatics and Telecommunications Engineering, vol. 388, pp. 216–229. Springer, Cham (2021). https://doi.org/10.1007/978-3-030-82565-2_18
15. Bocklisch, T., Faulkner, J., Pawlowski, N., Nichol, A.: Rasa: Open Source Language Understanding and Dialogue Management. Cornell University (2017). https://arxiv.org/abs/1712.05181
16. Reyes, R., Garza, D., Garrido, L., De la Cueva, V., Ramirez, J.: Methodology for the Implementation of Virtual Assistants for Education using Google Dialogflow. In: Martínez-Villaseñor, L., Batyrshin, I., Marín-Hernández, A. (eds.) Advances in Soft Computing. Lecture Notes in Computer Science (Lecture Notes in Artificial Intelligence), vol. 11835, pp. 440–451. Springer, Cham (2019). https://doi.org/10.1007/978-3-030-33749-0_35
17. Setiaji, H., Paputungan, I.V.: Design of Telegram Bots for Campus Information Sharing. In: ICITDA. Yogyakarta, Indonesia (2017)
18. Wu, Y., et al.: Google's Neural Machine Translation System: Bridging the Gap between Human and Machine Translation (2016). https://arxiv.org/abs/1609.08144
19. Aiken, M.: An Updated Evaluation of Google Translate Accuracy. Stud. Linguist. Literat. **3**, 253 (2019)

20. Hasan, K.S.: Automatic Keyphrase Extraction: A Survey of the State of the Art. In: ACL. Baltimore, Maryland, USA (2014)
21. Alami Merrouni, Z., Frikh, B., Ouhbi, B.: Automatic Keyphrase Extraction: A Survey and Trends. JIIS **54**, 391–424 (2020)
22. Chandrasekaran, D., Mago, V.: Evolution of Semantic Similarity - A Survey. (2021). https://arxiv.org/abs/2004.13820
23. Honnibal, M., Montani, I. (n.d.). spaCy. https://spacy.io
24. de Sousa Borges, S., Durelli, V.H.S., Reis, H.M., Isotani, S.: A Systematic Mapping on Gamification Applied to Education. In: SAC. New York, NY, USA (2014)
25. Luo, H., Robinson, A., Park, J.-Y.: Peer Grading in a MOOC: Reliability, Validity, and Perceived Effects. Onl. Learn. Off. J. Onl. Learn. Consort. **18**, 1–14 (2014). https://doi.org/10.24059/olj.v18i2.429.
26. Wölfel, M., Schlippe, T., Stitz, A.: Voice Driven Type Design. In: SpeD. Bucharest, Romania (2015)
27. Schlippe, T., Alessai, S., El-Taweel, G., Wölfel, M., Zaghouani, W.: Visualizing Voice Characteristics with Type Design in Closed Captions for Arabic. In: Cyberworlds. Caen, France (2020)

Approaching Adaptive Support for Self-regulated Learning

Niels Seidel(✉), Heike Karolyi, Marc Burchart, and Claudia de Witt

FernUniversität in Hagen, Research Cluster, "Digitalization, Diversity and Lifelong Learning – Consequences for Higher Education" (D^2L^2), Universitätsstr. 11, 58084 Hagen, Germany

{niels.seidel,marc.burchart,claudia.deWitt}@fernuni-hagen.de

Abstract. Students in distance education are expected to self-regulate their learning process. In this paper, we present a semester planning tool that enables learners to plan, monitor, and reflect on learning tasks over a semester. In this way, Zimmerman's model of self-regulated learning (SRL) could be mapped to an application that can be used in the established Moodle LMS. Furthermore, we introduce the Adaptation Rule Interface for implementing rule-based context-sensitive adaptations in Moodle. As a baseline for adaptive SRL support, we conducted two field studies (N = 157 and N = 93). In both studies, monitoring was used most frequently, while planning was mainly done at the beginning of the semester. Participants preferred to adjust provided milestones, instead of creating their own. Reflection played a minor role. The participants who used the tool extensively reported a strong affinity for time management and content elaboration in the LIST-K inventory. Using cluster analysis, participants with high learning activity and high SRL activity were grouped into one cluster, while the majority of participants combined into a second cluster. Through this empirical investigation of SRL processes, it is possible to design effective adaptive support for self-regulated learning.

Keywords: Self-regulated learning · Adaptive learning environments · Learning analytics

1 Introduction

Academic success depends on the ability to organize and structure one's own learning process. Thus, the aspect of self-regulated learning (SRL) is one of the decisive conditions for success, not only in traditional face-to-face studies, but also in online and distance learning. In terms of semester courses, this means considering all learning resources offered, completing exercises, clarifying gaps in understanding, reviewing content, and preparing for the exam. For a course that can span six months, metacognitive demands become a challenge, especially for first-year students. In distance learning, students often seem to get off to a good start in the semester and are dedicated learners until life circumstances or job responsibilities become time competition alongside their studies. At

D. Guralnick et al. (Eds.): TLIC 2021, LNNS 349, pp. 409–424, 2022.
https://doi.org/10.1007/978-3-030-90677-1_39

the FernUniversität in Hagen, this affects about 80 % of students. Necessary prioritization contributes to a loss of focus on the learning process and to course dropouts after an initially good performance. Helping students to be continuously on their path or to quickly get back on track — and, thus, to give them educational opportunities — results in an approach to support the metacognition of the learning process. In order to support students in self-regulation, learning analytics instruments are needed to track learning behavior and progress.

With this paper, we make four contributions to the research about SRL in digital learning environments (LE). (i) We present the semester planning tool for an LMS, which supports learners in resource-based and metacognitive learning tasks, and specifically addresses planning, monitoring, and self-reflection. (ii) We present results of two field studies of distance learning courses over a period of one semester. (iii) We examine and correlate data usage of the planning tool with behavioral data of the LMS and responses on the Learning Strategies of University Students (LIST-K) questionnaire. (iv) Finally, we present a concept and prototype of a rule-based and context-sensitive adaptive learning environment that aims at supporting learners in a LMS on a course-level.

The rest of the article is structured as follows. Section 2 provides an overview of the SRL model and supporting tools that assist student self-regulation, as well as methods on how to trace self-regulating behavior itself and in relation to other learning activities. We also show that adaptive systems have so far hardly addressed SRL. In Sect. 3, we present the design and architecture of a tool that enables learners to plan individual resource-related and activity-related learning tasks over the course of a semester. This tool helps learners to keep track of their learning progress through learning analytics dashboards and to reflect on goal achievements. We also report on the status of our developments of adaptive support for course-based support for SRL. In Sect. 4, we present a field study in which we evaluated the semester planning tool as a baseline for adaptive SRL support in two distance learning courses over the period of one semester. The article concludes in Sect. 5, with a conclusion and outlook.

2 SRL Models and Approaches for Adaptive Learning

2.1 Self-regulated Learning

SRL describes abilities and skills or the proactive action of a person to achieve knowledge based on learning- and experience-dependent (self-) set goals and expectations by making use of resources (e.g., from contexts, strategies, interactions, collaborations, experiences, skills, motivation, and reflection). This includes to independently plan, control, and adapt this process according to reflexively gained knowledge. In the context of distance learning, this adaptation of goals or their fulfillment processing, primarily considers motivational factors, time management, learning strategies, and contextual information [1–6]. Designing adaptations for SRL focuses on supporting awareness to the individual achievement process, performance and providing knowledge about learning strategies in order to foster self-monitoring perspectives for students. Self-monitoring, as well as the influencing characteristic of learning motivation, determine the self-control of the learning process [7, 10]. Models of SRL are oriented either to the process [1, 8, 9] or as a shell or layer structure [3]. The well-known process-based model described by Zimmerman [1], who referred to the work of Bandura [2] and Pintrich [8], are most often used as a basis for SRL research. The process-oriented model by Schmitz and Wiese [9] emerged from learning diary studies. This model differs in its terminology but only slightly in structure and scope. Landmann and Schmitz [10] provide an additional perspective; they describe a systematization of different levels of regulation according to goal, pattern, strategy, and execution.

Early process-oriented models of self-regulation, such as that of Wiener in 1948 [11], followed a basic cybernetic-like pattern. Recent process-based models of self-regulation of Schmitz & Wiese and Zimmerman [4, 9] consider three main action phases in an iterative process. While Zimmerman [4] locates self-monitoring and self-checking in the performance phase, Pintrich [8] attributes more importance to these two process steps and separates them into a four-phase process model. Planning, monitoring, control, and reaction/reflection form the regulatory aspects and are differentiated by the four areas of regulation: cognition, motivation/affect, behavior and context. This matrix results in a 16-part categorization scheme for the phases and domains of self-regulation. Pintrich et al. [12] derived from this matrix the Motivated Strategies for Learning Questionnaire (MSLQ), an instrument for assessing self-regulation in the learning process. For German usage, there is a tested version of this known as the "Learning Strategies of University Students" (LIST) [6], which can also be used as a short version LIST-K [13]. Research on MOOC learners indicates that self-regulation in online learning environments primarily requires support in goal setting, strategic planning, time management, self-evaluation, and note-taking [14, 15].

A number of tools can be found in the literature that assists learners in self-regulating their learning processes like *MetaTutor* [16], *eLDA* [17], *MyLearningMentor* [18], *Video-Mapper* [19], *Learning Tracker* [20] or *Note my progress* [21]. To the best of our knowledge, there is no mature tool that supports students in SRL over longer period of time within an online learning environment like Moodle. Therefore, we focused on a holistic approach that not only focus on micro-events but on a whole-semester course including all virtual course-related resources. It was our aim to create a theory-based system that is able to detect, track, model, and foster learners' SRL activities.

2.2 Adaptive Support

One of the challenges for SRL support is to consider individual learning progress, learning styles, and the potential for SRL support. Our main interest is to support students in their self-regulation process in such a way that they expand their skills. It is not the goal to compensate for partial steps of the SRL or to take over steps. Rather, adaptations should be used to create awareness for metacognitive, cognitive, or resource-related strategies of SRL and to support the self-responsible regulation of the learning process of a learner.

Since students cannot provide information about this on an ongoing basis, data-driven processes are needed. Beheshitha et al. [22], Cicchinelli et al. [23], and Lau et al. [24] examined SRL processes to cluster learner types according to learning strategies [22], to their activity level [23] and to the combination of learning content and goals [24]. Given the extensive body of research on adaptive and personalized LE, we extracted different approaches to design and implement such systems from systematic reviews of literature [25–29] and systems [30, 31]. In addition, systematic literature reviews can be found on specific aspects of adaptive LEs such as the learning style [32, 33] and the learner model [34, 35]. Grubišić et al. specifically address the requirements for adaptive courseware [26]. Only a few reviews mentioned self-regulation as a target of adaptation [25, 34] only in one of the cited SRL publications represented in an adaptive system and empirically investigated. Souki et al. [36] developed a framework that allows students to specify and organize their individual learning path by defining objectives and by capturing and adopting learning behaviors. Three SRL phases, namely planning (forethought), performance control, reflection (self-evaluation) are supported by predefined strategies and by activities adapted to the learning preferences of the learners. Learning preferences are classified into four styles which correspond to four learner types: (i) activist who relies on his gained experiences, (ii) reflector who uses his reflective observations, (ii) theorist whose learning is based on the development of abstract concepts, and (iv) pragmatist who actively conducts experiments.

With the goal set of providing SRL support, the architecture of a system should also integrate and map the learning process in order to concretely capture SRL phases using trace data and to be able to relate them to learning activities, learning performance, and the monitoring of planning and self-reporting. The first step in this process involves identifying how learners plan, monitor, and reflect on their learning activities and how this can be traced in the learning environment. The goal of adaptations that support SRL includes not only awareness of the SRL phases but also support for learning strategies. Adaptations should therefore support learners in the learning environment in the application of concrete learning strategies and contribute to the expansion of the learning strategy repertoire. Approaching the adaptation for SRL support in our project the first sub-goal is therefore to identify groups of learners in order to plan adaptations according to their needs.

3 Semester Planning Tool

Based on Zimmerman's SRL model [1] the semester planning tool was iteratively designed, developed, tested, and evaluated with students in a design-based research process. The tool was developed for Moodle as so-called course format plugin using Vue.js on the client side. All its components (Fig. 4) are presented to the user on the course overview page.

3.1 Design of the Components

Planning of Individual Learning Activities. Planning process: A brief initial survey forms the basis for personalizing an initial semester plan. Based on a student's individual goal of (i) taking the exam in the course, (ii) learning according to personal interests, or (iii) just wanting to get an overview of the course content, a customized template for a semester planning is loaded. This planning takes into account the time of the survey as well as the specified planning rhythms (e.g. weekly, fortnightly, monthly). The planning templates are based on the experiences of the instructors from previous semesters. However, students are encouraged to create a semester plan on their own or to edit the provided plan. A plan is composed of several time-bound milestones. A milestone is preceded by a (self-) defined learning goal, for the achievement of which learning resources and activities from the current Moodle course can be assigned. Further learning resources that are not part of the Moodle course can be added, as well.

Fig. 1. List view of the semester planning using milestones. Each milestone is assigned to activities or resources that learners can check on completion.

Self-Monitoring. The monitoring tool is used to consistently support self-monitoring needs during learning activities throughout the semester and refers to the individual planning and the resource-related learning progress. Therefore, the milestones can be viewed in chronological order, displayed in a list (Fig. 1) and on a timeline (Fig. 2). Both representations show to what extent the resources and activities, assigned to a milestone, have been marked as completed by the learner. Incomplete milestones and those whose due date has been exceeded are highlighted in the planner itself and by badges attached to the resources on the course overview page. Automatic check-off was deliberately not implemented, as some Moodle activities do not allow differentiated completion tracking (e.g. for reading course texts) and student learners should themselves decide to mark the learning activity as completed in relation to their learning objective.

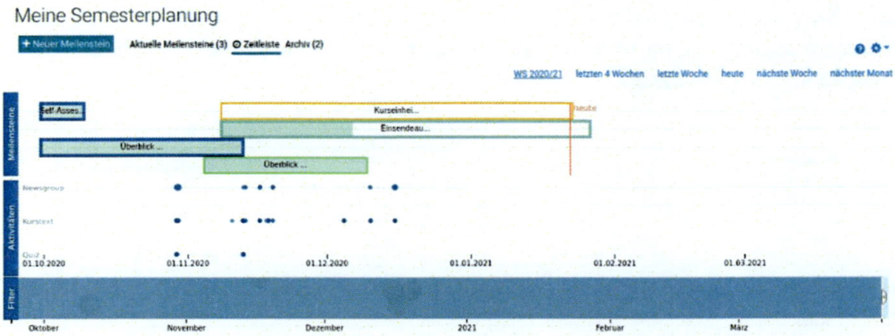

Fig. 2. Timeline view of milestones created since the beginning of the semester and worked on to different extents. The border color of the milestone boxes indicates the status and urgency. A blue border refers to reflected milestones, green borders show completed or ongoing milestones that are still in time, whereas orange borders signal a delay.

In addition to learning progress in relation to individual planning, a learner dashboard provides a resource-based representation of all learning activities offered in the course for each course section (Fig. 3). Each learning activity is represented by a blue box that automatically turns green as soon as the learner achieves the activity completion defined by the course tutors. The activities that have been considered in the individual planning are underlined by a gray bar. This enables learners to monitor their individual learning progress in relation to the course offering and their individual planning. By clicking on a box, a learner can navigate directly to the learning activity and start working.

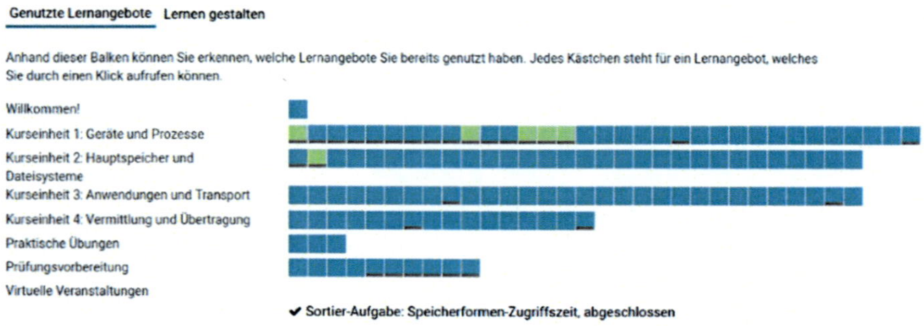

Fig. 3. Learning progress in relation to the completed activities offered in the course.

Reflection. The reflection of the learning process is supported by two instruments: firstly, by information and guidance on learning strategies and, secondly, by a guided self-reflection using reflection questions. In the learning environment, this becomes traceable by accessing and processing learning resources. On the basis of the trace data, only imprecise conclusions can be drawn about the learning strategies used. Nevertheless, an offer was created in the learner dashboard precisely for self-monitoring during the work phases in the Moodle course in order to reflect on or classify one's own learning strategies and, if necessary, to be made aware of alternative strategies. Crucial to this is knowledge of the sub-processes of self-monitoring and self-observation. According to Pintrich (2002), self-observation aims at the improvement of used techniques as well as at the progress of goal achievement. Thus, accessing information on the learning strategies in conjunction with the use of other resources can provide insight on how to regulate learning processes.

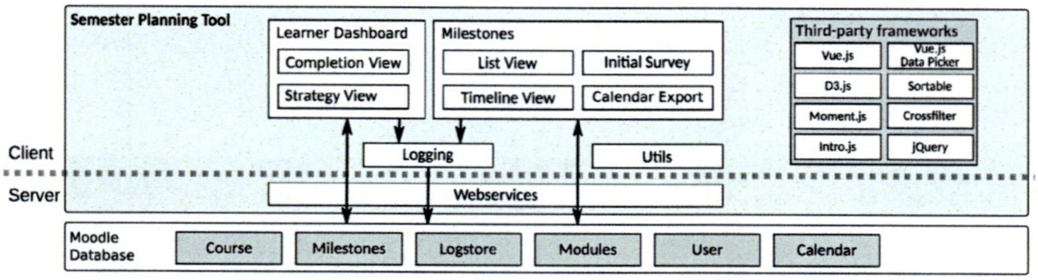

Fig. 4. System architecture of the semester planning tool.

After the completion of a milestone the learner undertakes a self-assessment through a guided self-reflection based on four questions. Self-reflection questions the learner about his or her performance through self-evaluation and causal attribution. Depending on the outcome of the reflection process, a reaction then occurs with affective, self-satisfaction expressions, such as a sense of pride positively reinforcing self-efficacy beliefs and accompanied by a corresponding adaptive response; this primarily affects self-motivational beliefs [1]. In order to support the reflection process, we related the goal achievement with the appropriateness of planning, used resources, and learning strategies. As part of the reflection process students are encouraged to review achievements related to milestones considering the individually applied cognitive learning strategies. This review aims to formulate a cue to improve SRL for the following milestone. Thus, a transfer of knowledge from self-regulation experiences from one cycle to the following process cycle is guaranteed. Technically, this is supported adaptively by displaying the information from the reflection directly in a memo field next to the information about the learning strategy. Thus, a learner can look up a suitable learning strategy for her or his hint.

3.2 Adaptive SRL Support

Regarding adaptive personalized learning environments, we distinguish six gradual scopes of adaptive support. The technically simplest form is limited to a single learning activity (e.g., computer-based adaptive tests) represented by one component of the online learning environment. Adaptive systems that include multiple learning activities represent the second level. In the third stage, learners experience adaptive support for all activities within a course of the online LE. Levels four through six span multiple courses, single degree programs, or even multiple consecutive or parallel degree programs. However, since the semester planning tool refers to an entire course and includes all activities offered in the course, the adaptations in this regard are classified at level three. In the following, we present Adaptation Rule Interface (ARI), a technical solution for adaptive support at the course level.

ARI is a mostly client-side implemented interface between the trace data and learner model on the one side and the perceivable adaptations in the browser view on the other side. It contains adaptations that a learner experiences visually and interactively. ARI is used to check the conditions that must be met for an adaptation rule to execute, using actual data from the learner model (Fig. 5), and to determine and initiate specific actors (Actors, Fig. 5), taking into account contextual information from the user session (Sensors, Fig. 5). Which input parameters are linked to which outputs is defined in adaptation rules (Rule Set, Fig. 5).

The definition of adaptation rules requires deep didactic knowledge for the definition of relevant metrics and the precise formulation of directives. The adaptation rules were initially formulated in natural language using fill-in parameters, independent of any specific subject or domain:

In a certain <situation> in a <period of time> characterized by the <key indicator> which is <like> a <value>, support the learner on the <adaptation level>, so that in

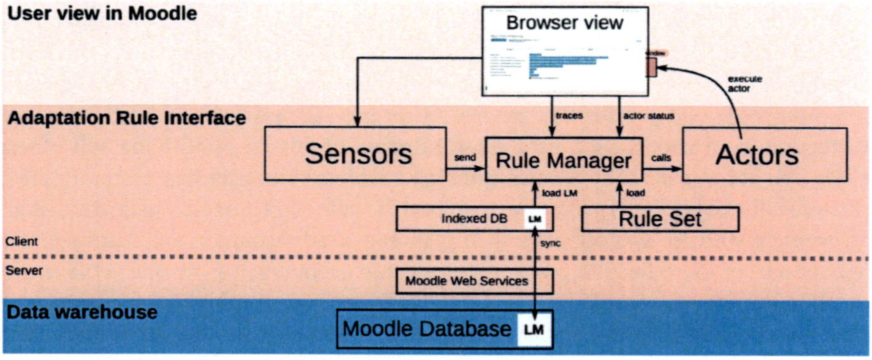

Fig. 5. ARI architecture.

the <area> the <action> is performed providing <information>. The elements are defined as follows:

- <situation> Description of the situation in which adaptive support becomes necessary
- <key indicator> A measurement contained in the learner model
- <period of time> Period to which the observation of the indicator refers
- <like> Operators like greater, less, equal, or unequal
- <value> A numeric or nominal value
- <adaptation level> Monitoring, Awareness, Recommendation, Intervention
- <area> Area of the learning environment where the adaptation should take place: user interface (e.g., emphasize certain materials), learning resources (e.g. substitute a quiz), or instructions (e.g., issue learning advice)
- <action> Type of action supported by ARI (e.g., popover notes, modal dialogs, badges, system notifications, chat messages, warning alerts, confirm dialogs, prompts)
- <information> Content or style information to be used by the action

The sensors determine, for example, the page currently being viewed in Moodle, the position on this page, or the respective user activity or inactivity. Actors can access Moodle-specific actions such as system notifications, messages, or even modal dialogs and provide them with specific information. Actors can also manipulate the layout of the page by specifying CSS, for example, to highlight elements or change their position and order.

Sensors and actors, as well as the rule set, and the learner model are designed modularly and can therefore be flexibly extended. The reaction and response behavior of the learners to the system's interventions are continuously captured and incorporated into the decision to trigger subsequent adaptations with the help of a reinforcement learning model. In this way, the adaptations desired by the user and provided by a so-called agent can be favored by rewards and the adaptations perceived as disturbing in the respective situation can be avoided (loss) in the future. The reinforcement learning agent is part of the Rule Manager (Fig. 5). As an example, to support monitoring, initial adaptive prompts have been implemented using ARI to alert learners about milestones that are due soon by a popover message, if it is unlikely to complete all assigned activities on time based on learning behavior demonstrated to date. In case the learner ignores this less intrusive notification more visually striking markers could be used (e.g., by using a signal color). If even these hints remain unnoticed and additional deviations are detected, more intrusive notifications such as modal windows or emails come into consideration of the Rule Manager.

With the knowledge of the technical possibilities to implement adaptations within a course, it is possible to define concrete use cases as adaptation rules. In this way, learners can be encouraged to use the semester planning tool in appropriate situations. In addition, the SRL phases can be addressed, such as planning for individual time management with regard to the division of learning material and adherence to deadlines. In monitoring, critical events or values can be highlighted and actions can be stimulated. With regard to reflection, recommendations for situation-dependent learning strategies can be based on behavioral data such as reading speed. In order to identify situations in which adaptation makes sense and to explore the field of possible starting points for adaptations, empirical baseline is needed.

4 Study

The aim of the study was to approach potentials for adapting SRL. Therefore, we captured the metacognitive behavior during SRL phases as a baseline using trace data that relate to learning activities and learning performances, as well as the observation of planning and self-reports. Hereby, we defined three research questions: RQ1 – How do learners use the tools to plan, monitor, and reflect on their learning activities? With this question, we want to identify relevant SRL support instruments. RQ2 – What is the relationship between tool usage and the learning strategies of students in higher education? This question helps us to find indicators that could moderate SRL activities enabled by the tool. RQ3 – What groups of learners can be identified in terms of course activities and tool usage? This question aims at differentiating groups with different needs regarding SRL support.

4.1 Methods

Participants and Design. The study was conducted in the compulsory course Operating Systems and Computer Networks of the distance learning B.Sc. Computer Science (hereafter referred to as CS course) and the Module 1D of the B.A. Educational Science (hereafter referred to as Edu course), which consists of three courses, in the winter semester 2020/2021. For each teaching course, a supplementary course was set up in a Moodle learning environment. The use of the learning environment was voluntary but was conditional on a two-step consent to use the platform and to participate in the study. The second informed consent could be withdrawn or granted again at any time, while the first consent was required for GDPR compliance. As an incentive for the student‘s participation, additional exercises such as self-tests, self-assessments, and assignments were offered, as were additional tools for semester planning and for reading the digital course texts. These differences in the learning offer are comparable to different didactic offers of tutors in face-to-face teaching. In any case, there are no disadvantages for the learners for the examination, since the course texts provided to all learners form the basis of the examination.

In total, 180 of the 534 CS course participants agreed to take part in the study and to use the Moodle instance set up for this purpose. By the end of the semester, the same number of active participants had been recorded. A total of 93 (52 %) of the 180 Moodle users participated in the survey at the beginning of the semester. These participants were between 21 and 60 years old (median 37 years). The sex of the participants was: 68 male, 24 female, and 1 diverse. In total, 80 participants have completed or started a university study program before; 29 are studying for the first time.

A total of 314 of the 563 students enrolled in the Edu course agreed to take part in the study and used the Moodle instance set up for this purpose. A total of 157 (50 %) of the users of the Moodle course took part in the survey at the beginning of the semester. These participants were between 21 and 61 years old (median 42 years): 138 female, 18 male, and 1 diverse participated. In total, 151 participants have completed or started a university study program earlier, just 6 are studying for the first time.

Material. Both Moodle courses offer learners comprehensive course texts (>150 pages) in a digital reading tool as well as various forms of self-testing tasks that can be assessed automatically by the system or by the learners themselves with the help of assessment criteria [37]. In the CS course, additional submission tasks are offered, and they are graded by reviewers. In the Edu course, comparable tasks are discussed in various discussion forums. In addition to the embedded course texts, further interactive supplementary material, such as videos and podcasts, links to further resources, riddles and interactive learning resources are available. The calendar provides the course-specific online lectures, seminars, and consultation hours. The semester planning tool appears at the top of the course overview page.

Data Collection. At the beginning of the semester, the students were mandatory surveyed about their socio-demographic background, and prior knowledge, as well as their university entrance qualification and past study programs. Part of the survey was also the short scale inventory "Learning Strategies of University Students" (LIST-K), which differentiates cognitive and metacognitive strategies as well as strategies of managing internal and external resources [13]. The LIST-K is composed of 39 items, grouped into 13 subscales with 3 items each. The students gave responses on a Likert-type scale, from 1 (very rarely) to 5 (very often). User interactions and user inputs within the Moodle environment have been captured in the database, especially in the standard log store. Customized Moodle plugins such as the semester planning tool and a plugin that supports the reading of long texts allowed more detailed and data-intensive logging capabilities.

Procedure. Both courses, and thus the field studies, began at approximately the same time on October 1, 2020. Between the 20th and 30th day, the survey had to be completed. Students were free to choose when to start with the course and when to engage with which tasks. The course ended after 182 days, and only data from the first 161 days could be considered for the following analysis. Results of the oral exam in the CS course are not expected until the course of the next six months; the written exam in the Edu course has not been graded at present. Due to this still incomplete data, we present only preliminary results.

4.2 Preliminary Results and Discussion

Survey Analysis. We grouped the responses of the LIST-K as proposed in the original paper [13]. The reliability of these constructs was validated with Cronbach's alpha, after inverting scores in reversed questions. In the Edu course, 5 items had to be excluded because of poor reliability with values below 0.6 for Cronbach's alpha. The remaining scores (Cronbach's alpha values M = 0.78, SD = 0.09 indicating almost good reliability) were averaged for each construct. In the CS-Course, one item had to be excluded. The remaining scores (Cronbach's alpha values M = 0.77, SD = 0.09 indicating almost good reliability) were averaged for each construct.

Usage of the Semester Planning Tool (RQ1). In preparation, the accumulated trace data events were assigned to the SRL phases planning (e.g., milestone_updated, milestone_resource_added), monitoring (e.g., dashboard_completion_item_hover), and reflection (e.g., reflection_completed, dashboard_strategy_strategy_click). In the Edu course 143 students made use of the tool. Figure 6 (right) shows the number of students using the tool over the course of the semester with regard to the SRL phases. Monitoring activities predominate and confirm the relevance of the learning analytics dashboards [38]. Planning activities were performed by nearly one-third of the students at the beginning of the semester, but this number decreased in the middle of the semester. The calendar weeks 51 and 52 in 2020, as well as the first week in 2021, showed a decrease in all activities due to holidays and vacation time. However, the participants preferred updating the provided milestone plan (350 times) compared to the creation of a new milestone (27 times). The opportunity to reflect on completed milestones with the help of a short structured questionnaire was used a total of only 71 times. Overall, the majority of activities in the reflection phase derive from retrieving information about learning strategies.

A total of 77 participants in the CS course made use of the semester planning tool. As shown in Fig. 6, left. across all weeks, most students prefer the monitoring features compared to planning activities that were only used by 14–19 students in the first 5 weeks. Only 5 persons created and 29 participants updated an existing milestone. The reflection questions were answered by 13 participants, while as many as 15 took advantage of the information on learning strategies.

Correlations[1] between Learning Strategies and Semester Planning Tool Usage (RQ2). The responses to the LIST-K with at least acceptable values for Cronbach's alpha have been examined for correlations with the activity measures related to the SRL phases and milestone usage as described in Table 1. Milestone updates and cumulative SRL activities correlate positively with statements about personal time management in the LIST-K (updates 0.19*, planning 0.19*, monitoring 0.18*, reflection 0.19*). Thus, the tool's offerings appeal to students who plan and also limit their learning time.

[1] Significance levels: *$p < 0.05$, **$p < 0.01$.

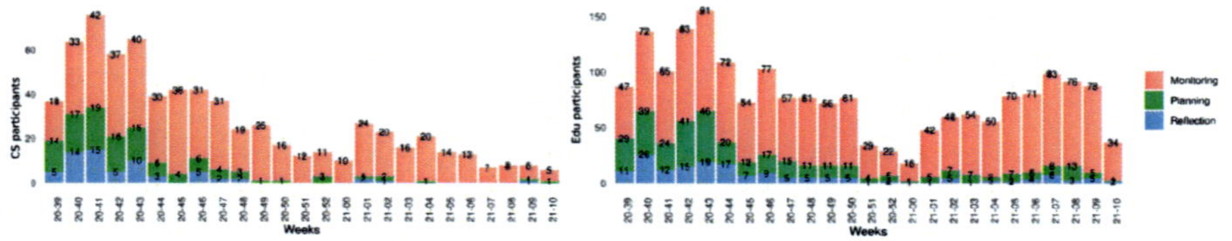

Fig. 6. Number of CS course students (left) and Ed course students (right) involved in the three SRL phases in the course of the semester.

Updating milestones also correlates positively (0.19*) with control strategies for closing knowledge gaps and the desire to test oneself.

In the CS course milestone creation correlated positively with the willingness to make an effort (0.21*) and the preparation of the learning workplace (0.22*). Monitoring correlates with the ability to concentrate (0.22*). Multiple correlations were found between elaboration ability and monitoring (0.24*), reflection (0.29**) as well as the creation (0.23*) and editing (0.22*) of milestones.

Clustering by Course Activities and Tool Usage (RQ3). Considering the first three attribute categories in Table 1 for cluster analysis using the K-Means algorithm we determined two clusters by calculating the silhouette measure (highest average silhouette width 0.76) for the Edu course. In total, 143 participants were assigned to the first and 14 to the second cluster. Both clusters are clearly distinguished across the 13 attributes. Participants in the first cluster are characterized by comparatively low usage of the semester planning tool (factor 2 to 6 lower) and less course-related activity (factor 1.6 to 8 lower). Across both clusters, only very few milestones have been created. Both clusters

Table 1. Learner model consisting of collected attributes and measures of learning behavior and survey responses.

Category	Attributes	Description
SRL phases	phase_planning, phase_monitoring, phase_reflection	Number of events assigned to each of the SRL phases aggregated from Moodle log store
Milestones	ms_created, ms_updated, ms_reflected, ms_resource_added	Number of events related to the use of milestone in the semester planning tool captured in the Moodle log store
Course activities	forum_create, forum_viewed, quiz_submitted, page_reading[2]	Number of selected course activities captured in the Moodle log store
Socio-demographic data	age, sex, university_entrance_qualification, part_time_study, profession	Survey responses about socio-demographic facts

[2] The page reading measure has been calculated by the number of individual hours in which at least one page scrolling event occurred in the log store.

showed considerable monitoring activities, although participants in cluster 2 used it 3.6 times more often. The CS course also shows two clusters (silhouette score 0.82). Ten participants were assigned to the first cluster and 83 to the second. Here, too, it can be seen that a small group of students who plan, monitor, and reflect on their learning more often read course texts (5.3 times more) and complete assessment tasks (4.2 times more).

5 Conclusion and Outlook

In this paper, we presented a Semester Planning Tool that enables learners in a Moodle course to plan resource-related and activity-related learning tasks over a semester in an individual way, to track their learning progress, and reflection on goal achievements. By this means Zimmerman's SRL model [1] could be mapped on an application to be used in widely used LMS.

Conducting field studies with courses in educational science (N = 157) and computer science (N = 93) could answer RQ1 regarding the usage of the tool regarding planning, monitoring, and reflecting activities. In both courses, monitoring was used most frequently, while planning was mainly done at the beginning of the semester. Reflection seems to be less frequent but requires more attention in the future, as it is the most important step of a successful SRL iteration. Participants preferred to adapt provided milestone plans. CS students were reluctant to use the tool. RQ2 aimed at the relation between tool usage and the self-reports about the LIST-K inventory. Positive correlations were found here, but these do not give a consistent picture across both courses. Moreover, several items of the LIST-K had to be excluded. However, items regarding time management and the ability to elaborate could be used as predictors for distinct SRL activities. In terms of RQ3, we found similar clusters of students in both courses. The first cluster contains the majority of students who made little use of the SRL support provided by the tool, but they were less active in the course as well. The second smaller cluster assembled participants who used the tool more intensively during their comparatively high rate of course activities. It remains an open question whether the tool has had a positive effect on learning activity or whether mainly the high-performing student can be attracted by tools to further improve their SRL capabilities.

However, these results show that the support of self-regulation in distance learning needs to address different types of learners in order to personalize assisting instruments. So far, it has only been possible to differentiate students according to the amount of activity in their courses. This distinction does not appear to be sufficient [36]. Across both courses, the learner dashboards gained a lot of attention and should therefore be used as an anchor for adaptively supporting SRL.

With the presented adaptation architecture and the learner model, we laid a foundation for an adaptive context-sensitive rule engine that provides guidance through awareness cues, hints, and instructions for better self-regulation.

The combination of self-reports and learning analytics data about SRL learning processes opens up new ways to investigate SRL processes in online and distance learning. A better understanding of SRL subsequently enables the design and implementation of personalized support for learning activities.

Acknowledgments. This research was supported by the Research Cluster "Digitalization, Diversity and Lifelong Learning – Consequences for Higher Education" (D^2L^2) of the FernUniversität in Hagen, Germany.

References

1. Zimmerman, B.: Self-efficacy and educational development. In: Self-Efficacy in Changing Societies, Bandura, A. (Hrsg.). Cambridge University Press, Cambridge, New York (1999)
2. Bandura, A.: Social cognitive theory of selfregulation. Organ. Behav. Human Decis. Process. **50**, 248–287 (1991)
3. Boekaerts, M.: Selfregulated learning: where we are today – theory, research, and practice. Int. J. Educ. Res. **31**(6), 445–457 (1999)
4. Zimmerman, B.J.: Attaining self-regulation. In: Handbook of Self-Regulation, pp. 13–39 (2000)
5. Zimmerman, B.J.: Theories of self-regulated learning and academic achievement: an overview and analysis. In: Zimmerman, B.J., Schunk, D.H. (Eds.). Self-Regulated Learning and Academic Achievement (2001)
6. Schiefele, U., Wild, K.P.: Lernstrategien im Studium: Ergebnisse zur Faktorenstruktur und Reliabilität eines neuen Fragebogens. Zeitsch. Different. Diagnost. Psychol. Jahrgang **15**(4), 185–200 (1994)
7. Schiefele, U., Pekrun, R.: Psychologische Modelle des fremdgesteuerten und selbstgesteuerten Lernens (1997)
8. Pintrich, P.: A conceptual framework for assessing motivation and self-regulated learning in college students. Educ. Psychol. Rev. **16**(4), 385–407 (2004)
9. Schmitz, B., Wiese, B.: Eine Prozessstudie selbstregulierten Lernverhaltens im Kontext aktueller emotionaler und motivationaler Faktoren. Zeitsch. Entwicklungspsychol. Pädagogis. Psychol. Jahrgang **31**(4), 157–170 (1999)
10. Landmann, M., Schmitz, B.: Welche Rolle spielt Self-Monitoring bei der Selbstregulation und wie kann man mit Hilfe von Tagebüchern die Selbstregulation fördern? In: Gläser Zikuda, M., Hascher, T. (Hrsg.). Lernprozesse dokumentieren, reflektieren und beurteilen. Lerntagebuch & Portfolio in Forschung und Praxis, pp. 149–169. Klinkhardt Verlag, Bad Heilbrunn (2007).
11. Wiener, N.: Cybercenetics: Control and Communication in the Animal and the Machine. MIT Press, Cambridge (1948)
12. Pintrich, P.R., Smith, D.A.F., García, T., McKeachie, W.J.: A manual for the use of the motivated strategies questionnaire (MSLQ). University of Michigan, National Center for Research to Improve Postsecondary Teaching and Learning, Ann Arbor (1991)
13. Klingsieck, K.B.: Kurz und knapp – die Kurzskala des Fragebogens "Lernstrategien im Studium" (LIST). Zeitsch. Pädagogis. Psychol. **32**(4), 249–259 (2018)
14. Milligan, C., Littlejohn, A.: How health professionals regulate their learning in massive open online courses. Internet High. Educ. **31**, 113–121 (2016)
15. Veletsianos, G., Reich, J., Pasquini, L.: The life between big data log events learners, strategies to overcome challenges in MOOCs. AERA Open **2**(3), 1–10 (2016)
16. Azevedo, R., et al.: MetaTutor: An Intelligent Multi-Agent Tutoring System Designed to Detect, Track, Model, and Foster Self-Regulated Learning (2012).
17. Onah, D.F.O., Sinclair, J.E.: A Multi-dimensional Investigation of Selfregulated Learning in a Blended Classroom Context: A Case Study on eLDa MOOC. Proceedings of the Advances in Intelligent Systems and Computing, pp. 63–85 (2017)

18. Alario-Hoyos, C., Estévez-Ayres, I., Sanagustín, M.P., Leony, D., Kloos, C.D.: MyLearningMentor: a mobile app to support learners participating in MOOCs. J. Univ. Comput. Sci. **21**(5), 735–753 (2015)
19. Fahmy, A., Chatti, M., Narek, D., Hendrik, T., Ulrik, S.: Video - mapper a video annotation tool to support collaborative learning. In: Proceedings of the 3rd EMOOCs, pp. 131–140 (2015)
20. Davis, D., Chen, G., Jivet, I., Hauff, C., Houben, G.: Encouraging metacognition & self-regulation in MOOCs through increased learner feedback. In: Proceedings of the CEUR Workshop, LAK 2016, pp. 17–22 (2016)
21. Pérez-Álvarez, R., Pérez-Sanagustín, M., Maldonado-Mahauad, J.J.: NoteMyProgress: supporting learners' self-regulated strategies in MOOCs. In: Lavoué, É., Drachsler, H., Verbert, K., Broisin, J., Pérez-Sanagustín, M. (eds.) EC-TEL 2017. LNCS, vol. 10474, pp. 517–520. Springer, Cham (2017). https://doi.org/10.1007/978-3-319-66610-5_53
22. Beheshitha, S.S., Gašević, D., Hatala, M.: A process mining approach to linking the study of aptitude and event facets of self-regulated learning. In: Proceedings of the Fifth International Conference on Learning Analytics & Knowledge, pp. 265–269. ACM, New York (2015)
23. Cicchinelli, A., et al.: Finding traces of self-regulated learning in activity streams. In: Proceedings of the 8th International Conference on Learning Analytics & Knowledge, pp. 191–200. ACM, New York (2018)
24. Lau, C., Sinclair, J., Taub, M., Azevedo, R., Jang, E.E.: Transitioning self-regulated learning profiles in hypermedia-learning environments. In: Proceedings of the Seventh International Learning Analytics & Knowledge Conference, pp. 198–202. ACM, New York (2017)
25. Vandewaetere, M., Desmet, P., Clarebout, G.: The contribution of learner characteristics in development of computer-based adaptive learning environments. Comput. Hum. Behav. **2**(27), 118–130 (2011)
26. Grubišić, A., Stankov, S., Žitko, B.: Adaptive courseware: a literature review. J. Univ. Comput. Sci. **21**(9), 1168–1209 (2015)
27. Xie, H., Chu, H.-C., Hwang, G.-J., Wang, C.-C.: Trends and development in technology-enhanced adaptive/personalized learning: a systematic review of journal publications from 2007 to 2017. Comput. Educ. **140**, 103599 (2019)
28. Talaghzi, J., Bennane, A., Himmi, M.M., Bellafkih, M., Benomar, A.: Online adaptive learning: a review of literature. In: Proceedings of the 13th International Conference on Intelligent Systems: Theories and Applications. Association for Computing Machinery, New York (2020)
29. Martin, F., Chen, Y., Moore, R.L., Westine, C.D.: Systematic review of adaptive learning research designs, context, strategies, and technologies from 2009 to 2018. Educ. Tech. Res. Dev. **68**(4), 1903–1929 (2020)
30. Newman, A., Gates, B., Fleming, B., Sarkisian, L.: Learning to Adapt 2.0: The Evolution of Adaptive Learning in Higher Education (2016). https://tytonpartners.com/library/learning-to-adapt-2-0-the-evolution-of-adaptive-learning-in-higher-education/
31. Osadcha, K., Osadchyi, V., Semerikov, S., Chemerys, H., Chorna, A.: The review of the adaptive learning systems for the formation of individual educational trajectory. In: CEUR Workshop Proceedings, 2016, vol. 2732, pp. 547–558 (2020)
32. Akbulut, Y., Cardak, C.S.: Adaptive educational hypermedia accommodating learning styles: a content analysis of publications from 2000 to 2011. Comput. Educ. **58**(2), 835–842 (2012)
33. Kumar, A., Singh, N., Ahuja, N.J.: Learning styles based adaptive intelligent tutoring systems: document analysis of articles published between 2001 and 2016. Int. J. Cognit. Res. Sci. Eng. Educ. (IJCRSEE) **5**, 83–98 (2017)
34. Afini Normadhi, N.B., Shuib, L., Md Nasir, H.N., Bimba, A., Idris, N., Balakrishnan, V.: Identification of personal traits in adaptive learning environment: systematic literature review. Comput. Educ. **130**, 168–190 (2019)

35. Nakic, J., Granic, A., Glavinic, V.: Anatomy of student models in adaptive learning systems: a systematic literature review of individual differences from 2001 to 2013. J. Educ. Comput. Res. **51**(4), 459–489 (2015)
36. Souki, A.M., Paraskeva, F., Alexiou, A., Papanikolaou, K.A.: Developing personalised e-courses: tailoring students' learning preferences to a model of self-regulated learning. Int. J. Learn. Technol. **10**(3), 188 (2015)
37. Haake, J.M., Seidel, N., Karolyi, H., Ma, L.: Self-Assessment mit High-Information Feedback. In: Zender, R., Ifenthaler, D., Leonhardt, T., Schumacher, C. (Hrsg.), DELFI 2020 – Die 18. Fachtagung Bildungstechnologien der Gesellschaft für Informatik e.V., pp. 145–150. Gesellschaft für Informatik e.V., Bonn (2020)
38. Schwendimann, B., et al.: Perceiving learning at a glance: a systematic literature review of learning dashboard research. IEEE Trans. Learn. Technol. **99**, 1 (2016)

Towards Personalized, Dialogue-Based System Supported Learning for MOOCs

Theresa Elfriede Isa Zobel(✉) and Christoph Meinel

Hasso Plattner Institute, Potsdam, Germany
{theresa.zobel,christoph.meinel}@hpi.de

Abstract. Since the start of the pandemic in March 2020 at the latest, e-learning has become a key factor. Many students and teachers had to shift from a traditional campus to online platforms and learn new skills in a very short time. Even before the pandemic, lack of technical skills, poor hardware, and unstable internet connections were not the only limitations of e-learning portals, such as Massive Open Online Course (MOOC) platforms. Many students complain about a lack of self-discipline and self-motivation when learning online. They often feel too stressed, which leads to more frustration. Lack of social interaction between students or the teacher, as well as lack of body language, complicate educational activities. Participants cannot ask teachers for assistance as easily as they can with traditional in-class learning. Alternatives to traditional campus facilities, such as libraries or career counseling, are also limited on e-learning platforms. To reduce or even solve the limitations of MOOC platforms, this paper presents a concept for personalized, automated dialogue-based systems. These systems will have various integrated scaffolding tools to encourage the users in their learning activities. Besides a library function and a quiz mode, the chatbot answers frequently asked questions, recommends tutorials, and helps the users with their learning paths. An integrated to-do list with reminders is designed to help students become more organized and complete coursework on time. Ultimately, such efforts to create more personalized e-learning tools and interactions could reduce drop-out rates and increase educational achievements and satisfaction among users of MOOC platforms.

Keywords: Chatbot · MOOC · Scaffolding tools

1 Introduction

E-learning has become an important educational method worldwide since the start of the pandemic in March 2020, at the latest. E-learning has several advantages over traditional classroom learning. This includes, most importantly, flexibility, as there are hardly any geographical or time restrictions for students. In terms of cost, e-learning is also usually much cheaper than traditional learning. In addition, knowledge and information can be conveyed more efficiently and broadly online, which makes e-learning very attractive for students. This is also reflected in the high number of new registrations on e-learning platforms, such as Massive Open Online Course (MOOC) platforms. Over 60 million

D. Guralnick et al. (Eds.): TLIC 2021, LNNS 349, pp. 425–435, 2022.
https://doi.org/10.1007/978-3-030-90677-1_40

users have registered and more than 2 million new courses have been added on MOOC platforms since the beginning of 2020 [30].

However, the transition from a traditional campus to online platforms can be difficult and stressful, as the user has to learn new skills in a short period [38]. Insufficient technical skills, poor hardware, and unstable internet connections are not the only limitations of e-learning platforms. Many students complain about a lack of self-discipline and self-motivation when learning online [11]. They often feel too stressed, which leads to more frustration. Lack of social interaction between students or the teacher, and lack of body language and emotions, complicate educational activities. Participants cannot ask instructors for assistance and guidance as comfortably as they can with traditional in-class learning. Alternatives to traditional campus facilities, such as libraries or career counseling, are also limited on e-learning platforms.

To reduce or even solve the limitations, a concept for automated, dialogue-based systems (chatbots) was developed that provides the learners with numerous essential scaffolding tools. A dialogue-based system is a conversational system that can interact with users in their natural language. Existing research in the field of chatbots for educational purposes includes, among others, the chatbot MOOCBuddy [17], which provides learners with personalized course recommendations, and Quiz-Bot [27], which helps students learn faculty knowledge using flashcards. This paper addresses an equally challenging research question: *What automated scaffolding tools could a personalized, dialogue-based system provide to support and encourage users in their learning activities?*

The systems presented in this paper are intended to have various integrated scaffolding tools. Thus, they should support, motivate and encourage users in their learning activities. In addition to a quiz mode, a library function should be integrated, which will suggest interesting literature to students and encourage them to engage in group discussions. Answers to frequently asked questions (FAQs), as well as recommendations of tutorials and support for the user's learning path, are further functionalities of the chatbot. An integrated to-do list with a reminder function is designed to help students become more organized and complete coursework on time, which should lead to less frustration and higher satisfaction.

In this context, the basics of e-learning and research in the field of dialogue-based systems are presented. The elaborated concept of personalized, dialog-based systems for MOOC platforms and their integrated scaffolding tools are introduced and described. Furthermore, possible implementation requirements, design and behavioral approaches, as well as future work and possible evaluation methods are explained.

2 Foundation

This section presents the basics of e-learning, its advantages and its limitations. Research in the field of dialogue-based systems and e-learning, along with MOOCs, is presented and discussed herein.

2.1 E-Learning

E-learning evolved from the concept of distance learning and is defined as self-directed, collaborative learning activities based on web-based technologies [5, 18, 34]. Among others, it used to support students who were unable to obtain higher education due to their geographical location. Online education was initially limited to technical subjects such as mathematics or engineering. Nowadays, the offer is much more extensive, and courses are also offered in many other fields, such as history or economics [35].

E-learning refers to learning activities that are largely conducted over the internet and thus differ from traditional in-class learning [2]. Communication between students and teachers, and among students, takes place via text-based forums, chats and video calls. Traditional classroom-style learning is always synchronous, resulting in real-time information exchange and communication [4]. Online learning can be both synchronous and asynchronous. In asynchronous learning, users do not have to be present at the same time, so communication takes place with a time delay.

E-learning can take place through many channels such as email, online research, or online discussions. For the purposes of this paper, e-learning will be limited to education-oriented online courses such as MOOCs.

MOOCs. MOOCs offer higher education and continuing education in the form of online education in a variety of fields, including medicine, machine learning, law, psychology, and many others [21]. Participation is often free and flexible, as learning usually takes place online with no time or location limits. In 2019, approximately 400 higher education institutions offered a total of 2.5k MOOCs. By the end of 2019, about 13.5k courses from more than 900 universities were available to about 110 million students worldwide [29]. In 2020, due to the pandemic, registrations increased enormously, resulting in a total of 180 million users worldwide and 2,800 course launches [30].

Benefits of E-Learning. The benefits of e-learning have already been discussed in many studies [3, 16, 20, 37]. The most important advantage over traditional learning is flexibility. Every user can access e-learning from any place and at any time and without great cost and effort. The students do not have to travel to a location, and e-learning is usually much cheaper than on-campus education, as there are no costs for premises. The user can also complete the course at their own pace, which can increase satisfaction and reduce stress. In addition, with the help of e-learning, a much larger and more efficient exchange of data and information is possible. Students can share information and discuss ideas online in the forums. This supports, especially, learners with language difficulties or anxieties to communicate more easily with other students.

Limitations of E-Learning. Based on *A Critical Literature Review on e-Learning Limitations* by Wong [38], the limitations of e-learning can be summarized in the following categories: technological limitations, personal limitations, limitations compared to traditional learning, design limitations and other limitations. Technological limitations are mainly related to missing or weak hardware — i.e., computer, smartphone, as well as a poor internet connection, which is essential for e-learning. In addition, students and instructors often feel overwhelmed with the skills and responsibilities related to the technology needed (personal limitation). The lack of self-discipline and lack of (self)-motivation is cited as the biggest problems students have with e-learning [4, 11]. The language barrier is a further personal limitation. Students with poor writing skills may be at a disadvantage, as most communication is written, which can lead to misunderstandings and problems. Another significant limitation of e-learning compared to traditional learning is the loss of physical interaction. Learners and teachers feel isolated. Body language and emotions, from which important cues can be derived, are completely absent. Seeking advice becomes more difficult as lecturers are almost only available by email [1]. Moreover, there is a lack of university facilities such as the library or career counseling. Poor course design leads to stress and frustration among participants. Course materials must be designed to be clearly understood by all participants, similar to traditional learning, as well as to allow easy access to further help and information (design limitations). One more limitation mentioned was constant accessibility. Emails and forum entries can be created at any time. This leads to a never-ending teaching and learning process, which puts a strain on both students and teachers.

2.2 Dialogue-Based Systems

Eliza is considered the first chatbot and a major innovation by Weizenbaum in the 1960s [36]. It was developed to mimic the reactions of a psychotherapist in a therapy session. Nevertheless, the field of dialogue-based systems (chatbots) has only become interesting in recent years with the further development of natural language processing and artificial intelligence. This results in faster and more efficient training of models and more accurate predictions [31]. In addition, the further development of social platforms, especially mobile messaging platforms, contributed to the adoption of chatbots. Mobile messaging platforms now have more than 3 billion users worldwide [32]. Users interact with the chatbot in natural language to obtain specific information or entertainment. Use cases for chatbots are diverse, the most popular being customer service, emotional and social support, and information and requests [8], as well as education.

One area of application for chatbots in education is language learning. Although dialogue-based systems as practicing partners have not yet reached their full potential, chatbots are considered a powerful and efficient tool [14]. Orlando et al. [24] describe and discuss an integrated system with natural language management to support and monitor students during their e-learning activities to find potential issues. T-Bot and Q-Bot [22] are two chatbots on online education platforms like Moodle that help students during their e-learning activities and support instructors in teaching activities whereby one bot takes on the role of a tutor (T-Bot) and the other the role of evaluating the students (Q-Bot). A similar chatbot is NLAST (Natural Language Assistant for Students), a dialogue-based system [13] that allows students to ask questions about tests and courses, obtain further material and view scored exams. Other papers describe customer support chatbots for universities [15, 25] and MOOC platforms [40] that answer FAQs for students and interested users. MOOCBuddy [17], a chatbot designed to encourage personalized learning, offers MOOC users recommendations based on their social media profile via Facebook Messenger. In addition, a dialogue-based system called Quiz-Bot [27] was designed and evaluated to help students learn faculty knowledge in science, safety and English vocabulary using flash-cards. Within the scope of the colMOOC [19] platform funded by the European Commission, there are many other interesting projects, among others, in the area of conversational agents for MOOCs [9, 10, 12].

3 Applications in the Area of Personalized Dialogue-Based System for E-Learning

To solve the previously mentioned limitations of e-learning, this section presents a concept for automated dialogue-based systems for MOOC platforms. It is not just an alternative source of customer service, but it also offers the user a great variety of automated scaffolding tools and support functionalities, some of which are based on artificial intelligence. The chatbot is constantly available for answering and fulfilling the request in dialogue-based conversations. The goal is to improve the e-learning experience, increase motivation, reduce the drop-out rate and encourage continuous-learning for students as well as the instructors.

Using the categories of limitations given by Wong [38], it is shown that especially students' limitations in e-learning, limitations compared to traditional on-campus learning and some other limitations can be reduced or even solved with the help of a dialogue-based system. In the following, the applications of the chatbot are present.

3.1 Recommendations and Learning Paths

Part of the personal limitations of students in e-learning, as well as the limitations compared to classroom-style education, is the lack of recommendations and guidance on professional development and advanced courses. The assistance that the chatbot can provide cannot be compared with in-person career counseling at the university's career center. Nevertheless, the automated dialogue-based system may give the user suggestions for further courses and possible learning paths.

These suggestions are often based on the courses and services already taken. However, course recommendations and learning paths can also be created based on users with similar interests. The prerequisite for both is that the user has already taken some courses and their interests are known to the chatbot. If this is not the case, the chatbot needs to collect data from the user to create recommendations. User data to create recommendations could be obtained, for example, via some multiple-choice questions in the dialog format. Additionally, continuous improvement of the chatbot is facilitated by user feedback on the recommendations and learning paths provided.

3.2 Library

In addition to the career center, another difference of the online campus compared to a traditional campus is the absence of a library. As a substitute, the chatbot would provide the user with a search engine to find current and interesting scientific papers, books or technical articles for a particular topic. Furthermore, students are invited and encouraged to share and recommend interesting articles and papers with other students. In specialized groups, students can further exchange and discuss specific papers or scientific fields. This is intended to promote social interaction and the transfer of information and knowledge between students. Social isolation and the loss of active exchange, such as after traditional on-campus lectures, are also counteracted.

3.3 To-Do List and Reminder

Another difficulty for the user is the efficient management of study time. To ameliorate this problem to a degree, the chatbot should provide the user with a to-do list containing a reminder function. General and course-specific to-do lists can be created. If a task on the list is due, the chatbot reminds the user on the website and, if desired, also by email. The user can request all or course-specific to-do lists from the automated dialogue-based system at any time. In addition, the user should be able to activate automatic reminders for coursework, homework or other important deadlines. This feature is designed to help students become more organized and complete coursework on time, resulting in less frustration and higher satisfaction with their educational activities.

3.4 Quiz

Another aspect of personal limitations is the lack of self-discipline in learning and coursework and the difficulty in motivating oneself. Some students need more encouragement and support than e-learning platforms can provide. A tool that encourages students and supports learning is a quiz [6, 27]. The quiz would be integrated into the automated, dialogue-based system with a learning and competition mode. In learning mode, students would practice with the chatbot, as well as request tips and further explanations. The difficulty of the quiz can either be set manually or automatically by the chatbot. In case of an automatic difficulty setting, the chatbot accesses previous quiz attempts as well as the user's course progress. Since many MOOC platforms have a course progress page, it would be possible to display the user's progress of the quiz in learning mode. Consequently, some users might feel encouraged to reach a 100% progress level on the quiz. Nevertheless, it is important that the users consider the quiz as a support tool that they enjoy using and do not feel forced to participate. This is why the focus of the integrated quiz is on practicing and motivating the user. Possible incentives should only be introduced at a moderate level.

3.5 Frequently Asked Questions (FAQs)

Many existing MOOC platforms already provide a page or section with answers to FAQs. In many cases, however, users do not find these pages and instead simply contact customer support with basic questions [40]. This results in unnecessary waiting time for the user and costly manual effort on the part of the instructors or platform operator. Therefore, the customer support domain should be an important part of the dialogue-based system. It includes answers to questions about the platform, pricing for courses and certificates, age restrictions, and course structure, as well as data privacy.

In addition, the chatbot will provide answers to personalized FAQs. These include questions about the user's current MOOC courses, such as information about course progress or downloading a certificate, or requests to change notification settings and update personal data.

3.6 Speech Support

Users also utilize mobile devices to attend MOOCs [7, 26]. Depending on the screen size of the device, typing is not always easy and spelling mistakes might occur. Speech support integrated into the dialogue system in the form of a dictation and output function could help ameliorate or even solve this problem. Furthermore, another limitation of e-learning is that students who have poor writing skills or are not proficient in the course or platform language are at a disadvantage. Thus, an integrated speech support tool could especially help these students to get better support and learning experience and be more successful.

3.7 Tutorials

For e-learning and taking part in MOOCs, specific new skills and responsibilities related to technology are needed. This is the case for both students, whose skillsets vary between demographic groups, and instructors. Instructors who have had little to no previous exposure to e-learning and mainly taught in traditional on-campus environments require guidance to find their way in the digital environment. The dialogue-based system should provide tutorials and instructions in text and video format for both user groups.

4 Implementation Requirements

A chatbot for MOOC platforms to answer technical FAQs to reduce expensive human customer support has already been implemented and initially evaluated [40]. The presented features will build on this and extend the concept. In the following, some details on the implementation of the automated, dialog-based system are discussed. Besides design decisions and characteristics that the chatbot should have, some technical implementation aspects are presented. Among other things, it is explained why users can communicate with the chatbot using a text-only input, as well as some action buttons.

4.1 Appearance and Behavior

In addition to the variety of tools that the dialogue-based system should offer, its character and appearance are also an influential aspect to convince the user of the chatbot and to encourage its use [33]. The chatbot should adopt a calm and friendly tone in all its messages. Since jokes might be understood differently by different cultures, they are kept to a minimum. Nevertheless, the chatbot should not come across as too boring or lecturing. It should be a friend or a tutor for the user. The user should enjoy communication with the chatbot, and the interaction should be satisfying. In all its messages, the dialog-based system should encourage the user to ask further questions if something is unclear and should not give the user the impression that the questions are annoying.

Furthermore, it is possible to give the chatbot a gender-neutral name and face to support an inclusive learning environment and motivate the user to engage with the chatbot and ask for help, as gender stereotypes also seem to apply to computers and therefore dialogue-based systems [23]. These aspects support the user's satisfaction and motivation to use the chatbot.

4.2 Technical Implementation

To avoid a user not being able to use the chatbot properly because, for example, they don't know how to start a conversation, some actions in the form of buttons should be directly available to the user. During an ongoing conversation, the action buttons are updated to match the dialog. For example, in the case of a decision question, the options "Yes", "No" and "I don't know" are displayed. These action buttons can support different user groups. They save users on mobile devices the inconvenience of typing on a small screen. These buttons also help users who are not too familiar with the platform or language to avoid misunderstandings and get help more easily. This reduces another personal limitation — namely, the pure text-based communication. To better handle requests to the chatbot and, in particular, to use the automated scaffolding tools, the chatbot should work in a specific course context.

The automated, dialogue-based system should be built to self-learn. This learning process should be two-fold through additions and exclusions. With the help of user feedback and ongoing research, the domain of the chatbot would constantly be updated to include additional tools and topics. For each topic, the test datasets and answer datasets would be expanded. At the same time, data that are incorrect or misleading would be removed from the datasets. Thus, by means of self-learning, the chatbot's knowledge and abilities are growing and the domain is kept clean from bad data.

5 Future Work

An important and difficult issue to solve in the future is the amount and type of help a user requires and whether a user wants to ask for help, or the system should determine that a user is struggling and consequently prompt the chatbot to intervene. It would be possible that the user adds their technical and domain understanding in the form of a competence level (e.g., novice, advanced beginner, proficient, skilled, expert) to their platform profile, so that the answers the chatbot provides are adapted to these specific levels. If the user needs more or less help than provided based on their chosen level, they can lower or raise the competence level of help.

After the implementation of the initial concept, the automated dialogue-based system should be tested with the help of user feedback and monitoring. The most crucial factor is to receive feedback from users. With the help of surveys on the platform, it would be possible for the students and instructors to anonymously evaluate the dialogue-based system. Among others, the following points could be considered: overall quality, performance, range and quality of tools offered, quality of answer, understanding of the question, design, and appearance of the chatbot. The user should also have the possibility to express wishes and suggestions in text form.

Furthermore, users are encouraged to evaluate their experience with the chatbot immediately following a conversation. The user is asked to rate on a scale whether their problem has been solved, whether they are satisfied with the solution or whether the tool fulfilled performance and quality expectations.

6 Conclusion

The paper presented a concept for automated, personalized, dialogue-based systems for e-learning portals such as MOOC platforms. These platforms have limitations for both learners and teachers that lead to dissatisfaction, stress and frustration. For this reason, automated scaffolding tools integrated into the chatbot have been introduced to reduce or even solve these limitations. The chatbot should always be available for the users and provide them with instructions, tutorials, answers to FAQs and recommendations. This should support the user to be more motivated and self-disciplined. Among other things, the integrated literature search for interesting research, papers and articles is intended to replace the lack of university facilities such as the library. In addition, it is possible to receive articles recommended by other students as well as to discuss relevant research with other students. This encourages social interaction between users and an active exchange of information and knowledge. To overcome students' isolation and demotivation, the chatbot also offers a quiz mode to support students' learning. It can be concluded that automated dialogue-based systems have the potential to support students in many different areas of their educational activities by creating a more personalized learning experience on MOOC platforms.

References

1. Abbasi, S., et al.: Perceptions of students regarding e-learning during covid-19 at a private medical college (2020). https://doi.org/10.12669/pjms.36.COVID19-S4.2766
2. Allen, I.E., Seaman, J.: Making the Grade: Online Education in the United States (2007)
3. Arkorful, V., Abaidoo, N.: The role of e-learning, advantages and disadvantages of its adoption in higher education. Int. J. Educ. Res. **2**(12), 397–410 (2014)
4. Bejerano, A.R.: Communication education the genesis and evolution of online degree programs: who are they for and what have we lost along the way? Commun. Educ. **57**(3), 408–414 (2008). https://doi.org/10.1080/03634520801993697
5. Bleimann, U.: Atlantis University: a new pedagogical approach beyond e-learning. Campus-Wide Inf. Syst. **21**(5), 191–195 (2004). https://doi.org/10.1108/10650740410567536
6. Bothe, M., et al.: On the acceptance and effects of recapping self-test questions in MOOCs. In: IEEE Global Engineering Education Conference, EDUCON, pp. 264–272. IEEE Computer Society (2020). https://doi.org/10.1109/EDUCON45650.2020.9125145
7. Bothe, M., Meinel, C.: Applied mobile-assisted seamless learning techniques in MOOCs. In: Calise, M., Kloos, C.D., Reich, J., Ruiperez-Valiente, J.A., Wirsing, M. (eds.) Digital Education: At the MOOC Crossroads Where the Interests of Academia and Business Converge: 6th European MOOCs Stakeholders Summit, EMOOCs 2019, Naples, Italy, May 20–22, 2019, Proceedings, pp. 21–30. Springer International Publishing, Cham (2019). https://doi.org/10.1007/978-3-030-19875-6_3
8. Brandtzaeg, P.B., Følstad, A.: Why people use chatbots. In: Kompatsiaris, I., et al. (eds.) INSCI 2017. LNCS, vol. 10673, pp. 377–392. Springer, Cham (2017). https://doi.org/10.1007/978-3-319-70284-1_30
9. Caballé, S., Conesa, J., Gañán, D.: Evaluation on using conversational pedagogical agents to support collaborative learning in MOOCs. In: Barolli, L., Takizawa, M., Yoshihisa, T., Amato, F., Ikeda, M. (eds.) 3PGCIC 2020. LNNS, vol. 158, pp. 199–210. Springer, Cham (2021). https://doi.org/10.1007/978-3-030-61105-7_20

10. Caballé, S., Conesa, J.: Conversational agents in support for collaborative learning in MOOCs: an analytical review. In: Xhafa, F., Barolli, L., Greguš, M. (eds.) INCoS 2018. LNDECT, vol. 23, pp. 384–394. Springer, Cham (2019). https://doi.org/10.1007/978-3-319-98557-2_35
11. Cappel, J.J., Hayen, R.L.: Evaluating E-learning: a case study. J. Comp. Inf. Syst. **44**, 49–56 (2016). https://doi.org/10.1080/08874417.2004.11647595
12. Demetriadis, S., et al.: Towards integrating conversational agents and learning analytics in MOOCs. In: Barolli, L., Xhafa, F., Javaid, N., Spaho, E., Kolici, V. (eds.) EIDWT 2018. LNDECT, vol. 17, pp. 1061–1072. Springer, Cham (2018). https://doi.org/10.1007/978-3-319-75928-9_98
13. Fonte, F.A.M., et al.: NLAST: A natural language assistant for students. IEEE Glob. Eng. Educ. Conf. EDUCON. 10–13-April, October 2017, pp. 709–713 (2016). https://doi.org/10.1109/EDUCON.2016.7474628
14. Fryer, L.K., et al.: Chatbot learning partners: Connecting learning experiences, interest and competence. Comput. Human Behav. **93**, 279–289 (2019). https://doi.org/10.1016/j.chb.2018.12.023
15. Granito, A., et al.: Adaptive feedback improving learningful conversations at workplace. J. E-Learning Knowl. Soc. **10**(1), 63–83 (2014). https://doi.org/10.20368/1971-8829/896
16. Hameed, S., et al.: Effective e-learning integration with traditional learning in a blended learning environment. In: Proceedings of the European Mediterranean Conference on Information System, EMCIS 2008, pp. 1–16 (2008)
17. Holotescu, C.: MOOCBuddy: a chatbot for personalized learning with MOOCs. Rochi – Int. Conf. Human-Computer Interact. **8**, 91–94 (2016)
18. Honey, P.: E-learning: a performance appraisal and some suggestions for improvement. Learn. Organ. **8**(5), 200–203 (2001). https://doi.org/10.1108/EUM0000000005913
19. Karakostas, A., et al.: ColMOOC-an innovative conversational agent platform to support MOOCs a technical evaluation. In: Proceedings - IEEE 20th International Conference on Advanced Learning Technologies, ICALT 2020, pp. 16–18. Institute of Electrical and Electronics Engineers Inc. (2020). https://doi.org/10.1109/ICALT49669.2020.00011
20. Klein, D., Ware, M.: E-learning: new opportunities in continuing professional development. Learn. Publ. **16**(1), 34–46 (2003). https://doi.org/10.1087/095315103320995078
21. Margaryan, A., et al.: Instructional quality of Massive Open Online Courses (MOOCs)
22. Mikic Fonte, F.A., et al.: T-Bot and Q-Bot: a couple of AIML-based Bots for tutoring courses and evaluating students Intelligent Tutoring Systems View project Formal Methods View project (2014). https://doi.org/10.1109/FIE.2008.4720469
23. Nass, C., et al.: Are machines gender neutral? Gender-stereotypic responses to computers with voices. J. Appl. Soc. Psychol. **27**(10), 864–876 (1997). https://doi.org/10.1111/j.1559-1816.1997.tb00275.x
24. Orlando, D., Pietro, G.F.: An integrated system, with natural language management, for the monitoring activities in e-learning environments. In: Proceedings of the - CISIS 2008 2nd International Conference on Complex, Intelligent, and Software Intensive Systems, pp. 965–970 (2008). https://doi.org/10.1109/CISIS.2008.137
25. Ranoliya, B.R., et al.: Chatbot for university related FAQs. In: 2017 International Conference on Advanced Computing and Communication Informatics, ICACCI 2017. 2017-Janua, September 2017, pp. 1525–1530 (2017). https://doi.org/10.1109/ICACCI.2017.8126057
26. Rohloff, T., et al.: Towards a better understanding of mobile learning in MOOCs. In: Proceedings of 2018 Learning With MOOCS, LWMOOCS 2018. pp. 1–4. Institute of Electrical and Electronics Engineers Inc. (2018). https://doi.org/10.1109/LWMOOCS.2018.8534685
27. Ruan, S., et al.: QuizBot: a dialogue-based adaptive learning system for factual knowledge. In: Proceedings of the Conference of Human Factors in Computing Systems, pp. 1–13 (2019). https://doi.org/10.1145/3290605.3300587

28. Sepehr, S., Head, M.: Competition as an element of gamification for learning: an exploratory longitudinal investigation. In: ACM International Conference Proceeding Series. pp. 2–9. Association for Computing Machinery, New York (2013). https://doi.org/10.1145/2583008.2583009
29. Shah, D.: By The Numbers: MOOCs in 2019 (2019). https://www.classcentral.com/report/mooc-stats-2019
30. Shah, D.: By The Numbers: MOOCs in 2020. https://www.classcentral.com/report/mooc-stats-2020
31. Shah, H., et al.: Can machines talk? Comparison of Eliza with modern dialogue systems. Comput. Human Behav. **58**, 278–295 (2016). https://doi.org/10.1016/j.chb.2016.01.004
32. Statista: Most popular Global Mobile Messenger Apps as of January 2021, Based on Number of Monthly Active Users. https://www.statista.com/statistics/258749/most-popular-global-mobile-messenger-apps/
33. Medhi Thies, I., Menon, N., Magapu, S., Subramony, M., O'Neill, J.: How do you want your Chatbot? An exploratory wizard-of-oz study with Young, Urban Indians. In: Bernhaupt, R., Dalvi, G., Joshi, A., Balkrishan, D.K., O'Neill, J., Winckler, M. (eds.) INTERACT 2017. LNCS, vol. 10513, pp. 441–459. Springer, Cham (2017). https://doi.org/10.1007/978-3-319-67744-6_28
34. Tsai, S., Machado, P.: E-learning, online learning, web-based learning, or distance learning : unveiling the ambiguity in current terminology. eLearn Mag. **7**, 3–5 (2002). https://doi.org/10.1145/566778.568597
35. Volery, T., Lord, D.: Critical success factors in online education. Int. J. Educ. Manag. **14**(5), 216–223 (2000). https://doi.org/10.1108/09513540010344731
36. Weizenbaum, J.: ELIZA-A computer program for the study of natural language communication between man and machine
37. Wentling, T.L., Waight, C., Gallagher, J., La Fleur, J., Wang, C., K.A.: E-learning : a review of literature (2000)
38. Wong, D.: A Critical Literature Review on e-Learning Limitations (2007)
39. Worm, B.S., Buch, S.V.: Does competition work as a motivating factor in E-learning? a randomized controlled trial. PLoS One. **9**(1), e85434 (2014). https://doi.org/10.1371/journal.pone.0085434
40. Zobel, T.E.I., et al.: Improving the scalability of MOOC platforms with automated, dialogue-based systems. In: 2020 IEEE Learning with MOOCS (LWMOOCS), pp. 42–46 (2020). https://doi.org/10.1109/LWMOOCS50143.2020.9234359

Author Index

D. Guralnick et al. (Eds.): TLIC 2021, LNNS 349, pp. 437–438, 2022.
https://doi.org/10.1007/978-3-030-90677-1